Social Media and the Transformation of Interaction in Society

John P. Sahlin
The George Washington University, USA

A volume in the Advances in Social Networking and Online Communities (ASNOC) Book Series

Information Science REFERENCE
An Imprint of IGI Global

Managing Director:	Lindsay Johnston
Managing Editor:	Keith Greenberg
Director of Intellectual Property & Contracts:	Jan Travers
Acquisitions Editor:	Kayla Wolfe
Production Editor:	Christina Henning
Cover Design:	Jason Mull

Published in the United States of America by
Information Science Reference (an imprint of IGI Global)
701 E. Chocolate Avenue
Hershey PA, USA 17033
Tel: 717-533-8845
Fax: 717-533-8661
E-mail: cust@igi-global.com
Web site: http://www.igi-global.com

Library of Congress Cataloging-in-Publication Data

Social media and the transformation of interaction in society / John P. Sahlin, editor.
pages cm
Includes bibliographical references and index.
ISBN 978-1-4666-8556-7 (hardcover) -- ISBN 978-1-4666-8557-4 (ebook) 1. Online social networks. 2. Social media. 3. Communication and technology. I. Sahlin, John P., 1971-
HM742.S628194 2015
302.30285--dc23
2015012012

This book is published in the IGI Global book series Advances in Social Networking and Online Communities (ASNOC) (ISSN: 2328-1405; eISSN: 2328-1413)

British Cataloguing in Publication Data
A Cataloguing in Publication record for this book is available from the British Library.

Advances in Social Networking and Online Communities (ASNOC) Book Series

Hakikur Rahman
University of Minho, Portugal

ISSN: 2328-1405
EISSN: 2328-1413

Mission

The advancements of internet technologies and the creation of various social networks provide a new channel of knowledge development processes that's dependent on social networking and online communities. This emerging concept of social innovation is comprised of ideas and strategies designed to improve society.

The **Advances in Social Networking and Online Communities** book series serves as a forum for scholars and practitioners to present comprehensive research on the social, cultural, organizational, and human issues related to the use of virtual communities and social networking. This series will provide an analytical approach to the holistic and newly emerging concepts of online knowledge communities and social networks.

Coverage

- Agent-Mediated Knowledge Management
- Meta-data Representation and Management (e.g., Semantic-Based Coordination Mechanisms, Use of Ontologies, etc.)
- Current State and Future Development of the Institutional Knowledge Management
- Knowledge as an Empirical Problem Solving Tool
- Organizational Knowledge Communication
- Information and Data Management
- Generation of Municipal Services in Multi-Channel Environments
- Local E-Government Interoperability and Security
- Knowledge as a Symbol/Model of Development
- Knowledge Management and Business Improvement

IGI Global is currently accepting manuscripts for publication within this series. To submit a proposal for a volume in this series, please contact our Acquisition Editors at Acquisitions@igi-global.com or visit: http://www.igi-global.com/publish/.

The Advances in Social Networking and Online Communities (ASNOC) Book Series (ISSN 2328-1405) is published by IGI Global, 701 E. Chocolate Avenue, Hershey, PA 17033-1240, USA, www.igi-global.com. This series is composed of titles available for purchase individually; each title is edited to be contextually exclusive from any other title within the series. For pricing and ordering information please visit http://www.igi-global.com/book-series/advances-social-networking-online-communities/37168. Postmaster: Send all address changes to above address.

Titles in this Series

For a list of additional titles in this series, please visit: www.igi-global.com

Cases on Strategic Social Media Utilization in the Nonprofit Sector
Hugo Asencio (California State University – Dominguez Hills, USA) and Rui Sun (California State University – Dominguez Hills, USA)
Information Science Reference • copyright 2015 • 370pp • H/C (ISBN: 9781466681880) • US $195.00 (our price)

Handbook of Research on Interactive Information Quality in Expanding Social Network Communications
Francisco V. Cipolla-Ficarra (Latin Association of Human-Computer Interaction, Spain & International Association of Interactive Communication, Italy)
Information Science Reference • copyright 2015 • 435pp • H/C (ISBN: 9781466673779) • US $255.00 (our price)

Implications of Social Media Use in Personal and Professional Settings
Vladlena Benson (Kingston Business School, Kingston University, UK) and Stephanie Morgan (Kingston Business School, Kingston University, UK)
Information Science Reference • copyright 2015 • 362pp • H/C (ISBN: 9781466674011) • US $195.00 (our price)

Identity and Leadership in Virtual Communities Establishing Credibility and Influence
Dona J. Hickey (University of Richmond, USA) and Joe Essid (University of Richmond, USA)
Information Science Reference • copyright 2014 • 321pp • H/C (ISBN: 9781466651500) • US $205.00 (our price)

Harnessing the Power of Social Media and Web Analytics
Anteneh Ayanso (Brock University, Canada) and Kaveepan Lertwachara (California Polytechnic State University, USA)
Information Science Reference • copyright 2014 • 305pp • H/C (ISBN: 9781466651944) • US $215.00 (our price)

Educational, Psychological, and Behavioral Considerations in Niche Online Communities
Vivek Venkatesh (Concordia University, Canada) Jason Wallin (University of Alberta, Canada) Juan Carlos Castro (Concordia University, Canada) and Jason Edward Lewis (Concordia University, Canada)
Information Science Reference • copyright 2014 • 465pp • H/C (ISBN: 9781466652064) • US $225.00 (our price)

Gender and Social Computing Interactions, Differences and Relationships
Celia Romm Livermore (Wayne State University, USA)
Information Science Reference • copyright 2012 • 343pp • H/C (ISBN: 9781609607593) • US $195.00 (our price)

Table of Contents

Detailed Table of Contents

Just as the three most important aspects of real estate are: Location, Location, Location, it can be said that the three most important aspects of seeking funds for major projects are Access, Access, Access. The advent of social media has opened a new avenue for artists and non-profits to fund their programs: social bootstrapping. This chapter will address the use of social media platforms to raise funds for major arts and nonprofit projects. This chapter will also consider the implications of social media to fund niche programs that may be considered "too risky" to fund under the traditional rules or those that simply don't have the access to major contributors. Pioneers in the arts and nonprofit industries have successfully adopted the best practices of the technology industry and used social media platforms to secure funding that would not traditionally be available.

Social media has become ubiquitous in the daily personal lives of students and teachers alike. But the question remains if social media should be integrated effectively in higher education or if it should be left out in the realm outside the classroom. This paper explores how students use social media in school, whether or not they find social media useful in the learning process and provides further discussion on the importance of adopting a social media strategy in the education sector. The authors facilitated a study in 2010 and a follow-up survey in 2011 to students taking the course International Media and Entertainment Management at NHTV Breda University of Applied Sciences in the Netherlands to assess how students are using social media in a learning environment. The study reveals a need for social media to be used as a learning tool in order to promote active participation through content creation and encourage a virtual space for dynamic dialogue which in turn helps link formal and informal learning connecting students, teachers, and colleagues around the world. But more complex issues such as privacy, copyright, policing and governance of social media needs to be addressed.

air traffic density. A prototype online social media 'radar' sensor system is developed and tested in a pilot study with a dataset of tweets gathered regarding the Occupy Wall Street movement. Further work is suggested which could provide anticipated crowd location, movement and intent in addition to size.

This chapter provides readers with an overview of how social media has enhanced large-scale natural disaster response at the Department of Homeland Security and its partners. The authors of this chapter present the history of the Federal Emergency Management Agency and how its successes and failures have shaped how the Department of Homeland Security has managed trends in increased community participation and information technology. Concepts from Systems Engineering frame the discussion around resilience engineering, network analysis, information systems, and human systems integration as they pertain to how social media can be integrated more effectively in large-scale disaster response. Examples of social media in disaster response are presented including a more in-depth case study on the use of social media during the 2012 Hurricane Sandy response. The chapter concludes with a proposed framework of a decision support system which integrates the benefits of social media while mitigating its risks.

The Internet is a really wide, intriguing world, difficult to penetrate in depth, rich with dissimulations, full of useful and also evil things, that are continuously changing. Cyberbullying represents an actual risk, especially for the online generation which is often unable to distinguish between virtual and real-reality. The first step for cyberbullying prevention is the knowledge acquisition of what cyberbullying is and how it occurs within a specific context. This is not an easy task, since cyberbullying is a complex and quite new phenomenon, so much that researchers' opinion is often divided on its definition and there isn't agreement on the extent of its diffusion. This article presents and comments some cyberbullying preventing tips, inspired by the "top ten tips" that Hinduja and Patchin suggested to educators engaged in cyberbullying prevention.

During the 21st century, businesses benefit from two key components of effective innovation: social media and collaboration. This chapter provides an in-depth analysis to study the interrelatedness of these components. Featured in this chapter are classroom exercises to deepen the learning of students. By providing multiple recent social media examples, the reality of the integral ways in which social media permeates our lives is delivered. Introduced in the chapter is a sixth discipline, an extension of Senge's five disciplines of collaboration. Through the presentation of a new leadership model based upon the six disciplines, the impact of social media is examined. The conclusion of the chapter contains definitions of the concepts introduced. The use of social media has been and still remains a strategically keen tool in business effectiveness.*

The semantic web, social media and the amount of user-generated content continues to grow at a staggering rate. Social Media significantly contributed to the information flow during the Arab Spring, the Occupy and Wall Street movement continue to maintain a global online presence using social media technology. But is the social media information explosion really a unique event in media history? How did story telling evolve into social media? In order to place social media in its historical context and anticipate digital native expectations, we explore the origins of narrative and storytelling from the perspective of a documentary producer. How did past media technologies prepare the way for social media? How do digital natives perceive the world via social media and what do they expect from today's documentary producer? What are the viewing habits of digital natives? What do previous 'information explosions' have in common with social and digital media? These are essential questions for media and documentary producers navigating their way through the vast maze of social media technology and the semantic web, in addition to traditional media.

The current chapter proposes media agent and multi-agent models aiming at improving mediated communication and information exchange in social networking. Great progress has been conducted during the last decades in Information and Communication Technologies, which is also reflected in social media. The proposed models exploit the latest media technologies for the augmentation of user-interaction and contribution experience in multiple levels. Features of the suggested agent and multi-agent approaches are discussed and elaborated through the prism of social computing, social media analytics and intelligence, resulting to a sophisticated communication mediator between users and social groups. In addition, enhanced user engagement and collaboration are considered in terms of rich media experience and augmented reality, semantic interaction services, intelligent content processing and management automation over interoperable multiplatform environments. Social media cooperation and integration is envisioned towards the realization of Web 3.0 and beyond, as the main chapter contribution.

Preface

Thank you for selecting this book. This work is intended to support those with an interest in the use of social media for uses other than mere marketing. Simply put, social media is only one of many tools for effecting communication. These channels can be used to heal or to harm, to comfort or to chastise, to sell or to buy, to connect or to isolate. This book investigates the uses and impacts of social media as we transform to a digital-first world.

A BRIEF HISTORY OF SOCIAL COMPUTING

Communication has always been the primary purpose of networked computing. When the Defense Advanced Research Project Agency (DARPA) created the first version of the Internet, it was meant to support near-instant communication among universities conducting research and development for military programs. The use of simple messaging was sufficient to meet the requirements of communication and information sharing. DARPAnet was never intended to become a platform for social interaction; it was certainly not meant as an avenue for commerce. In 1982 MIT drafted the first known rules of network etiquette (now commonly known as "netiquette"). This first netiquette guide was very clear regarding the purpose of DARPAnet: while using the platform to initiate greetings among colleagues and occasionally make arrangements for in-person meetings, the use of DARPAnet for the primary purpose of social interaction and commerce was not only rude but also illegal. How far the Internet has evolved from its original vision!

Asynchronous forms of communication such as Bulletin Board Services and later email were the standard for text-based communiques among colleagues and friends. As one of the first synchronous chat programs to reach wide adoption, AOL Instant Messenger opened the doors of instant digital communications to everyday users worldwide. In 1992, the Mosaic Browser allowed the Internet to become a myriad of words and images, driving companies to use the Internet as a primary channel for communicating with their customers.

When social media platforms such as MySpace, Facebook, and Twitter started, they sought to expand the communication channel of simple chat clients and rooms to support multimedia messaging. MySpace was meant as a platform of individual expression, enabling the non-technical inhabitants of the Internet community to have personal webpages rich with images, video, and sound. These platforms were not originally meant to change the world; they were meant to support limited use cases. In a 2014 interview with *Fast Company*, Twitter founder Jack Dorsey indicated that he did not have visions of grandeur for his invention; he used it primarily to make sure his mother knew he was eating well. Facebook founder

Mark Zuckerberg built his platform initially as an isolation platform – limiting membership to verified students at Harvard University.

Clearly these use cases expanded far beyond the original vision of their founders. Advances in communications technology drove further evolutions of social media platforms. The ubiquity of high-speed Internet access has transformed high definition multimedia from a luxury for the technocracy to a commodity available to the general public. Social media platforms evolved to support rich content, even providing video streaming and video chat capabilities. YouTube and Vimeo were developed to support the sharing of high quality digital video and opened a whole new line of communications between media companies and their audiences. It also led to the rise of a new brand of celebrity: the video blogger. Video blogging has launched the career of many media celebrities: Justin Beiber's success was originally attributed to YouTube. When the band Journey sought to replace Steve Perry, they turned to YouTube and found stunning videos of superfan Arnel Pineda singing. Journey hits in karaoke venues; Perry's vocal doppelganger soon became the new front man for his favorite band.

Mobility computing has overtaken desktop platforms as the primary onramp to the Information Superhighway. As users became unleashed from their desks and were more able to add the digital experience to their physical experience, social media platforms further evolved to align with the rise of mobile computing. Platforms such as Vine and Snapchat were born specifically to be used on mobile devices and provide instant bursts of rich multimedia content. By documenting the images of our experience in real time, we are more able to share experiences across the borders, shrinking the world in a way imagined only in the dreamy visions of Science Fiction writers the likes of Kurzweil and Clarke.

SCOPE

Innovation is more than merely using advanced technology never before seen in the marketplace. Innovation requires the purposeful and focused adaptation of technology in a new way to meet an extant business challenge. At times, new technology gives rise to new business challenges, and at others innovative minds find ways to use current technology in new ways to deliver new services. The social media innovators represented in these case studies reflect a broad spectrum of organizations with vastly different challenges.

The scope of this work is to focus on how organizations have used social media platforms in ways other than simple marketing to consumers – hundreds of books already exist on the topic of social marketing. Rather, this work provides case studies on how governments and non-profits have used social media to execute their missions, how social justice organizations use the platform to organize acts of digital and physical protests, and how educators use social media to enhance the student-teacher engagement model. Also, this book provides case studies discussing the darker side of social media, ranging from psychological impacts of social media to the effects of cyberbullying. While social media provides a powerful platform to effect globalization, it also serves to make those who would do harm more effective and efficient.

STRUCTURE

This book is organized in three sections, providing case studies from academic researchers across the globe.

Section 1: Changing the World through Microcommunication

Chapter 1: Social Bootstrapping: Microfunding Major Arts and Nonprofit Projects

This chapter is significant in that it offers a glimpse into how the entertainment and nonprofit worlds use the strengths of social media to start new projects that benefit society. Rather than focus on major benefactors to provide major gifts to a few projects, social media democratizes the arts and nonprofit arenas, allowing for more interesting projects that may not capture the attention of the super-rich but are worthwhile nonetheless. Through massive parallelism and microfunding transactions, projects such as *Allegiance*, George Takei's musical about the internment of American citizens during World War II whose only crime was being of Japanese descent. Without the benefit of social media, this dark yet important corner of American History may have died with the last generation to live the tale.

Chapter 2: Social Media – Changing the Way We Teach, Changing the Way We Learn

Social Media provides a new and exciting tool for educators. As a model of communication with students, social media builds new avenues into the daily lives of learners. It also destroys many barriers between educator and learner. In order for educators to take full advantage of social media, they must learn new pedagogical techniques to tailor their message to a new medium. While social media has many strengths in terms of instant feedback and removes many barriers to communication, it also provides new challenges for educators in terms of privacy and filtering. Should an educator "friend" a student? Should students be allowed to post directly to open discussion threads such as Twitter and Facebook? Is the "always on" nature of social media beneficial to learners or is it simply an extension of the classroom into every aspect of life?

Chapter 3: A Study of Two Microfinance Models and Their Suitability in Egypt

One of the most interesting applications of social media in the world of finance and social policy is the application of microtransactions to benefit the public. In this chapter, the authors compare two models where social media is used to multiply the generosity of everyday citizens to benefit developing economies. The authors focus on the economy of Egypt after the Arab Spring to provide financing for citizens to start their own businesses. Similar to how *Kickstarter* multiplies the small contributions of millions to fund ventures, these microfinance models focus on providing small business loans to those who would not qualify for business loans in their country. The results of the study are interesting in that the loan repayment rate of these microfinance loans outpace traditional business loans by a significant margin.

Section 2: Social Government – Communicating with the Constituency

Chapter 4: Globalisation, the Internet, and the Nation-State – A Critical Analysis

With every new advance in communications technology, our world appears to become smaller. News and culture from around the world are more readily available and access to information is now instant. As the use of social media and the Internet bring the world to our homes, it also raises a question: is it changing the nature of the nation-state? Political borders are becoming less effective as barriers to information. The only way to control information in the Internet Age is to disable it completely, resulting in the country's inability to participate on the world stage. This chapter investigates the impact of the Internet and Social Media on the nature and role of the nation-state as a provider of information and services to its citizen base. Is social media the harbinger of doom to the traditional nation-state? How are non-traditional actors using social media to increase their international influence? Non-state actors such as the Islamic State and Al Qaeda are using social media not only as an effective means of gaining influence, but are also using it as a weapon against its enemies, hacking into official social media platforms to expose the weakness of their foes on the international stage.

Chapter 5: Seeking an Online Social Media Radar

Grassroots political movements such as Occupy, Anonymous, and the Arab Spring all used social media to provide a call to action. These groups used social media effectively to coordinate the activity of millions of people across borders and in many cases change the world. When development plans in Istanbul became untenable, citizens turned to social media to coordinate massive rallies to protest government actions. The use of social media to organize political activity is so effective that Turkey banned access to Twitter in that country. Is it possible to mine social media activity to predict the next Arab Spring? The authors of this chapter investigate the potential of using data analysis to predict the effectiveness of social media calls to action. This model would help governments determine the degree of concern they give to some movements and prioritize activity. It would also help predict whether certain nonpolitical events are becoming so popular that additional crowd control may be necessary to protect the population.

Chapter 6: Large-Scale Disaster Response Management – Social Media and Homeland Security

Governments have turned to social media for a variety of interesting use cases. During the response to Hurricane Katrina, the U.S. military used public sources to help coordinate search and recovery activities. These ad-hoc experiences led to the use of social media platforms as a part of how the United States responds to major disasters. The flexible and evolutionary nature of social media provides the idea communications mechanism for disaster response. Telephone lines are down? Power out in a certain area? No problem; Twitter to the rescue. Wildfires in a region? Facebook and Google Earth can provide easy access to information. The U.S. Department of Homeland Security has become particularly adept at taking advantage of the power of social media to develop response plans for major disasters. This chapter reviews a number of use cases where The Department of Homeland Security applied innovative communication techniques to respond quickly and effectively in times of crisis.

Section 3: The Law of Unintended Consequences – Impacts of Social Media

Chapter 7: Cyberbullying Prevention – Some Preventing Tips

As a technology platform, social media is merely a tool for communication. It does not discriminate between beneficial and harmful communications. Just as a hammer can be wielded to build or to destroy, social media can be a tool for good or for ill. Technology provides a virtual mask behind which people can hide and say things they never would to another's face. Unlike face to face communications, where a person has to face the pain he or she causes, the inherent disconnect between sender and receiver in this form of communication allows a person to send a message and never have to see the implications of that message. With the increasing popularity of social media has come a new form of causing pain. Termed cyberbullying, this form of communication is particularly painful to its targets. Unlike typical schoolyard bullying, where the target can eventually escape his or her tormentor (at least for a period of time), the words in cyberspace never disappear. The target of cyberbullying is constantly reminded of the pain as others read and comment on the original painful post, causing the message to trend and reappear in the target's news feed. Ignoring cyberbullying becomes exponentially more difficult. This author discusses a number of tips to identify cyberbullying and offers some ideas on how to avoid it.

Chapter 8: Abuse of the Social Media Brain – Implications for Media Producers and Educators

As with any communication platform, social media can be a source of information or a cacophony of distraction. This author investigates the implications of the "always on" generation using social media in their personal and professional lives for everything from information to entertainment. How must educators and media professionals understand the benefits and drawbacks of social media to tailor their messages for this platform to avoid being lost in the noise?

Chapter 9: Business and Social Media – Collaboration for the Sixth Discipline

In his seminal work *The Sixth Discipline*, Peter Senge investigates how communication becomes the key to optimizing the supply chain in business. Communication is necessary to reduce waste in the system and ensure that the right products and services are available where they are needed, when they are needed. This chapter identifies how social media is bringing about a new form of collaboration to improve communication across the business lifecycle. With mobile ad-hoc networking, near field communications, wearable technology, and pico-processing, this evolution on the Sixth Discipline discusses how social media is rapidly changing the very nature of business operations itself.

Chapter 10: The Social Media "Information Explosion" Spectacle – Perspectives for Documentary Producers

In this chapter, the author considers how media producers can navigate the troubled waters of information to ensure their message is heard. As people are becoming more informed in real time, critical analysis is often reduced to a pithy comment or a sound bite. This information explosion started with the ubiquity of cable news networks and it accelerating with the rise of new social media platforms. Facebook rant that goes on for more than two sentences: TL;DR. Twitter restricts messages to 140 characters, but even that is too in-depth for our Short Attention Span Generation. Don't have time to watch a Ted Talk on YouTube? That's not a problem: Vine limits video feeds to six seconds. Instagram and Pinterest limit communications to a single image. The more available information becomes, the less able people are to absorb it and consider it without being enticed to jump to the next topic. This information explosion has significant implications to producers of educational media such as documentaries. How can one produce a serious and thoughtful analysis of a complex issue in the flashpoint world where even six seconds may be too long?

Chapter 11: Augmenting User interaction Experience through Embedded Multimodal Agents in Social Networking

Can social media be used to drive specific reactions in people? How can producers of a message use multimodal communications to enhance their words to spur others into action? These questions are considered by the authors of this chapter. By using multimodal agents in social media, users can provide an interactive feel to the social media exchange, reinforcing their message with images and video to encourage emotional responses. Studies indicate that the positive or negative nature of social media postings can affect one's own mood: if your friends are posting generally positive messages, you are likely to be in a better mood overall. If those posts are enhanced with multiple forms of media appropriate to the nature of the message, the e is even more powerful.

Chapter 12: The Increase in Immediate Reply Syndrome through the Popularization of SNSes

The art of letter writing may be dead. This is not because people no longer have time to write. It is because the mode of communication and its main purpose has shifted. One of the consequences of the rise of social media is the notion that a reply is expected immediately. Asynchronous communication is no longer acceptable. The Veruca Salt generation expects immediate feedback and gratification. If someone doesn't "like" your status or "pin" your creation, you begin to wonder why. Waiting more than a few seconds for a reply can be viewed as an affront. "He just commented on her status. She liked his picture. Aren't we friends anymore? I'd better post something else to make sure they see me." This need for immediate feedback is driving an increasing banality into the messages transmitted across social media. This need for constantly updating one's feed to ensure that others are replying has been popularized by the neo-punk band The Chainsmokers: But first, lemme take a #selfie.

The contributors of this work provided a wide variety of case studies investigating the alternate uses for social media and its implications for society and public policy. This research is invaluable in discussing the utility of social media as a platform for building communities in the digital age.

ALTERNATE USES OF SOCIAL MEDIA

Originally meant for fostering communications among friends, social media was quickly identified as an effective platform for developing communities. Never missing an opportunity to reach an audience of similar interests, marketers found ways to turn social media into a haven for targeted advertising. In the days of Internet banner ads, targeted marketing was limited to personalization through cookies or opt-in profiles within personalization websites. With social media, companies found ways to develop a brand relationship with their customers. The concept of viral marketing took off: accelerating word of mouth advertising through social media channels. The inventors of the GoPro camera knew well the value of viral marketing, encouraging their customers to post videos of their extreme experiences on social media and co-opting these personal videos into their official marketing campaigns.

Nonprofit organizations realized the value of viral messaging and memetic transfer of content to raise social awareness of issues and raise funds for worthy causes. 2014 was a banner year for Social Media as an agent for change and social justice. The ALS Foundation's viral fundraising campaign, the #IceBucket Challenge raised a record-breaking amount of donations in just six months. Hewlett Packard encouraged its employees through the #MakeItMatter and #MatterToAMillion social media campaigns, matching their charitable contributions up to a total of $1,000,000. Singer/Songwriter/Actress/Producer Cyndi Lauper founded the #GiveADamn campaign to raise awareness of issues faced by the Lesbian, Gay, Bisexual, and Transgender (LGBT) community, particularly teens.

Organizations also fond social media campaigns as an avenue for seeking justice. In Santa Barbara, California, Elliot Rodger undertook a deadly rampage and posted his misogynistic rants justifying his bloody actions through his video blog rife with derision for women and a sense of entitlement to romantic interactions with women. Publication of his ravings sparked an online conversation about misogyny giving rise to the hashtag #YesAllWomen striving to raise awareness of the universality of sexual harassment and a general culture of violence toward women. During the riots following the decision not to indict Officer Darren Wilson after the fatal shooting of Michael Brown, an unarmed teenager in Ferguson, Missouri, a new hashtag was born: #BlackLivesMatter. Just as #YesAllWomen resulted in a counter hashtag (#NotAllMen), many tried to correct users of #BlackLivesMatter by stating that #AllLivesMatter. When Lynn Fuller, a San Diego, California woman suffered 3rd degree burns over 75% of her body after her boyfriend set her bed on fire, her friends organized a social media campaign to help raise funds for her medical bills raining over $50,000 in just over a month – a grassroots effort that would not have been nearly as successful before social media.

Digital Activists found social media particularly useful to satisfy their aims. The hacktivist group Anonymous took pleasure in threatening to release private documents and personally identifiable information (an act known as "doxxing") through social media to apply pressure to achieve goals of social justice. When the Ku Klux Klan got involved in supporting the Ferguson Police Department during the riots, threatening to shoot violent rioters and looters, Anonymous stepped in by doxxing local members of the KKK unhooding educators, office workers, homemakers, and law enforcement officers. Social media has become such a powerful force in organizing for political activism, Turkey officially outlawed the use of Twitter, blocking access to Twitter from all Domain Name Service (DNS) servers in the country. Not to be outdone, Google responded to this official ban on free speech by publishing the IP Address of its public DNS servers (Keep Calm and 8.8.8.8, 8.8.4.4), implicitly offering to bypass the official DNS servers in Turkey and masking the original IP address and location of the source device.

As the Islamic State of Iraq and the Levant (ISIL) swept into Iraq and Syria, they used social media to spread news of their victories. ISIL also used social media effectively to transmit video messages recorded by hostages as well as executions. ISIL took credit as the Cyber Caliphate for hacking the Twitter and Instagram accounts of the United States Central Command (the joint command responsible for U.S. military operations in the Middle East) to demonstrate their ability to conduct operations in both the digital and the physical worlds. After numerous warnings via social media not to publish videos of executions, Anonymous stepped into the political fray once again by seizing dozens of social media accounts used by ISIL operatives for coordination of activities and distribution of their messages and videos.

CONCLUSION: THE FUTURE OF SOCIAL MEDIA

Predicting the future evolutions of social media is as difficult as predicting the future evolutions of humanity. The real difference is that the pace of this evolution is happening on Internet Time. That said, there are three distinct trends in the use of social media worth investigating (perhaps for a future edition of this work):

Advocacy and Activism

Advocacy groups will find ways to leverage the social media platform beyond simple awareness. Political activists such as Occupy and Anonymous have used social media to organize their efforts. Nonprofits and advocacy groups will continue to value targeted communications with like-minded people. The true value of social media is the ability to shift the control of the message from the organization to the community. By decentralizing the message, advocacy groups can better assess the need and focus their efforts to localities that need their help most. Integrating picture and video content with social media, activists have almost unfettered access to execute their mission and raising awareness. The ASPCA used social media to publish hidden videos exposing abuses by the farming industry

By placing the power of advocacy directly in the hands of those who need help, those who wish to persecute or otherwise take advantage of the weak lose much of their power and control. Allowing people to communicate their needs directly to organizations means that advocacy groups can deliver benefits exactly where they are needed. Combining this information with data analysis will help organizations provide justification for additional grants based on the true need of the community.

Microtransactions and Digital Payment

A primary benefit of advances in communications technology is that social media can move beyond simple messages and provide multimedia content. As trends toward microtransactions continues, digital payment directly between people will increase. Social media will be the platform of choice for transactions to raise money for social causes, personal transactions, and even small commercial transactions.

The implication of this is that social media is also likely to become the platform of choice for digital crime. The relative anonymity of digital communications provides an ideal environment for thieves and con artists. As our interactions become less personal and more digital, it is vital to maintain vigilance and common sense.

Digital Constituency

As governments seek to improve their services to their constituents, social media becomes increasingly important in strengthening the bond between government and the citizenry.

Governments will continue to evolve the trend of digital activism by soliciting direct support by citizens to provide assistance. Many government organizations have mobile apps for citizens to report public nuisance or safety issues such as potholes, abandoned vehicles, and graffiti. The County of San Diego, California took this a step further with the Pulse program, allowing citizens who are CPR-certified to volunteer as initial responders. When a 911 dispatcher sends official first responders to the scene of a heart attack, participants in the area also receive digital alerts in an attempt to get help to the scene in as little time as possible.

By providing direct channels of communications with constituents, government officials can better understand the needs of the citizens they represent. Governments will apply the lessons learned of corporate customer service organizations to improve the services they provide to the citizen base. Combining social media with data and sentiment analytics, governments can transcend mere hashtag trend analysis to a true understanding of how people feel about trending topics. This may have dramatic implications on the polling process. By including real time polling with social media, governments can get instant feedback on their performance. The question that remains is whether this will encourage our elected officials to lead or merely sway to the shifting winds of the electorate.

Acknowledgment

I would like to offer thanks to those who made this book possible.

To my wife Kristil, I extend my eternal gratitude; taking on a book project while working full time and teaching has been quite the effort. Thank you for your patience and support.

To the members of the Editorial Advisory Board, I wish to thank you for your support and review of the intent of this work. I truly believe that with your guidance, this work will meet its original intent.

My chapter reviewers were critical to getting this work over the finish line. Your input was vital in ensuring that this work is of high academic and research quality.

And last, but certainly not least, I'd like to thank the authors who contributed cases studies to this book. Your research literally made this work possible.

Section 1

Changing the World through Microcommunication

Chapter 1
Social Bootstrapping:
Microfunding Major Projects in the Arts and Nonprofit Organizations

John P. Sahlin
The George Washington University, USA

ABSTRACT

Just as the three most important aspects of real estate are: Location, Location, Location, it can be said that the three most important aspects of seeking funds for major projects are Access, Access, Access. The advent of social media has opened a new avenue for artists and non-profits to fund their programs: social bootstrapping. This chapter will address the use of social media platforms to raise funds for major arts and nonprofit projects. This chapter will also consider the implications of social media to fund niche programs that may be considered "too risky" to fund under the traditional rules or those that simply don't have the access to major contributors. Pioneers in the arts and nonprofit industries have successfully adopted the best practices of the technology industry and used social media platforms to secure funding that would not traditionally be available.

BACKGROUND

Traditional methods for funding major arts and non-profit projects generally require major capital investments from corporations, or at the least a major benefactor. Because the super-rich establish endowments like The Bill and Marilyn Gates Foundation, major projects can be executed with a single stroke of the pen. Even the U.S. National Endowment for the Arts has the ability to fund major projects, though resources are scarce in today's tight budgets and pressure by Congress to limit investment in the arts in favor of defense and health care programs; despite a slight increase from 2013, the 2014 U.S. Budget reduced funding for the arts by 28% from its peak 2009 funding (National Endowment for the Arts, 2014).

The main problem with seeking funding for major projects from a single source is that artists with a vision or people trying to start a new public service program may not have access to major benefactors; competition is high for finite resources and getting to the door is difficult. Nonprofit service organizations face this same problem. As the competition for public funds increases dramatically, so too does the need

DOI: 10.4018/978-1-4666-8556-7.ch001

to serve the community. Well-known nonprofits can get the attention of major benefactors or government agencies, and even cooperate with large organizations to access funds via major corporate or government fundraising efforts. One example of this approach is the Combined Federal Campaign (CFC). The CFC program is managed by the United Way, and allows U.S. Federal Employees (and contractors) to provide payroll deduction contributions to a variety of non-profit organizations that have been selected by the United Way to participate (Office of Personnel and Management, 2014). In this model, the United Way uses its well-established brand name to act as an intermediary between smaller non-profits and the U.S. Government. However, even getting access to the United Way program can be all but impossible for new or very small non-profits. Because of a lack of access to major resources, artists with a strong, unique voice or nonprofits that service a niche need may fail to complete their vision.

LITERATURE REVIEW

Microfunding: Not a New Concept

The concept of microfunding may be the answer to those programs that cannot garner the attention of major benefactors. Microfunding is the process of collecting large sums of money for a specific cause or project through a large number of donations in very small increments. The success of microfunding depends entirely on the network available. A local bake sale or car wash may be sufficient to help buy uniforms for the school band, but in order to launch a major project via microfunding, one needs to reach as many people as possible.

Capturing the attention of millions of people to contribute to a cause generally requires easy presence. One method of establishing this easy presence is to use the ubiquity of the Internet and mobile technology to access people directly. Mobile is particularly important to microfunding, as over 40% of time reviewing media is spent on mobile devices (Meeker, 2014). Current research in the field of computer-mediated microfunding suggests that people are far more likely to donate to worthy projects when the donations are made convenient and if the requesters have established a personal connection to the contributor (Gerber, Hui, & Kuo, 2013). Said another way, the likelihood of successful fundraising is dependent on the ability to make people care and make it easy for people to contribute.

Microfunding presents a secondary benefit to the general public: it allows people to feel they have contributed to a cause or project that is greater than themselves. As described in research published by the Poznan University of Technology, large-scale Non Governmental Organizations (NGOs) and organizers of charity projects have turned to computer-mediated microfunding as a source of income and increased awareness of their projects (Wojciechowski, 2009). Not only does microfunding remove the barrier to entry of participation by lowering the threshold of donations, they also allow people to participate in solving a global issue or major project. This approach also provides individuals with a sort of bragging rights to participation. Sometimes referred to as the "Sunday Effect," there is a distinct correlation between pro-social activity and public viewing of said behavior (Malhorta, 2010a). The idea that altruism can be social is an odd one; altruism by definition should not be dependent upon whether people's activities are being viewed by others. Despite this, research has shown that people have a tendency to be more active supporting projects when there is a sense of competition involved. Similar to a live auction, where participants become swept by the concept of winning the bid, project contributors are more likely to participate in fundraising efforts if they receive public recognition (Malhorta, 2010).

Computer-mediated microfunding sites such as Kickstarter, IndieGoGo, GoFundMe, and others take advantage of the Sunday Effect by publishing the names of contributors (and in some cases, the amounts). Of course, some contributors wish to remain anonymous and these sites allow for this option. Kickstarter includes a variation of this theme by not only listing the names of contributors, but also providing tangible and intangible rewards to contributors based on funding level. In the case of creative projects such as movies and books, names of contributors at a certain level may be listed in the credits of the movie or even appear as characters in the project.

What is motivating participants to contribute to projects? Tangible benefits may include signed movie posters, letters of appreciation, or even invitations to special events. These tangible rewards allow contributors to extend the Sunday Effect beyond merely recognition on the funding site; contributors can bask in their pro-social activity long after the initial funding drive. Regardless of whether the contributors are motivated by altruistic means or are participating in order to be seen doing so, microfunding has been an effective source of funding for projects and establishing a social bootstrap community, particularly for the arts. Kickstarter, the leader in computer-mediated social bootstrapping, has registered more than $1,250,000,000 in pledged funding for projects, a total of $423,130,000 (or 34%) pledged to art, film, music, theatre, and dance projects (Kickstarter.com, 2014). However, other bootstrapping sites are more targeted toward establishing bootstrap communities to fund non-profit and arts projects. For example, IndieGoGo provided merely ¼ the funding of Kickstarter in 2013, but 76% of its funds to art, film, music, theatre, and dance projects (Holmes, 2103). Computer-mediated crowdfunding has been particularly beneficial to the arts and non-profit communities. In 2012, 49.4% of all funds sourced in this way were dedicated to social causes, film, performing arts, and music projects (Statista.com, 2014).

Social Bootstrapping: Establishing a Digital Collective

The concept of bootstrapping describes self-funding a project or company. This process of funding small-scale projects or even small companies is the heart of the entrepreneurial spirit. Unfortunately, bootstrapping is only effective if the person leading the effort has sufficient capital (or credit) to fund the venture until it becomes self-sustaining (i.e., profitable). When the entrepreneur does not have sufficient capital available, he or she needs to seek funding. The traditional approach toward funding major projects is to gain access to a major benefactor such as a venture capital firm, or a private investor/patron. There are a number of downsides to this approach. This first, as mentioned previously, is access. Getting an idea in front of a major investor is a daunting task, as the competition for these funds is extreme. Social bootstrapping, in comparison, is the process of leveraging the power of Metcalfe's Law and the social network to provide collective resources for a project that is important to a community.

The advent of computer-mediated microfunding sites discussed earlier has opened the door to a new source of funding not only for entrepreneurs, but also artists and Good Samaritans hoping to raise funds for their particular projects. While the rules for various sites vary, the three market leaders in non-equity based crowdfunding are Kickstarter, Indiegogo, and RocketHub (Barnett, 2013). This method of seeking small donations for projects directly from the Internet user community enables social bootstrapping in a way never before possible. Rather than hosting rallies and events, or collecting signatures in public forums, fundraisers can develop an online social community to bootstrap a project.

Just as access is important in the traditional model of fundraising, the key to social bootstrapping is access. In the case of computer-mediated crowdfunding, the direct access to individuals is provided by the crowdfunding platform powered by the Internet. However, capturing the attention of the Internet community to the point where they will contribute to a given project also requires access. In the traditional method of fundraising, access meant access to the one or two key people possessing significant resources. Computer-mediated crowdfunding turns this model upside-down; the key to access is having access to a wide network of individuals who care about the given project. By building a powerful online social community, one can take advantage of the concepts of social network Transmission and Binding. Transmission refers to the use of personal networks to communicate information that is mutually important to members of the network. Binding refers to the phenomenon of individual members of a given network taking collective action as a single unit. These two concepts of social network analysis are key to predicting the overall success of a network to establish a social community (Marin & Wellman, 2011). The theories of social network analysis presented by Marin and Wellman are supported by the reality seen in successful computer-mediated social bootstrapping efforts. In 2013, the majority of contributors of projects came from people within the direct personal network of the originator (Feldman, 2013). Social bootstrapping takes advantage of the exponential growth predicted by Reed (2001) to establish group-forming networks that take on a life of their own by generating new connections. By using social media to capture the power of group-forming networks, artists and aspiring Good Samaritans can establish a community much larger (and far more quickly) than by traditional means. A recent study by Kuppuswamy and Bayus (2014) demonstrated that crowdfunding project backers are most likely to contribute in the first or the last week of availability. Based on this, it is vital to take advantage of a social network as soon as possible. Social bootstrapping is the process of building these communities in Internet speed.

ENTR'ACTE: SOCIAL BOOTSTRAPPING AND THE ENTERTAINMENT INDUSTRY

Attention Marshmallows: Veronica Mars is not the only Crowdfunded Movie

Perhaps the most well-known example of using social bootstrapping to fund a movie is the *Veronica Mars* movie. *Veronica Mars* was a popular television show following a teenage private eye (in training). After three seasons, *Veronica Mars* was cancelled, with what fans felt were too many loose ends. Rob Thomas (Executive Producer of *Veronica Mars*) failed to convince a major production company to fund a *Veronica Mars* movie. In 2013, the Thomas turned to Kickstarter to secure funding from Veronica Mars fans (affectionately known as "Marshmallows") (Chin, Jones, McNutt, & Pebler, 2014). After only 2 days the *Veronica Mars* project raised over $3,000,000 and ended with a total of $5,702,153 (Widjaya, 2013; Thomas, 2103). The movie was released in 2014. While this is a high-profile example of a Kickstarter success story where fans funded a major motion picture, Veronica Mars is not the only movie project seeking funding directly from fans rather than from "professional" producers.

According to the statistics portal Statista.com (2014), 11.9% of all crowdfunded projects go toward film and performing arts projects. Many of these projects are for short films by aspiring directors or film students. Notable exceptions including *Veronica Mars* comprise the majority of funding, the rest allocated to smaller independent films. Celebrities such as Don Cheadle and Rob Zombie have also launched crowdsourcing efforts to fund independent film projects (IndieGoGo, 2014; Grow, 2014). Arguably, celebrities with as much gravitas as Rob Zombie and Don Cheadle could easily fund projects through the traditional process. Social bootstrapping allows these producers the ability to generate marketing buzz for their projects. Research has shown that the box office success of film projects has a strong positive correlation to the amount of activity about the project in social media (Mestyán, Yasseri, & Kertész, 2013).

Current celebrities have a large social following, which helps spread the word for their projects and generate funding via computer-mediated crowdfunding sites. A strong social media network is vital to the success of crowdsourcing projects: for every order of magnitude increase in Facebook friends, the probability of success increases significantly (10 friends = ~9%; 100 friends = ~20%; 1,000 friends =~40%) (Fundable.com, 2014). How then, would small independent filmmakers get their projects funded via crowdsourcing? When independent filmmakers do not have the personal social network of a Don Cheadle, Rob Zombie, or Rob Thomas, they must rely on the power of group-forming networks to help them expand their social bootstrapping community. Often, small projects gain the attention of one group-forming network node to give their project a viral growth boost. The Upworthy social activism site has been particularly helpful in launching successful social bootstrapping efforts. For example, the Kickstarter project to launch a documentary about the first major female director Alice-Guy Blanche had only reached half of its $200,000 goal after four weeks. Within one week of being posted on Upworthy's website (and associated social media feeds), the project exceeded its goal (Renninger, 2013). Social media aggregators such as Upworthy and Buzzfeed have been highly influential in social bootstrapping efforts for those without the direct connections. They have helped prove Reed's theories on group-forming social networks. The correlative success of social bootstrapping for movie projects compared to the social networks of their creators has ben confirmed not only in American films, but in the European independent film industry as well (Sørensen, 2103).

The Social Road: Taking Japanese Internment Camps to Broadway

The 1959 Broadway meta-comedy *The Producers* introduces the audience to Max Bialystock, a less than honest producer. Bialystock's expertise is wooing rich elderly women into funding his ventures, gaining funding for his projects one lady at a time (IBDB.com, 2014). This tongue-in-cheek look at how theatrical productions can be funded strikes a chord: many people who are not "in the inner circle" want to be a part of the theatre. Actor, writer, comedian, and social activist George Takei proved this by taking his recent collaboration *Allegiance* to Broadway in 2014.

Allegiance is the story of the internment of Japanese-Americans during World War II. During this a dark part of U.S. History that is often overlooked, the U.S. Government sent hundreds of citizens to camps simply because they were of Japanese descent. More than a social issue, this project was very personal to Takei; at the age of five he was sent to an internment camp with his family, where they lived for four years. Their sole crime was "looking like the people from Pearl Harbor" (Takei, 2014).

This very personal and dark topic gave Broadway investors hesitation. *Alleigiance* had all the makings of a Broadway show: a headliner cast including Lea Salonga and Telly Leung, and receiving standing ovations and two extensions during their opening in the Old Globe Theatre in San Diego. The chief concerns investors had was that the American public (specifically, theatre-goers) may not resonate with this story enough to turn a profit with the investments necessary for a Broadway production (Savage, 2014). Often, investors shy away from important projects that may seem too "heavy" for mainstream audiences because they increase risk (Kennedy, 2014).

Allegiance's lead producer Lorenzo Thione knew the value social media could play in allaying these concerns. As the former co-founder of search engine Powerset (now a Microsoft holding) understood how to make technology drive business value (Cox, 2014). Thione knew that Takei could use his celebrity and status as an outspoken activist for LGBT rights to raise awareness (and funding) for Allegiance. Takei immediately launched a social media campaign first to improve his social profile, and then to improve social awareness for his personal project. In 2012, Takei launched a successful Kickstarter campaign to raise sufficient funds to complete *Allegiance*, and bring an all-star cast to San Diego. As Thione and Takei took the next step to a Broadway opening, they had a revolutionary idea for Broadway: sell priority access (and receive a download of the cast album) to purchase tickets when available. This Kickstarter-esque rewards approach was new to Broadway, but it served two critical purposes. First, it established to investors that fans were definitely willing to buy tickets to this "risky" venture. Secondly, it helped generate a media buzz that is critical in the early stages of development on the Broadway scene (Savage, 2014, Takei, 2014).

This model of proving interest with grass-roots efforts to secure funding is not new to the software industry. Software companies have been using concepts such as providing free downloads of software with the ability to upgrade to a fully-functional version for a fee, to demonstrate a viable user base in order to secure funding for over two decades, and is a particularly popular strategy for launching independent games (Kerr, Whitson, Harvey, Shepherd, & O'Donnell, 2013). Thione's background in the software industry paid off in this venture. Thione and Takei adopted a business strategy of demonstrating audience interest; their efforts and creativity paid off and Allegiance opens on Broadway in the 2014 season. They have proven that this model of rewards to establish interest and ability to succeed in the digital marketplace as identified by Gerber, Hui, and Kuo (2013) is viable not only for tech projects but also for mainstream Broadway productions.

Social Television: Reinventing (Saving?) an Industry

The television industry has long experimented with methods for incorporating audience feedback into the results of programming. These efforts ranged form performing before live audiences in the 1950s to weaving live Twitter conversations into programming. The goal of these efforts is to increase the emotional investment viewers have in the programming: if the audience has a say in the results, they will be more attached to the program.

Fox Television struck ratings gold when they co-opted the talent show format into *American Idol*. In this program, the viewers had the ability to vote on their favorite contestant to determine who would advance to the next round in the competition. When American Idol premiered in 2002, it instantly became a hit and in 2004 earned the top rating spot in the U.S., which it commanded for over eight years (Keveney, 2014). Other reality/competition shows took a cue from American Idol's success – adopting this viewer feedback in a variety of methods. Lifetime Network's *Project Runway* began to invite contestants from past seasons as determined by web votes and now includes a weekly Fan Favorite competition where designers can win $1,500 each week and compete for a grand prize of $10,000 based solely on votes form the audience (MyLifetime.com, 2014). BBC and NBC selected leads in major Broadway and West End productions using this same talent search / viewer feedback model with programs like *How Do you Solve a Problem like Maria?*, and *I'd Do Anything* (BBC, 2014; BBC 2014a).

This model has had mixed overall results. The perceived curse of American Idol is that while the winner (determined by viewer votes) receives a recording contract, he or she often does not enjoy great success after the show. This is counterintuitive, as theoretically the winner should have the best chance for a career because he or she was deemed most popular. Samantha Barks was passed over by the audience for the lead role in the West End production of *Oliver!* as the runner-up in BBC's *I'd Do Anything*. The winner (who won despite not being Sir Andrew Lloyd Weber's choice for the role) went on to star in *Oliver!* on West End, but has had a relatively quiet career (BBC, 2014a). In the same year, Samantha Barks received the London Critics Circle Film Award for Young British Performer of the Year and the Hollywood Film Festival Spotlight Award, as well as receiving several other nominations for her on-screen portrayal of Éponine in the film *Les Misérables* (IMDB, 2014).

The PBS program *Reading Rainbow* ran for over 20 years with the goal of introducing young viewers to the endless possibilities that are available through reading. The format remained largely unchanged for two decades, opening with its star LeVar Burton hosting "reading time" with a group of children. Within minutes, the show transforms from a hosted story time to an enactment of the book. A constant message in Reading Rainbow was that when a child reads, he or she enters a new world without limits. "I can be anything, take a look, it's in a book. A Reading Rainbow" (Lybio.net, 2014). When PBS announced that it would cancel *Reading Rainbow*, LeVar Burton launched what may be the most successful Kickstarter campaign in history, raising over $2,000,000 in fewer than 24 hours, and raising a total of $5,408,916 toward its original goal of $1,000,000 (Burton, 2014). This dramatic success underscores the power of Reed's group-forming networks and social bootstrapping to raise funds for projects that are important to a community.

Celebrities have transformed the television industry by incorporating audience feedback and developing a social community. In 2014, actor Wil Wheaton launched *The Wil Wheaton Project* on SyFy, primarily based on his social network community developed on Facebook and Twitter. Wheaton's social network is extensive, including 163,000 followers on Facebook and 2,670,000 on Twitter (Facebook, 2014; twitter.com, 2014c). Wheaton leveraged this network to build an impressive presence for his SyFy program, which has over 15,000 Twitter followers after only ten episodes (twitter.com, 2104d). Wheaton and SyFy encourage viewers to join the social community by monitoring and responding to posts using the hashtag #wilwproject. Further, Wheaton takes the idea of building a social community around his show to the next level by live-tweeting other broadcasts on the SyFy channel, including his most recent social conversation involving live tweets of *Sharknado 2* (syfy.com, 2014). Another example of using social media to transform television by engaging directly with viewers is Chris Hardwick's *@Midnight* on Comedy Central. As the title suggests, *@Midnight* is designed specifically to integrate with social media. This program invites celebrities to play a question-and-answer game show that includes hashtags meant to engage with the social media community, including inviting the guests (and its 272,000 Twitter followers) to come up with names for #DepressingDisney titles or #BadHogwartsClasses (twitter.com, 2014a). By engaging directly with the audience via social media, shows such as *@Midnight* and *The Wil Wheaton Project* quickly developed a strong cult-like following.

Wheaton builds his social community by including people with common interest, including science fiction and fantasy game play. For the past six years, Wheaton has sponsored a variety show coincident with San Diego Comic Con called w00tstock. Sometimes referred to as "Nerd Church," Wheaton and partner Adam Savage (co-host of *Mythbusters*) attracts comedians, musicians, and actors to conduct a Vaudeville-style live show. Wheaton encourages the audience to record the sessions and post them on social media, provided they are posted with free-to-share rights (w00tstock.net, 2014) Wheaton further builds on his social community by expanding his on-screen personality with his private interests in home brewing, collaborating with the Stone Brewing Company in Escondito, CA to develop a co-branded beer called w00tstout, drawing collaborators including Fark.com founder Drew Curtis and comedienne Aisha

Tyler (Hernandez, 2014). By combining these personal interests with professional presence, celebrities are able to expand their social networks to provide crossover awareness of television projects. In an era of declining television viewership in favor of Internet programming, these crossover projects are critical to help bootstrap new media projects.

SOCIAL GOOD: BOOTSTRAPPING TO SUPPORT NONPROFITS

Microfunding is not a new concept to non-profits. The United Nations International Children's Endangerment Fund (UNICEF) was launched in 1946 (UNICEF.org, 2012). UNICEF encouraged children to collect spare change from houses while enjoying trick-or-treat festivities during Halloween. The power of a little spare change multiplied by millions of contributors cannot be ignored. In 2013, the Ronald McDonald House collected over $11,000,000 through donations of spare change at McDonald's restaurants worldwide and small-increment Internet donations (RMHC.org, 2014). The Salvation Army has had great success in localized fundraising by deploying "bell ringers" to storefronts during the Christmas season for 120 years (Salvation Army, 2013).

As mobile phones and Internet access became ubiquitous, it was a logical to incorporate technology into the microfunding process. In the mid-2000s, direct mobile donation services appeared on the scene. At first, organizations were concerned that participation would be low out of fear of scams. The American Red Cross was an early adopter of this process, launching the Text 2HELP campaign to raise funds for victims of natural disasters. In order to combat the fear of scams, the American Red Cross teamed with digital identify provider Verisign to confirm the validity of the Text 2HELP program texts. In 2008, Text 2HELP raised $190,000 through nearly 40,000 individual text messages (Khan, 2008). Merely two years later, the Red Cross parlayed on this program's success and raised over $32,000,000 in text donations of $10 each to support victims of a devastating earthquake in Haiti (Hamblen, 2010). Red Cross teamed with mobile donation provider mGive to make donations simple and convenient. By texting "Haiti" to a predetermined SMS account, phone users could donate $10 to be billed directly to their phone service.

Nonprofit organizations have taken to social media to build upon their current networks and raise overall awareness of an issue. As mentioned previously, social awareness sites such as Upworthy.com are often vital in raising awareness of key social issues. The most common usage of social media activity by nonprofit organizations was to raise general awareness of social issues; actual fundraising took a back seat to every other activity identified in recent studies conducted by Nonprofit Technology Network and the Case Foundation (Common Knowledge, 2012; Sharma, 2014). Not surprisingly, the elements of a sound social media strategy mirror the best practices of social bootstrapping for any purpose: access and engagement stems from a clear message that individuals internalize and pass on to their networks (Lasica, 2012). By presenting a clear message that resonates with individuals, nonprofits take advantage of Reed's group-forming networks to build momentum and awareness that is limited only by the speed of light and the reach of the Internet.

Alex's Lemonade Stand is an excellent example of how a nonprofit can use social media to reach a global audience by presenting a clear and personal message. Alexandra Scott was born in 1996 and within a year was diagnosed with cancer. When she was receiving hospital treatments at the age of four, she told her parents she wanted to open a lemonade stand to raise money for her hospital. She and her brother raised over $2,000 in one day with the help of her local community. When this story was covered in news stories across the country, other people opened lemonade stands to raise money for this cause. After Alex's death in 2004, her parents founded the Alex's Lemonade Stand Foundation (ALSF) to establish a centralized national effort to raise money to fund research to combat childhood cancer. ALSF launched a targeted social media campaign to raise awareness and suggest grassroots effort to help with the campaign. Since the first lemonade stand in 2000, ALSF has raised over $80,000,000 (alexslemonade.org, 2014).

Celebrities have also made use of social networking to raise awareness of social issues they hold dear. Actor Mark Ruffalo's Twitter feed often includes comments warning his followers of the dangers of fracking and issuing a call to action toward a more green future (Viskontas & Mooney, 2014). John Cusack has been under fire for his Twitter campaign against war, which has been confused with anti-Semitism as he tweeted against the Israeli occupation of Gaza (Kanazi, 2014). Anne and Wil Wheaton have often posted openly about their struggles with anxiety and depression in order to help destigmatize mental health issues, most recently including a tweet by Anne Wheaton from a train while on their way to Comic Con San Diego "There are so many ways to get help for depression. Talk to friends, doctor, call hotlines, go to free clinics. Seek help. You deserve it" (twitter.com, 2014b), this in response to the fact that their train to San Diego was delayed because of an apparent suicide by jumping in front of a train in Ventura, CA (KCLU.org, 2014).

FUTURE RESEARCH

This chapter has demonstrated that the keys to a successful social bootstrapping campaign are:

- An established social network
- A clear message with a defined social strategy
- A story that can be personalized by the viewer
- Convenience in options for donation
- Tangible and intangible benefits to contributors (including bragging rights)

Future research on this subject should be conducted to compare the rates of speed to funding non-profit and arts projects to a variety of variables associated with the project, including (but not limited to) the social network of the sponsor, whether the sponsor is a registered nonprofit, success of precious social bootstrapping campaigns, and access to traditional channels of funding. This line of research will help determine whether computer-mediated crowdsourcing is an effective method for bootstrapping a campaign, or has become an online marketing tool for those who already have the needed access for traditional funding.

CONCLUSION AND RECOMMENDATIONS

This chapter has investigated the various methods nonprofits and artists have used to leverage computer-mediated social networking platforms to bootstrap their projects. This social bootstrapping has been an effective source of raising general awareness of social and political issues as well as garnering interest in arts projects. However, use of social media to bootstrap a project is no guarantee of success. In order to accomplish one's goals, the project sponsor must define a clear message that can be internalized by others. By capturing the attention of individual social networks, nonprofits and artists can launch a virtual call to action to expand the project's social network exponentially, following Reed's Law of expansion. By turning the social network into a group-forming network, nonprofits and artists can quickly raise both awareness and funding for issues they hold dear. While early social bootstrapping efforts have proven successful, it remains to be seen whether this platform will continue to level the playing field for entrepreneurs, independent artists, and Good Samaritans or will be overrun by the influential, wealthy few.

REFERENCES

Alexslemonadestand.org. (2014). Our mission & history. Retrieved from http://www.alexslemonade.org/about/our-mission-history

Army, S. (2013, December 4). Day one down & four remain for world record! Retrieved from http://blog.salvationarmyusa.org/2013/12/04/day-one-down-four-remain-for-world-record/

Barnett, C. (2013, May 8). Top 10 crowdfunding sites for fundraising. *Forbes*. Retrieved from http://www.forbes.com/sites/chancebarnett/2013/05 /08/top-10-crowdfunding-sites-for-fundraising/

BBC.com. (2014). I'd do anything archive. Retrieved from http://www.bbc.co.uk/oliver/archive.shtml

BBC.com. (2014). How do you solve a problem like Maria? Retrieved from http://www.bbc.co.uk/maria/about/index.shtml

Burton, L. (2014). Bring Reading Rainbow back for every child, everyone! Retrieved from https://www.kickstarter.com/projects/readingrainbow/bring-reading-rainbow-back-for-every-child-everywh

Chin, B., Jones, B., McNutt, M., & Pebler, L. (2014). Veronica Mars Kickstarter and crowd funding. In M. Stanfil & M. Condis (Eds.), *Fandom and/as Labor. Transformative Works and Cultures 15* [Special Issue]., doi:10.3983/twc.2014.0519

Cox, G. (2014, March 11). George Takei uses social media following to tout Broadway-bound tuner. *Variety*. Retrieved from http://variety.com/2014/legit/news/star-trek-veteran-george-takei-uses-his-social-media-following-to-tout-broadway-bound-tuner-1201129693/

Facebook.com. (2014). Wil Wheaton. Retrieved from https://www.facebook.com/itswilwheaton?fref=ts

Feldman, A. (2013, November 2). Indiegogo vs. RocketHub: Which crowdfunding site should you choose? Retrieved from http://www.crowdfundinsider.com/2013/11/25715-indiegogo-vs-rockethub-crowdfunding-site-choose/

Fundable.com. (2014). Crowdfunding statistics. Retrieved from http://www.fundable.com/crowdfunding101/crowdfunding-statistics

Grow, K. (2014, July 31). Rob Zombie talks insane clowns and crowdfunding his new film: 31. *Rolling Stone*. Retrieved from http://www.rollingstone.com/music/news/rob-zombie-talks-insane-clowns-and-crowdfunding-his-new-film-31-20140731

Hamblin, M. (2010, February 13). Text 'Haiti' earthquake relief effort raises $32M for Red Cross. *Computerworld*. Retrieved from http://www.computerworld.com/s/article/9156678/Text_Haiti_earthquake_relief_effort_raises_32M_for_Red_Cross

Hernandez, B. (2013). What the w00t?: Drew Curtis/Wil Wheaton/Greg Koch Stone Farking Wheaton w00tstout. Retrieved from http://blog.stonebrew.com/index.php/w00tstout/

Holmes, D. (2013, October 14). Infographic: Kickstarter vs. Indiegogo. *Pando Daily*. Retrieved from http://pandodaily.com/2013/10/14/infographic-kickstarter-vs-indiegogo/

IBDB.com. (2104). The Producers. Retrieved June 30, 2014: http://www.ibdb.com/show.php?id=10510

Indiegogo.com. (2014). "Join MILES AHEAD – A Don Cheadle film." Retrieved July 5, 2014: https://www.indiegogo.com/projects/join-miles-ahead-a-don-cheadle-film

Kanazi, R. (2014, July 31). "Tweet and delete: On Gaza, celebrity courage – and cowardice – over social media." *Salon*. Retrieved July 31, 2014: http://www.salon.com/2014/07/31/tweet_and_delete_on_gaza_celebrity_courage_and_cowardice_over_social_media/

KCLU.org. (2014, July 23). "Woman struck, killed by passenger train on South Coast." Retrieved July 23, 2014: http://www.kclu.org/2014/07/23/woman-struck-killed-by-passenger-train-on-south-coast/

Kennedy, M. (2014, February 24). Backers of 'Allegiance' embrace innovation. *Associated Press*. Retrieved from http://news.yahoo.com/backers-39-allegiance-39-embrace-innovation-121731702.html

Kerr, A., Whitson, J., Harvey, A., Shepherd, T., & O'Donnell, C. (2013). Strategies and tactics for promoting indie game design. *Selected Papers of Internet Research*, North America, 3. Retrieved from http://spir.aoir.org/index.php/spir/article/view/874

Keveney, B. (2012). 5 reasons 'American Idol' may go on nearly forever – or not. Retrieved from http://usatoday30.usatoday.com/life/television/news/story/2012-01-14/American-Idol-endurance/52625742/1

Khan, M. (2008, December 4). Red Cross attracts $190k in pledges via Text 2HELP program. *Mobile Marketer*. Available at: http://www.mobilemarketer.com/ cms/news/messaging/2226.html

Kickstarter.com. (2014). Wojciechowski s. Retrieved from https://www.kickstarter.com/help/stats

Knowledge, C. (2012). *4th Annual Nonprofit Social Network Benchmark Report*. Retrieved from http://www.nten.org/sites/default/files/2012_nonprofit_social_networking_benchmark_report_final.pdf

Kuppuswamy, V., & Bayus, B. (2014, January 29). Crowdfunding creative ideas: The dynamics of project backers in Kickstarter. UNC Kenan-Flagler Research Paper No. 2013-15. Retrieved from http://ssrn.com/abstract=2234765 or10.2139/ssrn.2234765

Lasica, J. (2012, February 15). The 7 elements of a strategic social media plan. Retrieved from http://www.socialbrite.org/2012/02/15/the-7-elements-of-a-social-media-strategic-plan/

Lybio.net. (2014). Reading Rainbow theme song lyrics. Retrieved from http://lybio.net/tag/reading-rainbow-theme-song-lyrics/

Malhorta, D. (2010). The desire to win: The effects of competitive arousal on motivation and behavior. *Organizational Behavior and Human Decision Processes*, *111*(2), 139–146. doi:10.1016/j.obhdp.2009.11.005

Malhorta, D. (2010, April). (When) are religious people nicer? Religious salience and the "Sunday Effect" on pro-social behavior. *Judgment and Decision Making*, *5*(2), 138–143.

Marin, A., & Wellman, B. (2011). Social network analysis: An introduction. In J. Scott & P. Carrington (Eds.), *The Sage Handbook of Social Network Analysis* (pp. 11–25). Thousand Oaks, CA: SAGE Publications, Inc.

Mestyán, M., Yasseri, T., & Kertész, J. (2013, August 21). Early prediction of movie box office success based on Wikipedia activity big data. *PLoS ONE*, *8*(8), e71226. doi:10.1371/journal.pone.0071226 PMID:23990938

MyLifetime.com. (2014). Vote for your favorite designer! Retrieved from http://www.mylifetime.com/shows/project-runway/season-13/fan-favorite

National Endowment for the Arts. (2014). National Endowment for the Arts Appropriations. Retrieved from http://arts.gov/open-government/national-endowment-arts-appropriations-history

Office of Personnel and Management. (2014). Combined Federal Campaign. Retrieved http://www.opm.gov/combined-federal-campaign/

Reed, D. (2001, February). The law of the pack. *Harvard Business Review*. Retrieved from http://hbr.org/2001/02/the-law-of-the-pack/ar/1

Renninger, B. J. (2013, September 19). Why Getting on Upworthy is the New Goal for Kickstarter Campaigns. Retrieved from http://www.indiewire.com/article/why-getting-on-upworthy-is-the-new-goal-for-all-kickstarter-campaigns

RMHC.org. (2014). Tax returns & audited financials. Retrieved from http://www.rmhc.org/tax-returns-and-audited-financials

Savage, K. (2014, March 2). Fans pledge their allegiance to Allegiance (a new musical). Retrieved from http://blog-ken-savage.com/facebook/fans-pledge-their-allegiance-to-allegiance-a-new-musical/

Sharma, R. (2014, March 13). How nonprofits use social media to engage with their communities. *Nonprofit Quarterly*. Retrieved from https://nonprofitquarterly.org/management/23837-how-nonprofits-use-social-media-to-engage-with-their-communities.html

Sørensen, I. (2012, September). Crowdsourcing and outsourcing: The impact of online funding and distribution on the documentary film industry. *Media Culture & Society*, *34*(6), 726–743. doi:10.1177/0163443712449499

Statista.com. (2014). Most active crowdfunding categories of 2012. Retrieved from http://www.statista.com/statistics/269975/most-active-crowdfunding-categories-of-2012/

SyFy.com. (2104). The Wil Wheaton Project. Retrieved from http://www.syfy.com/wilwheaton/about

Takei, G. (2013, March 28). *My Broadway debut at age 75.* TEDxBroadway Presentation delivered to TEDxBroadway. Retrieved from http://blog.ted.com/2013/03/28/george-takei-on-star-trek-musicals-and-japanese-american-internment/

Thomas, R. (2103). The Veronica Mars movie project. Retrieved from https://www.kickstarter.com/projects/559914737/the-veronica-mars-movie-project

Twitter.com. (2014). @midnight. Retrieved from https://twitter.com/midnight

Twitter.com. (2014). Anne Wheaton. Retrieved from https://twitter.com/AnneWheaton/status/492020385487917056

Twitter.com. (2014). Wil Wheaton. Retrieved from https://twitter.com/wilw

Twitter.com. (2014). The Wil Wheaton Project. Retrieved from https://twitter.com/wilwproject

UNICEF.org. (2012, September 12). Milestones: Year by year through UNICEF's history. Retrieved from http://www.unicef.org/about/history/index_milestones.html

Viskontas, I., & Mooney, C. (2014, January 3). Mark Ruffalo wants you to imagine a 100 percent clean energy future. *Mother Jones.* Retrieved from http://www.motherjones.com/environment/2014/01/inquiring-minds-mark-ruffalo-fracking-solar

w00stock.net. (2014). w00tstock: A night of geeks + music. Retrieved from http://w00tstock.net/

Wojciechowski, A. (2009). Models of charity donations and project funding in social networks. In R. Meersman, P. Herrero, & T. Dillon (Eds.), *On the Move to Meaningful Internet Systems: Proceedings of OTM 2009 Workshops* (pp. 454 – 463). doi:10.1007/978-3-642-05290-3_58

ADDITIONAL READING

Indiegogo. (2014). Retrieved from http://www.indiegogo.com

Kickstarter. (2014). Retrieved from http://www.kickstarter.com

RocketHub. (2014). Retrieved from http://www.rockethub.com

Scott, J., & Carrington, P. (Eds.). (2011). *The Sage Handbook of Social Network Analysis.* Thousand Oaks, CA: SAGE Publications, Inc.

KEY TERMS AND DEFINITIONS

Crowdfunding: A subset of crowdsourcing in which community contributors are asked to provide funding for projects executed by the sponsoring organization.

Crowdsourcing: Method of using economies of scale to develop large communities of interest to execute a project or secure funding to achieve a specific, temporary goal of either collective action or crowdfunding.

Group-Forming Networks: Social networking nodes that expand individual social networks by linking groups with common interests.

Hashtag: Metadata associated with social media posts to allow for easy tracking and searching for common themes, subjects and keywords to effect data analysis.

Microfunding: Another term for crowdfunding, specifically where contributors are allowed to provide funding in small increments, reducing the barrier to entry for participants.

Reed's Law of Expansion: The utility of large networks (particularly social networks) increases exponentially with the size of the network.

Social Bootstrapping: Using social networks (specifically, computer-mediated networks) to start new projects or business ventures with limited extant infrastructure.

Chapter 2
Social Media:
Changing the Way We Teach and Changing the Way We Learn

Arleen Cuevas
NHTV Breda University of Applied Sciences, The Netherlands

Fritz Kohle
NHTV Breda University of Applied Sciences, The Netherlands

ABSTRACT

Social media has become ubiquitous in the daily personal lives of students and teachers alike. But the question remains if social media should be integrated effectively in higher education or if it should be left out in the realm outside the classroom. This paper explores how students use social media in school, whether or not they find social media useful in the learning process and provides further discussion on the importance of adopting a social media strategy in the education sector. The authors facilitated a study in 2010 and a follow-up survey in 2011 to students taking the course International Media and Entertainment Management at NHTV Breda University of Applied Sciences in the Netherlands to assess how students are using social media in a learning environment. The study reveals a need for social media to be used as a learning tool in order to promote active participation through content creation and encourage a virtual space for dynamic dialogue which in turn helps link formal and informal learning connecting students, teachers, and colleagues around the world. But more complex issues such as privacy, copyright, policing and governance of social media needs to be addressed.

1 INTRODUCTION

This study explores how students use social media at school, whether or not they find social media useful and discusses the need in creating an appropriate response to social media for the educational sector. Over a short period of the time, social media has become ubiquitous and the question remains if educators should adapt or ignore its use in the sphere of the learning environment. According to Zhang

DOI: 10.4018/978-1-4666-8556-7.ch002

and Yang (Yang, X. & Zhang S. 2010) use of social media is exponentially growing and there is a real need for teachers and students to understand and engage with social media in the educational sector.

For the purpose of this study we define social media as "a group of Internet-based applications that builds on the ideological and technological foundations of Web 2.0, which allows the creation and exchange of user generated content". (Kaplan, A. & Haenlein, M. 2010) The most popular types of social media include social networks such as Facebook and Myspace, file sharing sites such as Youtube, Vimeo and Flicker, blogging sites such as Blogger and Tumblr and also microblogging sites such as Twitter.

In our previous paper (Cuevas, A. & Kohle, F. 2010), we found out that majority of students are already well-adept with the use of social media sites. Almost all respondents had an active account in at least one type of social media, such as Facebook, Youtube or Twitter and they log into their accounts at least one time a week. Students in general had a positive experience with the integration of these social media in producing the multimedia projects in the course, although there were still students who would rather use email instead of making face to face contact with regards to asking questions or being more critical to peer assessments.

According to Siemens and Weller (Siemens, G. & Weller, M. 2011), the main benefits of social media for students is that it encourages interactive participation through peer to peer dialogue, promotes sharing of resources through content creation and provides a virtual space for collaboration and instant communication. Social media also enables students to build up "social capital" wherein resources coming from relationships and closed communities are enriched (Ellison, N. B., Steinfield, C., & Lampe, C. 2007).

But with these benefits, there are also challenges that come along with it. There is a need for training for faculty to use it and provide the same set of standard criteria across all courses. Instances such as students knowing more about social media and how it works should be avoided.

Many schools also do not have or are in the process of developing and implementing appropriate guidelines with regards to social media applications. The lack of guidelines in many universities often results in a fragmented faculty body – some engaging in social media and others ignoring it. Situations such as students uploading copyrighted university material or photos and videos of teachers behaving badly can be avoided with application of rules for the use of social media. The lack of these regulations can also potentially result in negative publicity for the educational institution as well as the student.

There are a variety of topics to be discussed such as: how to structure and deconstruct learning spaces, applying content control between the educator and the student, blending of informal and formal learning and how to address the issue of privacy and copyright in social media (Siemens, G. & Weller, M. 2011).

2 RELATED LITERATURE

The social constructivism theory is based on how socialization and interaction with other people can help students learn and construct their own knowledge and personal learning processes (Lin, Q. 2011). Social media can then be seen as a set of constructivist tools that can facilitate participative learning models that is being introduced in higher education today. (Espuny, C., Gonzalez, J., Lleixa M. & Gisbert, M. 2011)

There has been a lot of discussion about the perceived advantages and disadvantages of using social media in the education sector. According to Weller and Dalziel (Weller, M.J. & Dalziel, J. 2007), social media clashes with the education system because the education sector depends on the hierarchichal struc-

ture of professors, lecturers and students and is very dependendent on the quality assurance of content. This means there is a top-down approach in formal assessment and reviews of the work of students. The traditional education structure also emphasizes on the role of the "expert educator" as the right person who has the right amount of knowledge and skills to teach a certain course or class.

At the 2010 Ed-Media Conference, Selwyn (Selwyn, N. 2010) states that advocates of the traditional education structure have reasoned that teaching a set curriculum, enforcing attendance and being told what learn and providing fixed assessments on a fixed term offers a tried and tested measure of how students learn compared to the perceived potential of social media's use for learning. The one-way traditional informationa flow is still predominant in today's educational sector.

Social Media does not aim to replace traditional learning, but can supplement the landscape of traditional learning by adding more tools at its disposal. The type of learning and the assessment used in traditional education needs to evolve in order to adapt to an increasingly networked world. Instead of just using a whiteboard and a powerpoint presentation for a lecture, interactive materials can come from online videos, photos, blogs, podcasts or even skype sessions that can be found online (Bishop, M. & Yocom, J. 2011). Interactive dialogue can also come from discussion boards, online forums, virtual meetings, and recorded text messaging transcripts on social media sites. Students can upload their work and projects online which allows transparency wherein other students can see, review and learn from the work of others. This also allows educators easier access in assessing the material online and enables them to give feedback in an instant. The use of social media can also increase interactive discussions and allow collaborations with students in different universities or countries without the limitations of physical space. Because of these changes, the types of learning assesments also need to adapt to the type of learning tools that a course utilizes.

Social media also transforms the role of the educator from being a content provider to a content interpreter or a skills developer (Weller, M.J. & Dalziel, J. 2007). The learning environment will become more fluid, with students being able to make feedback and exchanges with regards to study tools. Educators in return can provide guidance in order for students to find, interpret and use the right content and information. Even if there are students who might have technical skills and know-how, there is an enormous amount of information out there and students need guidance into thinking more critically by finding the right resources and finding expert information from all the rubbish that is available in the web.

Social media's use in different applications rapidly expanded because making and uploading content became simplier, making digital publishing more accessible to non-techies. There is no need anymore to learn extensive html codes or protocols and no need to learn complicated web design softwares (Alexander, B. 2008).

Future discussions need to address issues arising from the loss of privacy, copyright issues, and policing and governance of the open web. The very nature of social media also remains very open and volatile, where it is always in "beta mode" where revisions and additions regularly take place which makes it hard to have a fixed structure or assesesment. (Hemmi, A., Bayne, S. & Land, R. 2009)

For all the talk about the openness of social media to be used in the classroom, the reality shows that it is influenced by the commercial sector as well. Social Media sites such as Facebook or Myspace is driven by ad revenues and possibly sharing personal information of its users to third party sites and companies. Some universities have experimented into making their own personal social media sites for education purposes. (Lea, B. L., Clayton, M., Draude, B., & Barlow, S. 2001 & Silius K., Miilumäki,

T., Huhtamäki J., Tebest T., Meriläinen J. & Pohjolainen, S. 2010) But this post several challenges in terms of costs in building a new platform and also the investment of a time and motivation of a student or teacher to learn or instruct using this new application.

The discussion about the importance of social media is based on the "ethical question of what counts as worthwhile learning and worthwhile education". (Standish, P. 2008)

3 BACKGROUND OF THE STUDY

The case study examines the use of social media at NHTV, University of Applied Sciences, Breda, Netherlands. 141 students participated in a survey as part of the 55 hour multimedia video project. For the purpose of this paper, a multimedia video project is defined as an audio-visual project that students created using a video camera and video-editing software in a computer. This is then combined with creating a film poster that corresponds to the theme of the said audio-visual project. The 55 hour video project makes extensive use of social media. 30 student group video projects are screened at the end of the project, posters are created using photoshop and uploaded in youtube. Afterwards students create a Facebook account in order to vote on each project, delivery of the final project takes place on Youtube.

In our last paper (Cuevas, A. & Kohle, F. 2010) we reported that this project remains very popular among our students as it increases curricular transparency by recording accurately when students upload their project. It also enables them to provide instant feedback and allowed them to share their learning experience with friends and family within and outside the classroom setting.

In this study we wanted to examine the students social media behaviour before and after joining NHTV and look into more detail how students perceive the use of social media as part of their education. The 55 hour project takes place in three batches within a two month period and each time we asked the students to participate in our survey.

According to Bishop and Yocom (Bishop, M. & Yocom, J. 2011), integration of social media to students doing multimedia projects such as audio-visual projects can enrich creative and collaborative social interactions. It can also make it simplier for educators and students to upload their work to be screened, critiqued, assessed and appreciated by other students or viewers within and outside the classroom environment.

3.1 Goals and Limitations of the Study

The goal of this study is to further understand how students are using social media in education and if they find it beneficial to utilize in a learning environment. The other goal is to possibly contribute to the development of best practices and adopt comprehensive guidelines regarding the application of social media in the educational sector.

The limitations of this study is the representation of sample size of students from NHTV Breda University of the Netherlands which represents conditions in European universities wherein there is a wide and free access of Internet use compared to other countries.

4 RESEARCH APPROACH TAKEN

The survey consisted of a set of 16 questions in total. There were 12 multiple choice questions and four free form questions which allowed the participants the opportunity to give individual answers. The participants (n=141) are students enrolled in an International Media and Entertainment Management course. The survey was conducted during their participation for the 55-hour multimedia video project. The sample was representative of the selected multicultural university located in North Brabant in the Netherlands. All 141 questionnaires were filled in properly, none were disqualified. Qualitative data was collected from the survey and the outcome was discussed in relation to similar studies providing relevant data and conclusions. The primary statistical software tools used were Microsoft Excel and SPSS.

5 FINDINGS AND DISCUSSION

5.1 Survey

Multiple-choice questions were given to three batches of our 55 hour project students totalling 141 students. Among all gender and age groups Facebook is the most popular platform followed by Youtube. We noticed that use of social media increased significantly for most students after enrolment and did not find any significant gender related differences.

When it came to communicating good or bad news the use face to face contact was still the preferred option with social media being the least favourite option.

Both male and female are equally concerned about privacy issues when using social media followed by freedom of speech. The difficulty and complex task to regulate postings, i.e. who can see what in on FB, YT or blog and the potential of social disruption on campus because of inappropriate content online are of concern to male and female students.

5.2 Discussion

Social media has become ubiquitous in the educational sector. Employees and students alike increasingly use social media to communicate and disseminate information in and outside the class room.

The surveys results from 2010 and 2011 show that students are increasingly familiar with the use of social media before and after enrolment at the University. Use of social media increases significantly after enrolment which is not surprising considering the new social bonds are created in the educational sector, either peer to peer or with faculty members. We also find that there are subtle differences regarding a student's willingness to engage social media: Students below the age of 20 are more likely to have multiple accounts in different types of social media and is more regular in their weekly usage, while older students tend to be a little more reluctant to engage on multiple levels. There is a noticeable increase in the use of Facebook groups to facilitate instant feedback and communication among peers and to share information about courses.

Some students limit their use of social media because of privacy matters. There was also a concern of separating their personal life with school life and some students do not want to combine both and share information. And because the social web is free and unregulated, there is always the possibility that inappropriate information disseminated via social media can cause social disruption on campus, in terms of putting false or wrong information as anybody can post or upload materials online.

During this research, students had the tendency to manage group projects virtually via email or social media and avoiding face to face communication. Davis and Zigur (Davis, A. and Zigurs, I. 2008) reports that the inappropriate application of virtual teams resulted in less integrated products, lack of ownership, lack of team bonding, polarisation of some team members and trust problems. On the other hand, a study conducted by Fortune, Spielman and Pangelin (Fortune M., Spielman M. & Pangelinan D., 2011) found no significant difference in the learning process between an online group and a face to face group during a study for a hospitality and recreation course. Both groups used social media such as facebook and twitter but stated that their online interaction was not greater that their face to face interaction thus supplementing the needed communication for the course.

This study examined how formal and informal learning can be connected using participatory media. Increasingly students and teachers alike are sharing multimedia material online in the class room as our 55hour multimedia project demonstrates. The 55hour project demonstrates the full potential social media can have regarding communication and peer feedback in a classroom setting. The students in our survey are skilled users of mobile technology, always ready to send and receive messages or post videos online. The challenge for media courses is not only to turn student users into skilled content creators but also to keep up to date with new developments. Social media tends to be dealt more as a symptom rather than a tool capable of synchronising cultural, political and sociological paradigm shifts such as the Arab Spring and the Occupy Wall street protests. Using social media in the class room encourages a dynamic dialogue, helps link formal and informal learning, which in turn connects students, teachers, colleagues, family and friends around the globe.

Web 2.0 technologies and social media are increasingly impacting the way teachers and students communicate with each other. Some educators are enthusiastically communicating with students via social media, while others find engaging in social media a difficult challenge – a situation any older traditional university educators would probably recognise. Bull et. al. (Bull, G., Thompson, A., Searson, M., Garofalo, J., Park, J., Young, C., & Lee, J, 2008) defines these set of faculty members as "digital immigrants" to contrast it with the younger net generation of students called "digital natives". Digital natives were born using this technology and find it easy and intuitive to use, while some of the older generation are still struggling to adapt to these new online tools.

"Digital immigrants" such as older faculty members need to engage with social media in a proactive and more structured way. A social media policy on its own is not sufficient to deal with the challenges ahead as there needs to be more training and resources allocated to be better integrated into the curriculum. A need for social media guidelines and "ethos" still needs to be developed and implemented.

6 CONCLUSION

The increasing number of research regarding social media and its use in different areas, most especially in education, gives proof that it can lead to a significant change in how we structure learning spheres in the future. It has the potential to change the traditional relationship and between teacher and student, thus giving more control and guidance to motivated students which can results in a more satisfactory learning experience. Dabbagh and Kitsantas (Dabbagh,N. & Kitsantas, A. 2011) propose that the use of social media can support self-regulated learning in the development of Personal Learning Environments (PLE) to integrate formal and informal learning. But on the other hand, not all students are capable of self-direction and knowledge management, so there is a need to put some kind of structure to the unstructured realm of social media if it is going to be utilized for education purposes.

A growing number of educators are using the advantages social media offers in the classroom to engage into a more dynamic dialogue with students and other faculty members. Implementing clear social media guidelines is not sufficient to prepare students and faculty for the challenges inherent with social media. Establishing classes instructing students and faculty alike in the appropriate use of social media offers a low-cost marketing platform making the school and its students more transparent to all stakeholders and reduces the risk of inappropriate content being released by faculty or students online. Attempts to ignore social media are bound to end in failure or worse in a publicity disaster if inappropriate content attracts hundreds or perhaps even millions of viewers on a site such as Youtube. This potential exists in an unregulated web, thus further strengthening the case for a formal introduction for students and faculty to social media early on in higher education.

There is a need to develop best practices, and compare what other universities have done for technical support and framework. There is also a need for further studies to different branching topics of social media. Foroughi (Dabbagh,N. & Kitsantas, A. 2011), proposes a more rigorous in depth research, detailing the macro, mezzo and micro levels of how social media can be developed and implemented by students, faculty and other stakeholders to fully understand its potential and enable to make best practices in the field of higher education.

There are still a lot of questions that needs to be answered: Educators need to evolve in order to keep up with the fast paced technologies afforded by the Internet. Ultimately, the use of social media makes us rethink of the whole educational process in order to enable students adapt to a future where everything rapidly evolves.

Conflict of interests: the authors of this paper declare that their research has been entirely self-funded.

REFERENCES

Alexander, B. (2008). Social Media and Higher Education – Scholarship, The Wave of the Future in the Digital Age. In R. Katz (Ed.), *From The Tower and the Cloud: Higher Education in the Age of Cloud Computing*. Educause 2008. Retrieved from http://www.educause.edu/thetowerandthecloud

Bishop, M., & Yocom, J. (2011). Education, Multimedia and Social Media: Are We Ready for Showtime? In *27th Annual Conference on Distance Teaching and Learning*. Retrieved from http:www/uwex.edu/disted/conference

Bull, G., Thompson, A., Searson, M., Garofalo, J., Park, J., Young, C., & Lee, J. (2008). Connecting Informal and Formal Learning: Experiences in the Age of Participatory Media. *Contemporary Issues in Technology & Teacher Education*, *8*(2). Retrieved from http://www.citejournal.org

Cuevas, A., & Kohle, F. (2010). A Case study in using youtube and factbook as social media Tools in enhancing student centred learning and engagement. In *ICERI2010 Proceedings* (pp. 3596-3601). ICERI.

Dabbagh, N., & Kitsantas, A. (2011). Personal Learning Environments, Social Media and Self Regulated Learning: A Natural Formula for Connecting Formal and Informal Learning. *The Internet and Higher Education, 15*(1), 3–8. doi:10.1016/j.iheduc.2011.06.002

Davis, A., & Zigurs, I. (2008). Teaching and Learning about Virtual Collaboration: What We Know and Need to Know. In *AMCIS 2008 Proceedings*. Paper 168. Retrieved from http://aisel.aisnet.org

Ellison, N. B., Steinfield, C., & Lampe, C. (2007). The benefits of Facebook "friends": Social capital and college students' use of online social network sites. *Journal of Computer-Mediated Communication, 12*(4), article 1. Retrieved from http://jcmc.indiana.edu

Espuny, C., Gonzalez, J., Lleixa, M., & Gisbert, M. (2011). University Students' Attitudes Towards and Expectations of the Educational Use of Social Networks. *Revista de Universidad y Sociedad del Conocimiento, 8*(1), 186-199. Retrieved from http://rusc.uoc.edu/

Foroughi, A. (2011). A Research Framework for Evaluating the Effectiveness of Implementations of Social Media in Higher Education. *Online Journal for Workforce Education and Development, 5*(1). Retrieved from http://opensiuc.lib.siu.edu

Fortune, M., Spielman, M., & Pangelinan, D. (2011). Student's Perception of Online and Face to Face Learning and Social Media in Hospitality, Recreation and Tourism. *Merlot Journal of Online Learning and Teaching, 7*(1). Retrieved from http://jolt.merlot.org

Hemmi, A., Bayne, S. & Land, R. (2009). The Appropriation and Repurposing of Social Technologies in Higher Education. *Journal of Computer Assisted Learning, 2*, 19–30. doi: 10.1111/j.1365-2729.2008.00306.x

Kaplan, A., & Haenlein, M. (2010). Users of the World unite! The challenges and opportunities of social media. *Business Horizons, 53*(1), 59–68. doi:10.1016/j.bushor.2009.09.003

Lea, B. L., Clayton, M., Draude, B., & Barlow, S. (2001). The Impact of Technology on Teaching and Learning Faculty Attitudes. *EDUCAUSE Quarterly, 2*(2), 69–71. Retrieved from http://net.educause.edu/

Lin, Q. (2011). The Role of Web-Based Activities in Mediating Student Interaction and Engagement in Four Teacher Education Classes. *Merlot Journal of Online Teaching and Learning, 7*(1). Retrieved from http://jolt.merlot.org

Selwyn, N. (2010). *The Educational Significance of Social Media – A Critical Perspective*. Keynote debate at Ed-Media Conference 2010, Toronto, 28th June to 2nd July. Toronto: Selwyn. Retrieved from http://www.scribd.com

Siemens, G., & Weller, M. (Eds.). (2011). The Impact of Social Networks on Teaching and Learning. *Revisita de Universidad y Sociedad del Conocimiento, 8*(1), 164-170. Retrieved from http://oro.open.ac.uk

Silius, K., Miilumäki, T., Huhtamäki, J., Tebest, T., Meriläinen, J., & Pohjolainen, S. (2010). Social Media Enhanced Learning in Higher Education. *IEEE Educon Education Engineering 2010 – The Future of Global Learning Engineering Education.* Retrieved from http://www.doaj.org

Standish, P. (2008). Preface. *Journal of Philosophy of Education, 42*(3-4), 3–4, 349–353. doi:10.1111/j.1467-9752.2008.00659.x

Weller, M. J., & Dalziel, J. (2007). On-line Teaching: Suggestions for Instructors. In L. Cameron & J. Dalziel (Eds.), *Proceedings of the 2nd International LAMS Conference 2007: Practical Benefits of Learning Design* (pp. 76-82). Sydney: LAMS Foundation. Retrieved from http://lamsfoundation.org

Yang, X., & Zhang, S. (2010). Use, Challenges and Potential of Social Media in Higher Education: Evidence from a Case Study. In C. Wankel (Ed.), *Cutting-edge Social Media Approaches to Business Education: Teaching with LinkedIn, Facebook, Twitter, Second Life, Blogs, Information Age* (pp. 217–240). Charlotte, NC: Publishing.

Chapter 3
A Study of Two Microfinance Models and Their Suitability for Egypt

Ahmed Youssry
Coleman University, USA

Brett Winklehake
San Diego Mesa College, USA

Jaime A. Lobera
Coleman University, USA

ABSTRACT

Developing countries around the world strive to implement one of the several current models of micro-finance. This study focuses on two models: Grameen Bank, which is considered the change factor for the microfinance field, and Kiva.org, an organization that understood the importance of the Internet and crowdfunding to create a different model of microfinance. The purpose of the study is to analyze these two models and determine which would be more suitable for application in Egypt. This study provides a strengths, weaknesses, opportunities, and threats (SWOT analysis), a financial analysis, and a structural analysis, as well as historical background for both organizations along with a scan for the political, economic, social, and technological infrastructure in Egypt to determine the most suitable microfinance model.

INTRODUCTION

Currently, poverty and unemployment are the two main obstacles facing developing countries. According to Giugale and Mubarak (1996), more than 23% of Egyptians live below the poverty line, and unemployment is almost 30%. The country is trying to get through the bottleneck of the political situation that has existed since the 2011 revolution, and efforts are ongoing to solve the problem of employment.

DOI: 10.4018/978-1-4666-8556-7.ch003

One of the most effective solutions other countries have undertaken to deal with employment issues is implementing microfinance.

Microfinance institutions around the world and specifically in developing countries have helped thousands of people to cross the poverty line. These institutions help the poor through different financial services including microloans, microinsurance, and microsaving. Given the current economic conditions and the number of microenterprises that exist in Egypt that might need support, the country should view microfinance as a major economic tool. Many African countries and others in similar situations have tried different microfinance solutions.

Statement of the Problem

Poverty and unemployment in Egypt are among the main issues facing the country after the revolution in 2011 (Giugale & Mubarak, 1996). Microfinance institutions are one of the solutions for these issues, yet in Egypt, microfinance institutions are still underdeveloped and have dependency on external fund sources which has led to a very slight impact on the society. Small projects that help the poor and unemployed segment of the society usually require limited amount of fund that can be supplied through microfinance institutions.

Purpose of the Study and Thesis Questions

Microfinance has become a solution in several countries in the help of elevating poverty and decrease unemployment. The purpose of this study is to analyze two microfinance models that were implemented successfully in other locations. Study the current environment in Egypt based on four aspects; political, economic, social and technological, and recommend one of these two models to be implemented in Egypt.

RQ1: "How did Grameen Bank achieve success in Bangladesh?"
RQ2: "How did Kiva.org manage to use online microfinance and crowdfunding concepts so successfully?"
RQ3: "Which microfinance model out of Grameen Bank and Kiva would be more successful in Egypt?"

Significance of the Study

The results of this research would help social entrepreneurs in Egypt identifying the most appropriate model of microfinance between the two models studied to be implemented in Egypt. The study would also help researchers and students that are welling to investigate more in microfinance and crowdfunding. Arab spring revolutions have given hope to all young generations that they could make their countries a better place and they can elevate poverty and fight unemployment. This study could be a base for starting these battles with a national project.

Methodology

The method followed in this paper to answer the research questions is based on the findings from scholarly references, dissertations, academic research and articles. The conceptual framework of the study is explained on the first section of the paper in order to engage the audience on the main concepts behind the two models of microfinance selected for comparison. The ground work in this paper includes a study of Egypt's context and current practices in microfinance and non-profit organizations. Furthermore the ground work includes the study of Grameen Bank and Kiva.org. The research questions are answered based on the research done in the ground work section.

CONCEPTUAL FRAMEWORK

The conceptual framework in this paper includes an overview of the concepts related to thesis topic and research questions. The first part in this section is highlighting the concepts related to microfinance. These concepts include the explanation of microfinance and the different sub concepts developed through it; microcredit, micro insurance, and micro saving. Moreover the history of microfinance, the profitability of microfinance organization, and the key principles of microfinance are presented.

The second part of the conceptual framework is a focus on crowdfunding and its related concepts. A brief about the history of crowdunding and the misconceptions that several people get deceived by is also discussed in this section. An elaboration of the mechanisms of crowdfunding is presented in this section as well. Lastly, the pros and cons of crowdfunding based on the research conducted will be the last concept in this section.

Microfinance

Microfinance is considered to be one of the most efficient approaches in providing different financial services to the poor (Akram & Hussain, 2011). These services include microcredit, microinsurance, microsavings, transactional services, and financial consultations. Many microfinance institutions limit their role to microcredit, assuming that it is the only benefit they can provide to micro entrepreneurs (Akram & Hussain, 2011). The microfinance industry includes a wide range of organizations that includes nonprofit organizations, credit unions, cooperative societies, some commercial banks, and some non-banking financial institutions (Akram & Hussain, 2011). Microfinance is the a smaller scale concept of banking as it includes most of the services that banks offer to people, yet on a much smaller scale for people that usually doesn't possess collaterals. The main service of microfinance is the micro loans offered without collateral to cover it.

In certain developing countries, such as Egypt, governments have tried to use microfinance as an alternative to solve infrastructure problems such as poverty, lack of education, unemployment, and victims of natural disasters (Marinescu, 2009). The governments in these countries consider microfinance to be a new solution, yet it usually fails because the main objective in the above-mentioned cases is giving credit to people who are incapable financially, which usually requires a 98% performance rate (Marinescu, 2009). In order to achieve this rate, the government needs to ensure that each transaction performed is monitored, followed up on, and checked case-by-case. The Monitoring of case by case would cost the government more money than the low interest that they would usually announce they can offer. In other words to monitor the performance rate it would cost more money than the return from the interest (Marinescu, 2009).

Microcredit

The Microcredit Summit Campaign (2005) described microcredit as tool or program that provides poor, unemployed people who are looking to improve their financial status the opportunity of starting small projects or becoming micro entrepreneurs. This happens by facilitating microloans without the proven financial records or credit history that most regular financial institutions require. These loans are usually much smaller than the regular loans given to businesses and to people that have collateral to cover the loan they are asking for. Microcredit is the main service that the microfinance institutions offer to assist the poor segment of the society, as it's the concept of loaning money for starting micro projects.

Microinsurance

This is the protection of low-income people from possible risks and hazards in exchange for a specific regular payment amount that suits the type of risk and likelihood of occurrence (Churchill, 2006). The difference between regular insurance and microinsurance services is that the latter are aimed at the poor (Churchill, 2006). A larger difference is that with regular insurance a person usually has a stable or predictable income, whereas microinsurance is targeted to the low-income population that does not have a stable income (Churchill, 2006).

Microsaving

This can be defined as the usual concept of saving, but on a micro scale, meaning that the same concepts that apply to household savings apply to microsaving (Browning & Lusardi, 1996). Saving is keeping an amount of income in order to use it at another time (Browning & Lusardi, 1996). The reasons for saving could be building a reservoir against unexpected events, anticipating a future relationship between needs and withholding, increasing financial safety with interest and appreciation, enjoying slowly increasing purchasing power, increasing the ability to take a risk on a project, and accumulating enough income to satisfy other needs. The difference between regular saving and microsaving is the amount and predictability of income that might be saved (Browning & Lusardi, 1996).

Profitability of Microfinance Institutions

In 2005, a microbanking report showed that microfinance institutions always consider their priority to be profitability, which sometimes makes the decision-making process more biased toward turning a profit than helping the poor (Marinescu, 2009). The initial efforts of the first 63 microfinance organizations around the world have resulted in profitability after regulation of inflation rates and help received from governments or international organizations (Marinescu, 2009). This profit can be up to 2.5% of the organization's total income (Marinescu, 2009).

The main sources for financial revenue for microfinance institutions are; interest charges, and fees and commissions (Guntz, 2011). Other profits that the micro finance institutions make are generated form; the net gain on financial assets, and the revenue from insurance of micro projects (Guntz, 2011). The fact that microfinance institutions are profitable or in other words, generate revenues doesn't hide the important role they play in helping small and micro projects.

History of Microfinance

Although not referred to as such, the concept of microfinance has existed since early civilization, due to differences in wealth between societal classes (Helms, 2006). In early times, microfinance took the form of informal credit practices and savings between individuals and groups. In the middle Ages, specifically in 1462, an Italian monk made the first formal attempt at microfinance in the form of a pawnshop that offered moneylending (Helms, 2006). In the 16th century, Pope Leon X allowed pawn shops to request a certain amount of money in return for their services to cover expenses (Helms, 2006). In the 18th century, the writer Jonathan Swift founded the Irish Loan Fund System to give microloans to farmers who were below the poverty line and had no guarantees or assets. The system was successful to the extent that at some point this type of lending was almost 20% of the household income in Ireland (Helms, 2006).

Friedrich Raiffeisen and his supporters from Germany developed the concept of a financial cooperative in the 19th century, which expanded rapidly throughout the developed world. In the 20th century, specifically between 1950 and 1970, a great effort was made to increase access to agricultural loans and credit using farmers' cooperatives at lower interest rates than the market (Helms, 2006). Those efforts failed due to the high expense of operations that exceeded the rates. It was not until the early 1970s when alternative-banking efforts became part of experimental programs to give small loans to poor women to start modest businesses that microcredit succeeded. Grameen Bank in Bangladesh, ACCION International in Latin America, and the Self-Employed Women's Association Bank in India were the pioneers of the microcredit movement (Helms, 2006).

In the 1980s, microcredit developed through institutions. The concept of cost-recovery interest rates and high repayment was introduced by Bank Rakayat in Indonesia in order to help micro lenders achieve long-term sustainability and increase the number of their clients (History, 2013). Since the 1990s, when the term "microfinance" began to replace the term "microcredit," most of the microfinance institutions tried to adopt a commercialization strategy that would turn them into profitable organizations to appeal more to clients. As the World Bank became aware of microfinance as a solution for poverty, they initiated the Consultative Group to Assist the Poor (CGAP) in 1995 (History, 2013). In the early 2000s with the rise of Internet technology, online transactions became an essential component of microfinance. Some organizations started to apply the model of online crowdfunding to microfinance to help poor people, and Kiva.org became prominent (History, 2013).

Consultative Group to Assist the Poor

The main idea of CGAP was to help any person access a financial service needed to improve his or her situation. CGAP created groundbreaking solutions for all the possible financial problems faced in poor communities through intensive and continual research, the group's active involvement with service providers on all levels, and funders and policy makers. CGAP was founded in 1995 by the World Bank as a response to the rising microfinance movement of the 1970s and 1980s (About CGAP, 2013). Currently, the organization acts as a reference source providing solutions and research for old and new microfinance institutions. CGAP's research database is considered one of the largest and most reliable sources for microfinance information (About CGAP, 2013).

Key Principles of Microfinance

In their 30th summit, June 10, 2004, G8, the group of the largest eight economies worldwide, recognized the effort done by CGAP to develop microfinance and came up with a statement by the end of the summit to formalize the key principles of microfinance. One of the principles was that people under the poverty line not only need loans in terms of financial services but also other financial tools such as savings, money transfers, and insurance. Microfinance is a very effective tool in the fight against poverty as poor households use microfinance services to support their families. Microfinance is about building a financial system that does not just offer a temporary solution and the concept should be included in countries' financial planning (G8 Summit Final Report, 2004).

The statement included other principles such as that microfinance institutions must support their operations through adequate fees that are not a burden on the poor. Microcredit is always a better solution in the guarantee of the looping of funds. The job of government is to facilitate financial services but not directly provide the services. Donor funds are a method of enhancing a microfinance institution but should not compete with it. Microfinance institutions should focus on management as means of success and finally, institutions that offer microfinance should always be able to measure their own performances in order to advance (G8 Summit Final Report, 2004).

Crowdfunding

Crowdfunding is one of the most effective methods of fund raising. Some experts prefer to say that it is a collective effort by an individual or a group of people to gather the support and the capital required to implement, improve, and enhance an idea or project through explaining and presenting a project to the public, which is usually done online (Prive, 2012). Crowdfunding depends on the support and funds of regular people, not traditional banking types such as bankers, venture capitalists, or angel capitalists.

History of Crowdfunding

The rise of crowdfunding in the last decade did not just happen to be a sudden phenomenon without previous indicators that there was something huge coming up, there were indicators and milestones that the crowdfunding concept went through (History of Crowdfunding, 2013). Crowdfunding began in 1996 when Andy Klein, the founder and owner of Spring Street Brewery, posted his company's public offering online Web to allow people to own stock directly (History of Crowdfunding, 2013) It wasn't until that year when, Mark Perlmutter lunched the first direct IPO platform to allow the public to put investments in small public offerings. Later, in 1998, Silicon Valley venture capitalist Tim Draper built on that success and founded a venture capital company based on public investing in early technology deals (History of Crowdfunding, 2013).

Michael Sullivan first used the term "crowdfunding" eight years later on his blog, fundavlog. The real watershed for small business investment in general, including crowdfunding, was when the Nobel Prize was awarded to Muhammad Yunus, the founder of Grameen Bank in 2007, for raising $6.8 billion through microloans to aid the poor. Following that, the crowdfunding platform started to branch out to different fields. In 2008, IndieGoGo began as a crowdfunding platform for independent films. Kickstarter, the famous crowdfunding platform, was formed in 2009 to provide an opportunity for all types of entrepreneurs and business seekers. That same year, crowdfunding grew by 74%, which brought the practice to the attention of politicians and economists (History of Crowdfunding, 2014).

Crowdfunding is currently in its early stages, yet it is growing enormously with the rise of the Internet over the last decade (Prive, 2012). There are different fields and projects that utilize the idea of crowdfunding including the music business, start-up businesses, charities, political campaigns, and software development, among others. The main idea is the concept of "it is not what you do, but why you do it." (Prive, 2012).The success of crowdfunding is due to the idea that people support a certain project because they believe in the cause or because they think that the creator is authentic and that his or her contribution might make a significant difference in bringing this idea to reality (Prive, 2012).

Misconceptions about Crowdfunding

There are several myths and misconceptions about crowdfunding among user platforms, which started at the beginning of the 1990s with the rise of the Internet (Howes, 2013). Some examples of these misconceptions include the idea that crowdfunding is only useful for start-ups and small business, which is not necessarily true (Howes, 2013). Crowdfunding can be a great tool for large, established companies and some people launch crowdfunding campaigns on their own Websites although exposure will not be as widespread as when an already-established platform is used. Another misconception is that some small business owners think that crowdfunding can maximize exposure by launching the same campaign on different platforms, which is not effective because the audience is essentially the same (Howes, 2013).

Crowdfunding for Non-Profits

The same concept that applies to for-profit start-ups or businesses also applies to nonprofit organizations (Howes, 2013). Crowdfunding for nonprofits is concerned with a project's cause and the story behind the cause drives the investors or, in this case, the loaners. The idea has also been used to create awareness campaigns and even to fund an election. In the U.S. in particular, politicians have used the concept to support their campaigns. The most famous example was the last presidential race, in which Barak Obama managed to raise almost $214 million from small donors, which changed the history of political campaigns (Crowdfunding for Nonprofits, 2011). There are several known nonprofit organizations that use the concept of crowdfunding to support their causes and raise funds (Crowdfunding for Nonprofits, 2011).

Crowdfunding Mechanisms

There are several mechanisms for crowdfunding through the use of an online platform or other kinds of platforms. According to Gueyie, Manos, and Yaron (2013), crowdfunding mechanisms can be split into three main categories: through donations, which is the concept of giving up a certain amount of material, whether money or some other valuable entity, to another person without the expectation of any return. The second category consists of an active investor, in which the crowd funder usually makes the decision or at least gives advice to the person or the project he is funding. The third and last type consists of a passive investor, who usually invests or loans money without trying to get involved. (Gueyie, Manos, & Yaron, 2013).

Pros and Cons of Crowdfunding

There are several advantages and disadvantages for using a crowdfunding platform to fund a project regardless of the scale of the project. An entrepreneur needs to take various factors into consideration when considering funding his project through a crowdfunding platform. Out of these factors are the pros and cons of crowdfunding.

Pros

- Crowdfunding helps in reducing the amount of time spent on the fundraising process and gives the entrepreneur more time to focus on other aspects of his future business (Dupree, 2012)
- Any person could be part of the funding process since there is no minimum level for the contribution to a certain project. This helps in expanding the outreach and the base of funders which also help the entrepreneur in getting more reviews on his project (Dupree, 2012).
- Investors can track the progress of the project they have chosen to fund, which make the business more transparent (Rogers, 2012)
- Crowdfunding platform are usually an online tool which make it easier for both entrepreneurs and investors to stay connected (Rogers, 2012)

Cons

- The exposure of ideas through the crowdfunding platforms makes the process of copying ideas easier and effortless
- Crowdfunding platforms mostly are online which means that there are a large portion of possible entrepreneurs with ideas that can't use it as they don't have access to the internet specially in some of the developing countries (Rogers,2012).
- Some project could be a fraud due to the lack of filtration process on the existing crowdfunding platforms (Dupree, 2012).

GROUND WORK

The main research in this thesis paper is done in this ground work section. The ground work consist of three main parts; Egypt context, Grameen Bank, and Kiva.org. The first part is Egypt context, which is a general overview of Egypt with an analysis of the political, economic, social, and technological environment. Additionally, a review of the current microfinance institutions in Egypt with the problems facing them is also presented in this part.

The second part of the ground work summarizes Grameen Bank as an organization. In addition to that it shows the socio-economic environment that lead to formation of Grameen Bank and highlights the relationship between the Bank and its employees. This part concludes with a SWOT analysis for Grameen Bank. The third and final part of the ground work is the synopsis of Kiva.org and their core principles. This part also concludes with a SWOT analysis for Kiva.org

Egypt Context

Egypt is considered the heart of the Middle East in terms of geography, as it is in the northeast of Africa, with Sinai situated in northwest Asia (Country Profile, 2014). Egypt shares borders with Libya, Sudan, Jordan, and Palestine. To the north of Egypt is the Mediterranean Sea, and to the east is the Red Sea. The Nile River, which is considered the source of water and agriculture, runs through the country from south to north. The surface area of Egypt is approximately 1 million square kilometers, of which less than only 5% is populated and more than 95% is desert (Country Profile, 2014).

Egypt is currently the largest populated Arab country, with 83,667 million people and continues to grow at a rate of 2.17% according to the Central Agency for Public Mobilization and Statistics (2013). Almost more than 98% percent of the population lives on the Nile strip, which runs from the south to the north. According to Brown (2004), life expectancy, which is the amount of years people expect life to be better after, in Egypt has increased from 55.3 years in 1980 to 67 years in the year 2001. In the same period, population density has increased from 44 people per km^2 to 69, which explains the previous increase. (Brown, 2004)

PEST Analysis for Egypt

Political

Because the political environment in Egypt has been very unstable since 2011, this section will be divided into two parts. In the first section, the political situation before the revolution in 2011 will be discussed and then the situation after 2011. Before the revolution, most Egyptians did not take part in the political process, as Hosni Mubarak's regime stayed in power for over 30 years (Osman, 2010). During this time, Mubarak made sure that his was the only party in power.

Egyptians in this era had a lack of awareness of rights and responsibilities, and there was little or no hope of change (Osman, 2010). Politics was not a part of the country's education and it was forbidden to practice any sort of politics in schools and universities (Osman, 2010). The country was ruled by the police state, in which everyone lived in fear. Starting in 2006, the return of Mohamed El Baradi raised hopes and initiated some constitutional changes; however, there was no active or effective movement in the streets (Osman, 2010).

On January 25, 2011, a few weeks after demonstrations had begun in Tunisia, a demonstration started in Egypt's Tahrir Square (Gamel, 2013). Many events happened during the first 18 days of the revolution, and eventually Mubarak had to resign his position and the Egyptian army took control of the country for a transitional period. Shortly after the fall of the old regime, the situation changed and an internal fight for power caused a period of political instability (Gamel, 2013). Currently, the situation is complicated because the army has remained in power after a short period of rule by the Muslim Brotherhood, which is considered by many people to be a terrorist organization (Gamel, 2013).

A new election for the presidency has been announced to take place in few months and there are two potential candidates that are running up (Gamel, 2013). The first candidate is the former head of the Egyptian army, and the second candidate has been in the civic society for 40 years (Gamel, 2013). To sum up, the political situation is extremely unstable, with the government changing every few months, and this has led to fear within the public, increasing prices for basic goods and services, and no real solutions for the country's long-term problems (Gamel, 2013).

One other aspect of the current political situation in Egypt that affects nonprofit organizations in Egypt is the recent shutdown of several nonprofit organizations and a halt to funding others under the claim that they were receiving money from foreign entities. Egyptian media has focused on this incident only from the government point of view as if these organizations were in Egypt to spy or to fund vandalism. This action caused many international nonprofit organizations to be reluctant to operate in Egypt (Kirkpatrick & Myers, 2011).

Economic

In this section, several ways will be discussed to assess the economy of countries. After the revolution in Egypt, the economy was hit hard due to the street demonstrations and a lack of tourism. The Egyptian economy is built on several components—returns from the Suez Canal, tourism, oil, and some agricultural and industrial production—which were all affected after the revolution. According to the United Nations Situation Analysis for Egypt (Handoussa, 2010), the country was making great progress until the year 2009, when inflation reached almost 16%.

One of the main economic and social issues facing Egypt is poverty. Poverty rates are different, based on geographical areas, with poverty in rural areas being 67% higher than in the urban areas. More than 23% of Egyptians live below the poverty line, and 7% live in absolute poverty (Guigale & Mubarak, 1996). Approximately six million people are trying to survive on a quarter of a dollar per day or less (Guigale & Mubarak, 1996). The unemployment rate is estimated to be around 21%, and 85% of that is composed of young people, aged 19 to 29 (Guigale & Mubarak, 1996).

The informal economy in Egypt is a very important factor in employing people, as almost 60% of the workforce either work for a private company or operate a small business in a local market. This informal economy provides a large variety of goods and services. The government tried to support this movement so that the private sector could help solve the problem, but because there are no formal rules yet it could not moderate and aid the process (Drolet, 2005).

Social

Egypt has a very unique social composition, in which there is a clear distinction between different social levels. According to Amin (2001), the social mobility in Egypt since the 1950s has progressed for different reasons and along different pathways, starting from the social era in the 1950s and 1960s, passing through the era of capitalism, and ending with globalization in the new millennium (Amin, 2001). In Egypt, two important cultural aspects are a large and lively social network and the fact that it is a religious society. These social aspects need to be taken into consideration when initiating a new project as they could be the difference between success and failure.

There are several longstanding social problems such as late marriage, illiteracy, unemployment, and sexual harassment, among others. Illiteracy represents a fundamental problem in Egypt, as approximately

39% of the population cannot read or write (El-Baz, 1997). One of the main reasons behind this is the lower social value of women's education, which leads to a high dropout rate, especially in the rural areas (El-Baz, 1997).

Technological

According to Hafheinz (2007), Egypt has the largest and fastest-growing population of Internet users around the Arab world, yet this is dependent on individuals' social level and gender. Egypt has infrastructure for technology in general and for the Internet in particular, yet this infrastructure is not sufficient in rural areas, just in the main cities. In the 2011 revolution, social media played a very important role and it was considered the catalyst for motivating people to take to the streets after they shared news of the demonstrations online (Iskander, 2011).

Recently in Egypt, young entrepreneurs have developed many companies that depend on social media and the Internet (Iskander, 2011). Venture capitalists in Egypt support the movement toward building Websites, mobile applications, and Internet-based services, yet these small businesses and entrepreneurs focus on a small segment of the society who have continuous online access and can actually benefit from these services. Technology and the Internet could be used more than at present to help all Egyptians (Iskander, 2011).

Microfinance Institutions in Egypt

According to Waly (n.a) the first attempt to start microfinancing in Egypt was by the Alexandria Business Association and the National Bank of Development in the form of microloans to small enterprises with the support of the U.S. Agency for International Development (USAID). In 1997, the Principle Bank of Development and Agricultural Credit developed an initiative called "The Productive Families Project," which was supported by the Ministry of Social Solidarity (Planet Finance, 2008).Currently there are 15 different microfinance institutions in Egypt, although only eight are fully operational. These organizations receive funding from USAID, but other external sources of funding include the Ford Foundation, the United Nations Development Program (UNDP), Save the Children, and others (Planet Finance, 2008).

All of the microfinance institutions in Egypt are currently operating under law number 141, which was introduced in 2004 and regulates microfinance activities in the country. Waly (n.a) claims that currently there are more than 1.13 million active clients who are borrowing around 1.8 billion EGP (equivalent to 266 million USD). The average loan size is estimated to be 1600 EGP (230 USD) and the penetration rate is 6%.

Problems that Microfinance Institutions Face in Egypt

According to research and the survey done by Blackwood (2012), the creation of social entrepreneurship and microfinance institutions in Egypt has not been easy. There are several issues that confront most of the current organizations and that will affect any future organizations. At the top of the list are problems of funding and government restrictions. Other organizations that Blackwood surveyed indicated that a lack of partnership was also a major problem, and 100% of the organizations surveyed responded that their start-up funds came through external funding sources. Any nonprofit organization would face these

Table 1. Loans vs. Borrowers in Microfinance Institutions (MFIs) in Egypt (Planet Finance, 2008)

MFI's most recent data	Report Date	265.9m Loans (USD)	1m Borrowers	0 Deposits (USD)	0 Depositors
ABA	2012	60,929,049	237,371	-	-
ABWA	2012	5,281,522	-	-	-
Al Tadamun	2012	5,206,214	40,823	-	-
ASBA	2011	61,854,750	225,289	0	0
BDC	2005	35,347,222	93,516	-	-
CEOSS	2012	8,774,845	46,542	-	-
DBACD	2012	26,589,887	107,608	-	-
ESED	2011	16,300,861	70,640	0	0
FMF	2012-09-30	5,014,048	18,654	-	-
Future	2013-09-30	1,419,373	10,451	0	0
Lead Foundation	2012	21,659,657	141,299	0	0
NSBA	2008	667,285	5,055	0	0
RADE	2012-06-30	2,248,960	12,735	-	-
SBACD	2012	12,602,047	-	0	-
SCDA	2009	784,561	1,951	0	0
YMMA	2006	1,267,649	11,856	-	0

problems when starting up, but there are always ways to solve these problems, as this research is trying to point out (Blackwood, 2012).

Grameen Bank

In 1983, microfinance institutions took a leap forward with the founding of Grameen Bank, a rural banking institution that was based on a microfinance model to aid the poor in Bangladesh. Grameen Bank was different in that it was aimed at women who owned less than half an acre of land or who owned assets of less than one acre of land (Khanker, Khalily, & Khan, 1995). The loans did not need collateral, which was phenomenal at that time.

The idea, started by Dr. Muhammad Yunus, was initially against all the rules that he had been teaching as an economics professor at Chittagong University. It took a great deal of research for Dr. Yunus to decide that he wanted to be teaching about the real economy in Bangladesh and then try to fix what he saw around him. During his study, Dr. Yunus came up with a list of 42 people that needed a small amount of money (not exceeding $27 each), to improve their financial situation (Yunus, 1999).The main motive behind Dr. Yunus' idea was that banks would not give poor people loans without collateral. In the beginning, he started loaning to the people on his list from his own money, and from funds contributed by his research group colleagues who believed in his ideas (Yunus, 1999).

However, Dr. Yunus was still concerned with the rest of the poor population, so he tried to convince the local university bank to loan to the poor with no collateral, but they refused. He then took a loan under his name and started loaning to the poor women in the village close to the university, and after they repaid their loans, he tried to put together a case study and resubmit his request to the bank, but the bank still refused (Yunus, 1999). Dr. Yunus did not give up at this point but instead moved to another village and tried to approach the local bank—but this bank also refused (Yunus, 1999).

Although Dr. Yunus eventually expanded to over 100 villages, he still couldn't convince any bank to lend money to the poor. He then decided to create his own bank. He went through several negotiations with the government, and in 1983 he reached an agreement to establish Grameen Bank or, as it is translated from Bangali, "Village Bank." The main concept was to lend to the poorest of the poor who did not have any sort of collateral (Yunus, 1999).

Grameen Bank's Credit Delivery Model

Grameen Bank operates with a different business model than other banks, similar to its different types of customers (Khandker, 1996). The Grameen model was based on a concept of groups of five, with different groups for men and women. The initial focus was on the women's groups, to help unemployed, low-income women in Bangladesh. The loans were received by individuals within a group, yet the entire group was accountable for repaying the loan.

If one member of the group failed to repay, then the entire group was affected and no other group member could qualify for more credit. This group pressure guaranteed discipline between members. On the other hand, in order for Grameen to protect the groups against anti-group actions that might affect one individual, the bank organized every six to eight groups into a larger community called "The Center" (Khandker, 1996).

Today, all of the lending and repaying transactions are made within the Center meetings. The two- tier structure helps to relieve the social pressure created by the Grameen Bank credit model. Each center is assisted by a bank representative to facilitate transactions, and each group of five elects a representative who acts as chairperson for a year to conduct meetings and take care of the paperwork (Khandker, 1996). The amounts that the bank usually lends to an individual are small, less than $100 per year, and he or she must repay the loan in 50 installments during the year. The loans are given to the individual to support the activity that the group of five has agreed upon (Khandker, 1996).

Power of Context

In his book *The Tipping Point*, Malcolm Gladwell discusses the concept of Grameen Bank's credit-delivery model. He examined similar projects that were dependent on peer pressure and group decision-making, and he demonstrated the importance of how groups change the decisions of individuals. Indecisive individuals within a group usually follow the logic of the group in order for them to reach a decision. Group numbers could differ from one application to the other, yet the effect of the group and peer pressure remains the same. (Gladwell, 2000)

Gladwell also discussed another decisive factor in any social epidemic, which is the power of context. The facts that Bangladesh was undergoing a tough economic period and that Dr. Yunus had just returned in 1974 were key factors in the success of Grameen Bank. Other factors include the place, timing, and the people involved in developing the idea of Grameen Bank, which are all considered part of the context. Gladwell claims that in order for any phenomena to become a social epidemic, it has to pass three laws—the third being the power of context (Gladwell, 2000).

Table 2. Distribution Pattern of Rural Landownership in Bangladesh from 1979 to 1983/84 (Grunewald, & Baron, 2011, Table 2)

Size of Land Owned	1979 Land Occupancy Survey		1983/84 Agricultural Census		Annual Rate of Growth 1979-84
	Number of Households	Share of Households	Number of Households	Share of Households	
{Acres}	{Thousands}	{Percent}	{Thousands}	{Percent}	{Percent}
None	1,979	15.4	1,198	8.7	-10.5
Less Than 0.50	4,286	33.3	5,201	37.6	4.4
0.50-0.99	1,516	11.8	1,660	12.0	2.0
1.00-2.49	2,660	20.7	2,979	21.6	2.5
2.50-4.99	1,436	11.2	1,598	11.6	2.4
5.00-7.49	501	3.9	650	4.7	6.0
7.50-14.99	374	2.9	415	3	2.4
15.00 or more	114	.9	117	.8	.6
Total	12,866	100	13,818	100	1.6

Socio-Economic Environment

Grameen Bank was established in Bangladesh in 1983. Bangladesh is a country of almost 60,000 mi^2 that borders India and was formerly a part of Pakistan until it achieved independence in 1971(Grunewald, & Baron, 2011). The country's population is 162,221,000 people, which is considered the seventh-largest populated country in the world (Grunewald, & Baron, 2011). The GDP in Bangladesh is $624 per person, which is relatively low in comparison with other countries. The currency is Taka and as of September 2010, the exchange rate was 69.64 Takas to one U.S. dollar. The major issues in Bangladesh are health and education. The literacy rate in the country is very low: 54.7% of the population is illiterate, and out of that, 81.9% are women (Grunewald, & Baron, 2011).

Relationship between Grameen Bank and Its Employees

Grameen Bank's employees are its main assets. The bank hires highly motivated individuals who are loyal to the cause of the bank and committed to helping bring the poor to a higher level of self-sufficiency. It has a very selective hiring process for potential employees, and applicants must go through several interviews in order to be recognized for a position (Dowla, 2006).

The recruited applicants must undergo an intensive training program. According to Woolcock (1998), this training program is very difficult, and almost half of the trainees leave before the end. The bank maintains a high level of trust and confidence in its employees, who have a deep sense of pride and compassion for the poor after finishing the training program. This trust can be felt and seen in the different branches of the bank, where employees' make decisions without checking with their supervisors (Dowla, 2006).

Grameen Bank also has a very unique system of evaluation to motivate employees and push them to achieve more success in helping the poor. This system is known as the Five-Star System, in which the bank ranks the different branches according to colored stars. A green star, which is the lowest-ranked star, is given to a branch if it achieves a 100% repayment rate (Mainsah, Heuer, Karla, and Zhang, 2004). The blue star, second in ranking from below, is awarded to the branches that manage to achieve a profit for the organization.

The violet star, middle-star rank, is given to the branches in which the amount of the deposits exceeds the loans. The brown star, the second-highest star, is given to the branches where the children in a borrower's family attend school. Finally, the red star, the highest- ranking star, is given to the branches that (1) have borrowers who are no longer considered to be in the poorest segment of society, and (2) have managed to fulfill Grameen Bank's ten requirements (Mainsah, Heuer, Karla, & Zhang, 2004).

To summarize, Grameen Bank enhances and builds relationships with employees' three different methods:

1. A rigorous recruiting system in order to acquire the most talented and loyal people to the organization and to the cause (Mainsah, Heuer, Karla, & Zhang, 2004).
2. Empowering the employees who are selected by giving them the power to make decisions without going back to their supervisors (Mainsah, Heuer, Karla, & Zhang, 2004).
3. Creating a unique system for evaluation—the Five-Star System—that promotes a healthy, competitive environment and keeps employees motivated and challenged (Mainsah, Heuer, Karla, & Zhang, 2004).

Thus, Grameen bank is keen to build relationships with employees through the ways mentioned above, which in turn benefits the organization and creates a sense of loyalty among the staff (Hossain, 2013).

Grameen Bank's Early Sources of Funds

Since its establishment in 1983, Grameen Bank has secured financing through domestic and international sources of funds at concessionary rates. Those sources have included the Central Bank of Bangladesh and the International Fund for Agricultural Development, among others. A portion from the bank's funds were secured through grants from organizations. In the early years, Grameen Bank had a total of 1532.7 million TK between 1983 and 1993 from the International Fund for Agricultural Development (Khandker, Khalily, & Khan, 1995). The bank also received approximately 200 million TK per year from the Bank of Bangladesh (Khandker, Khalily, & Khan, 1995). Starting in 1994, in order to sustain itself, the bank had to borrow 3250 million TK from commercial banks, which represented almost 46% of its resources (Khandker, Khalily, & Khan, 1995).

SWOT Analysis for Grameen Bank

Strengths

- Grameen Bank helps local communities alleviate poverty by providing small loans to individuals without collateral (Khandker, Khalily, & Khan, 1995).
- The idea of lending to groups of five helps in securing a social environment for the borrowers that helps them to achieve success (Khandker, 1996).
- The relationship between the bank and its employees enhances their ability and empowers them to make the right decision in aiding the poor (Dowla, 2006).

Weaknesses.

- Grameen Bank focuses on women in different societies, which might lead to a higher poverty level among men (Khandker, 1996).
- According to Bornstien (2011), it may be questionable whether the concepts of Grameen Bank help the poor to save money for the future and thus alleviate poverty.
- Funding of a similar foundation like Grameen Bank would require government support as well as external-funding support (Khandker, Khalily, & Khan, 1995).
- There are supply and demand constraints because Grameen Bank depends on its different branches (Khandker, 1996).

Opportunities.

- After Grameen Bank reached the point of self-sufficiency, it is easier to branch out to different countries around the world (Khandker, 1996).
- The Arab spring revolutions might encourage new governments to accept organizations such as Grameen Bank to operate within their countries (Osman, 2010).
- Public acceptance of the concept of microfinance might lead to more opportunities of funding of necessary (Dowla, 2006).
- Social media and technology represent an opportunity for Grameen Bank to expand and explore new ideas to help the poor (Dowla, 2006).

Threats

- The concept of the group of five might lead to conflict within the groups if one person fails to make a payment (Bornstien, 2011).
- New technology has made entry to microfinance much easier, which has created competitors to Grameen Bank. (Chang, Huang, & Wang, 2009).
- Peer-to-peer lending is an appealing concept that could be considered competition for the credit delivery model that Grameen Bank depends upon (Bornstien, 2011).

In order for a social entrepreneur to successfully implement the model of Grameen Bank, he needs to capitalize on strengths like the power of the idea itself and the lending group idea. In the same time avoiding weakness like the focus on women only or depending on the government support, would help the social entrepreneur in making the model successful.

Kiva.org

Kiva.org is an online platform that operates as a nonprofit organization that gives individuals the opportunity to participate in microfinance activity using small amounts of money directed to a small business or a micro entrepreneur (Chang, Huang, & Wang, 2009). In other words, Kiva.org tries to connect people who are looking to help other people in developing countries through micro lending. The process is done through an intermediary, a local microfinance institution, or an MFI, which identifies micro entrepreneurs and small businesses that need help to grow and inputs their data on the Website and facilitates communication between the lender and the borrower (Chang, Huang, & Wang, 2009).

The lending cycle starts when the lender chooses the entrepreneur or small business they want to fund. After the lenders make their decision, Kiva collects the loan amounts from all of the different lenders around the world and distributes the sums to each local MFI partner based on the projects that were selected in their areas (How Kiva Works, 2014).Then the local MFI partner distributes the total amount of money funded from each entrepreneur. The entrepreneur gives monthly payments back to the local partner that collected the money and transfers them to Kiva.org, which repays the lenders and completes the lending cycle. At this point, the lender can chose to take back his money from Kiva.org or relend the money to a different entrepreneur (How Kiva Works, 2014).

The idea for Kiva.org, an online microfinance platform, was initiated by an American couple, Matt and Jessica Flannery, during a visit to East Africa in 2004 and was founded a year later. Matt and Jessica met at business school at Stanford University while he was studying and she was a staff member in the public Management Program (Flannery, 2007). In 2004, Jessica had an opportunity to consult for a nonprofit organization that specialized in microfinance in East Africa.

She took the opportunity and few months later, Matt went on a month-long vacation from his work at TiVo. They came up with the idea for Kiva.org when they were visiting Uganda. According to Matt, when they were asked about their career goals, Matt replied that he wanted "to live in the San Francisco Bay Area and be an entrepreneur," while Jessica's answer was "to go to Africa and do microfinance." They combined their career goals by founding Kiva.org (Flannery, 2007).

Core Principles of Kiva

Matthew and Jessica's idea went through different phases before they arrived at their main concept for Kiva in 2005. Currently, Kiva's vision is: "Everyone connected, everyone empowered" (Flannery, & Shah, 2009). The organization's current mission is "to connect people through lending for the sake of alleviating poverty." In his study, Carlman identified seven core beliefs of Kiva.org, which include the power of microfinance, partner relationships, transparency, the spread of technology, lending as connecting, sustainability, and industry inclusion. The organization's employees and partners make sure to take these principles and guidelines into consideration before making any decision in the operation cycle (Carlman, 2010).

Kiva and Web 2.0

Kiva is considered to be a Web 2.0 application in microfinance. O'Reilly (2005) defines the Web 2.0 "network as a platform…making the most out of intrinsic advantage of that platform." Kiva is creating an atmosphere and a platform to connect people around the world for a bigger cause. In Web 2.0, the key to success is being user driven, and Kiva, according to Flannery (2007), the organization depends on lenders to bring in more lenders by word of mouth and reviews as the main tool for growth. Collaboration of different online tools, which is the essence of Web 2.0, is one of the elements that Kiva.org considers primary in operations whether it is PayPal, YouTube, Amazon, or another online tool (Carlman, 2010).

In their study, Ashta and Assadi (2008) claim that Kiva is not yet a perfect Web 2.0 model because the features and the technology are not the same for all types of users. The borrower end of the Website is limited since there is almost no interaction between the borrower and the lender except through the local partner. On the other hand, the lender side of the Website has many different technological features that enable the lender to make better choices and to select the entrepreneur or small business to which he wishes to lend money. Kiva is even planning to partner with existing Internet and Web 2.0 gurus such as Facebook in order to further enrich the lender experience (McCarthy, 2009).

Statistical Facts about Kiva.org

Kiva currently operates in over 73 countries around the world, although with limited access in some countries due to governmental regulations and a lack of local institutional support. (Statistics, 2014). There are over 1,083,984 lenders using the Kiva online platform to lend small or large amounts of money to people in other countries (Statistics, 2014). Kiva currently has 242 local field partners in the countries in which it operates and over 450 volunteers. The amount that has already been loaned is nearly $537,703,175, and Kiva stands as one of the most successful online platforms for microfinance (Statistics, 2014).

Not only are the amounts of loans and lenders astonishing but the repayment rate, which stands at 98.97%, is also amazing. The lenders at Kiva often lend more than once, and the average lending on the online platform is 10.15 times per lender, which shows how fast a lending cycle can occur. On the borrower side, Kiva has more than 1,256,171 borrowers who have been funded (Statistics, 2014).

Lending Teams

Using the Kiva.org online platform, users can choose to lend money individually and directly to particular entrepreneurs or they can choose to join a lending team (Kiva.org, 2014). Lending teams are self-organized groups that form a community of lenders who share lending goals and connect with one another. There are two types of teams—open teams and closed teams, and each type has its own advantages (Hartley, 2010).

For example, an open team does not usually have accountability, and any person or lender can join the group based on his interest. There could be a lending team composed of Egyptian natives with none of them born in Egypt, yet they share a common interest that enhances the in-group solidarity and usually has a big impact on the group lending in general. On the other hand, the closed-lending groups are usually very moderated and organized; members of closed teams are more cohesive and they share the same interests (Hartley, 2010).

There are many books and studies that discuss the effects of human cooperation individually and within a team or a group as well as studies that delve into the effects that similarities between members of a team might have on the lending behavior and solidarity of the members. Chen and Li (2008) have indicated that a group identity results from the mix of group members with similar interests and thoughts, which result in a 47% increase in charity concerns, and around a 93% decrease in envy because of a lower payoff (Chen, & Li, 2008)

SWOT Analysis for Kiva.org

Strengths

- Connecting lenders directly to borrowers in remote places, specifically in developing countries, has helped Kiva to raise the transparency level. Investors or lenders can choose the projects to which they would like to contribute and view the entire lending cycle through the platform. This transparency inspires trust between Kiva and the lenders and enhances opportunities for helping developing countries reduce poverty (Chen, & Li, 2008)
- The idea of loaning instead of donating money through the platform helps entrepreneurs in the long term to become independent and makes them appreciate the value of the money that they are borrowing (Chang, Huang, & Wang, 2009).
- Kiva serves as an alternative funding method for micro entrepreneurs in developing countries, which allows them to avoid local money lenders who may charge interest rates of up to 300% (Chang, Huang, & Wang, 2009).
- The marketing strategy for Kiva is based on cooperation with large online platforms, marketing, and advertising Websites such as Google, Facebook, Yahoo, etc. This strategy has enabled them to promote the awareness of poverty around the world.

Weaknesses

- Kiva only supports poverty in developing countries and does not create job opportunities for those who do not have entrepreneurial skills.
- Not controlling the field partners—the local microfinance institutions—leads to some of these local partners charging up to 21% interest rates on the loans provided by Kiva lenders (Chang, Huang, & Wang, 2009).
- Kiva depends solely on the local partners to provide business support to the micro- entrepreneurs, although they could provide other tools for them (Chang, Huang, & Wang, 2009).
- Kiva does not provide enough tools for borrowers as they do for lenders, as the organization's main focus is the lenders (Ashta, & Assadi, 2008).

Opportunities

- Kiva's platform is mainly online, so it is scalable and can reach countries in the developing world that do not have sufficient infrastructure.
- Kiva's media attention and its story have provided the opportunity to attract more lenders and investors, for which, for some time, there were not enough borrowers—in other words, the supply exceeded the demand (Uganda: A Little Goes All the Way, 2009).
- The uprising events around the world since 2011 have led developing nations to believe that different sources of funding for the poor may help economies get back on track.

Threats

- The growth in Kiva's operation and lender base may lead to an increased risk of scamming local microfinance institutions and scamming local entrepreneurs. This situation may occur if Kiva does not increase the amount of time and effort devoted to choosing local partners as well as the criteria for local projects and entrepreneurs (Chang, Huang, & Wang, 2009).
- Currently, lenders and lending teams are essential for Kiva and more research should be done to retain these lenders and convince them to continue their efforts.
- Resistant regimes in some developing countries could be a threat to the expansion of Kiva into new countries.
- The competition of new online platforms that offer the same services as Kiva might cause some lenders to go to competitors, which would provide local entrepreneurs with more options to demonstrate their projects.

For a social entrepreneur to successfully implement Kiva's model in Egypt, he needs to focus on solving the current issue that Kiva has like controlling the local partners by having certain criteria for selecting them and performing audits. Also the social entrepreneur would need to think of new tools for the borrowers in order to engage them more and in the same time avoid depending solely on local partners. Furthermore, the social entrepreneur will need to exploit the current strengths of Kiva like their marketing strategy and their concept of lending rather than donating.

DISCUSSION

In response to the first thesis question, "How did Grameen Bank achieve success in Bangladesh?" many factors led to the bank's success in Bangladesh in 1983. The first factor was the persistence by its founder Dr.Yunus and his group. When he was rejected multiple times by different banks in Bangladesh that refused to provide credit without collateral, he loaned his own money to groups. The second factor was the credit delivery model created by Grameen Bank, which depends on the creation of groups of five and starts lending to each of them on an alternating basis (Hartley, 2010). For instance, the second person in the group would not receive a loan until the first person made a first repayment (Hartley, 2010).

The credit delivery model also helped the bank to control repayments using peer and group pressure—if one individual within the group failed to make a repayment, the whole group would be suspended from further loans. The third reason behind the success of Grameen Bank is within the context of the country of Bangladesh: poverty rates were very high, particularly among women; Illiteracy rates reached almost 81% among women; and the government failed to fulfill the needs of the increasing population.

The fourth reason is the relationship created between Grameen Bank and its employees that enables an employee to make decisions without going back to their supervisors, and the trust created in making decisions that will help alleviate poverty. The fifth and final reason for Grameen Bank's success was the internal and external funds the bank received in the early days of operations, which almost closed the gap between the supply and the demand at that time. To sum up, Grameen Bank's success has been based on several factors, which have brought the concept of microfinance to a new level.

In response to the second research question, "How did Kiva.org manage to use online microfinance and crowdfunding concepts so successfully?," Kiva.org has managed to put together the concepts of microfinance and crowdfunding to come up with an online platform that aims to connect people and alleviate poverty around the world. Kiva made use of the Internet revolution and the Web 2.0 movement to attract lenders to their cause. They introduced peer-to-peer lending, which is based on an individual project or idea that the lender would like to support. The lending cycle that Kiva created makes the operation very transparent, which has also attracted more lenders.

Kiva.org has used social media and Internet tools to the fullest extent to market their platform and help people in developing countries. Using local partners to select local entrepreneurs reduced the amount of staff that Kiva needed to operate because it depends solely on input from local partners. Finally, the concept of lending teams has made it easier for lenders to find a community of people who share the same ideas and beliefs to invest in the same projects.

n response to the third research question, "Which microfinance model out of Grameen Bank and Kiva would be more successful in Egypt?" which is based on the research done on both models and also on Egypt's current political situation, it would be easier and more effective to implement the Kiva model. There are many factors behind the selection of Kiva: the first factor is the increased use of the Internet in Egypt and the increased use of social media and other components of Web 2.0.

The second reason is the limitations that the government creates for the startup of any nonprofit or microfinance organization. The third reason is the limited funds required for Kiva to start and to operate in Egypt's current economic situation. The fourth factor is the current microfinance institutions that exist in Egypt, could be potential partners for any created platform; and finally, the fifth factor is the advantages that any online platform would provide—flexibility, ease of use, quick access to funds, and the creation of communities.

CONCLUSION AND RECOMMENDATIONS

Microfinance and crowdfunding have become tools that could be used to alleviate poverty within societies. While microfinance has always been a way to provide different financial services to the poor, crowdfunding has recently become popular for the same reason. Both Grameen Bank and Kiva have managed through the years to become successful organizations through different strategies and different business models. Grameen Bank has developed the credit-delivery model that has proved to be a huge success in Bangladesh, while Kiva has used the concept of crowdfunding to create an online platform for microfinance to emphasis the concept of peer-to-peer lending.

The political, economic, social, and technical analysis of Egypt has shown that the use of microfinance would help the country's major problems such as poverty and unemployment. The fact that there are currently restrictions and limitations for the start-up of new microfinance organizations and that Egyptians are using the Internet in their everyday lives has led to the recommendation of using the model of Kiva.org for developing a microfinance presence in Egypt.

This conclusion was based on the current research and the concepts that are included in this paper. Future researchers may want to investigate the different steps needed to initiate the Kiva model in Egypt. Researchers could also investigate possible government support for an Internet-based microfinance organization. This future research will be important in supporting a social entrepreneur to take the initiative and use the results to implement the model. Other future research could include a study of the negative impact of microfinance and crowdfunding for social entrepreneurs in order to avoid these impacts.

REFERENCES

About C. G. A. P. (2013). *CGAP's organization*. Retrieved March 28, 2013, from CGAP.org: http://www.cgap.org/about

Akram, M., & Hussain, I. (2011). The role of microfinance in uplifting income level: A study of district Okara, Pakistan. *Interdisciplinary Journal of Contemporary Research in Business*, *3*, 83–90.

Amin, G. (2001). *Whatever happened to the Egyptians? changes in Egyptian society from 1950 to the present*. Cairo: American University in Cairo Press.

Ashta, A., & Assadi, D. (2008). *Do social cause and social technology meet? The impact of Web 2.0 technologies on peer-to-peer lending transactions*. Paper presented at the Asia Microfinance Forum, Hanoi.

Blackwood, C. (2012). Social entrepreneurship in Egypt: Turning revolutionary ambition into tangible results. *SIT Graduate Institute, 26*. Retrieved from http://digitalcollections.sit.edu/cgi/viewcontent.cgi?article=3594&context=capstones

Bornstein, D. (2011). Grameen Bank and the public good. *The New York Times*. Retrieved from http://opinionator.blogs.nytimes.com/2011/03/24/grameen-bank-and-the-public-good

Brown, L. (2004). *State of The World 2004. WorldWatch Institute Report on Progress Toward A Sustainable Society* (pp. 170–180). New York: W.W. Norton Company.

Browning, M., & Lusardi, A. (1996). Household saving: Micro theories and micro facts. *Journal of Economic Literature*, *34*(4), 1797–1855.

Carlman, A. (2010). *Development 2.0? The case of Kiva.org and online social lending for development*. Stellenbosch: Stellenbosch University.

Chang, A., Huang, F., & Wang, E. (2009). Strategies for efficiency expanding the operations of Kiva.org. *The American Economic Review*, *99*(1), 431–457. Retrieved from http://www.mcafee.cc/Classes/BEM106/Papers/2009/Kiva.pdf

Chen, Y., & Li, S. (2008). Group identity and social preferences. Retrieved from http://www.jstor.org/discover/10.2307/29730190?uid=3739560&uid=2&uid=4&uid=3739256&sid=21103613955097

Churchill, C. (2006). *Protecting the poor: A microinsurance compendium* (Vol. 1). Munich: Re Foundation.

Country Profile. (2014). Un.org.eg. United Nations. Retrieved from http://www.un.org.eg/UNInner1.aspx?pageID=3

Crowdfunding for Nonprofits. (2011). Greenmarketing.tv. Retrieved from http://www.greenmarketing.tv/nonprofit-crowdfunding/

Dowla, A. (2006). In credit we trust: Building social capital by Grameen Bank in Bangladesh. *Journal of Socio-Economics, 35*(1), 102–122. doi:10.1016/j.socec.2005.12.006

Drolet, J. (2005). Women and Microcredit: Towards an understanding of women's experiences in Cairo, Egypt. The School of Social Work, McGill University.

Dupree, S. (2012). Crowdfunding 101: Pros and Cons. Center of entrepreneurial studies, Stanford Business. Retrieved from, http://www.gsb.stanford.edu/ces/crowdfunding-101

El-Baz, S. (1997). *The impact of social and economic factors on women's group formation in Egypt.* Oxford, UK: Berg Press.

Flannery, M. (2007). Kiva and the birth of person-to-person microfinace. Retrieved from http://www.mitpressjournals.org/doi/pdf/10.1162/itgg.2007.2.1-2.31

Flannery, M., & Shah, P. (May 2009). *Kiva and Web 2.0.* Presentation given at Kiva Fellows Training.

G8 Summit. (2004, June 10). *G8 action plan: Applying the power of entrepreneurship to the eradication of poverty.* Retrieved from http://www.canadainternational.gc.ca/g8/summit-sommet/2004/action-plan-action_090604.aspx

Gamel, H. (August, 2013). What's going on in Egypt: A brief explanation. *Huffington Post.* Retrieved from http://www.huffingtonpost.com/2013/08/14/whats-going-on-in-egypt_n_3758890.html

Giugale, M., & Mubarak, H. (1996). *Private sector development in Egypt.* Cairo: American University in Cairo Press.

Gladwell, M. (2000). *The tipping point, how little things can make a big difference.* Retrieved from http://ebookbrowsee.net/the-tipping-point-how-little-things-can-malcolm-gladwell-pdf-d493731844

Grunewald, D., & Baron, P. (2011). A tale of two banks: Grameen Bank and SKS microfinance. *Journal of Leadership, Accountability & Ethics.* Retrieved from http://Web.a.ebscohost.com/ehost/pdfviewer/pdfviewer?sid=48b97718-0676-44a2-b169-9690e62a7ef7%40sessionmgr4003&vid=2&hid=4107

Gueyie, J., Manos, R., & Yaron, J. (2013). *Microfinance in developing countries: Issues, policies and performance evaluation.* Palgrave Macmillan. Retrieved from http://www.palgraveconnect.com/pc/doifinder/10.1057/9781137301925.0014

Gunts, S. (2011). *Sustainability and profitability of microfinance institutions.* Center for applied international finance and development. Retrieved from, http://www.th-nuernberg.de/fileadmin/Fachbereiche/bw/studienschwerpunkte/international_business/Master/CAIFD/ResearchPapers/SustainabilityAndProfitabilityOfMicrofinanceInstitutions_Guntz.pdf

Hafheinz, A. (2007). Arab internet use. In Arab media and political renewal (pp. 56–79). Academic Press.

Handoussa, H. (2010). *Situation analysis: key development challenges facing Egypt.* United Nations. Retrieved from http://www.un.org.eg/docs/101100%20SA%20Report%20final%20pdf%20version.pdf

Hartley, S. (March, 2010). *KIVA.org: Crowd-Sourced microfinance and cooperation in group lending.* Retrieved from http://dash.harvard.edu/bitstream/handle/1/3757699/Hartley_Kiva_DASH.pdf?sequence=2

Helms, B. (2006). *Access for all: building inclusive financial systems*. Consultative Group To Assist The Poor.

History. (2013). *KIVA.org*. Retrieved from http://www.kiva.org/about/historyhttp://www.kiva.org/about/history

History of Crowdfunding. (2013). *Crowdfunding World's Summit*. Retrieved from http://www.crowdfundingworldsummit.com/crowdfunding/

History of Crowdfunding. (2014). Retrieved from http://www.fundable.com/crowdfunding101/history-of-crowdfunding

Hossain, D. (January, 2013). *Social capital and microfinance: The case of Grameen Bank*. University of Dhaka.

How Kiva Works. (2014). Retrieved from http://www.kiva.org/about/how

Howes, L. (November, 2013). *5 Misconceptions about crowdfunding*. Analytical Ones. Retrieved from http://www.analyticalones.com/5-misconceptions-about-crowdfunding/

Iskander, E. (2011). Connecting the national and the virtual: Can facebook activism remain relevant after Egypt's January 25 uprising? *International Journal of Communication*, 13–15.

Khandker, S. (1996). Grameen Bank: Impact, costs, and program sustainability. *Asian Development Review*, 65–85. Retrieved from http://www.microfinancegateway.org/gm/document-1.9.28425/1835.pdf

Khandker, S., Khalily, B., & Khan, Z. (1995). *Grameen Bank: performance and sustainability*. Retrieved from http://books.google.com/books?id=rNN3Dfxcc3oC&printsec=frontcover#v=onepage&q&f=false

Kirkpatrick, D., & Myers, S. (2011). Egypt raids offices of nonprofits, 3 backed By US. *The New York Times*. Retrieved from http://www.nytimes.com/2011/12/30/world/middleeast/egypts-forces-raid-offices-of-us-and-other-civil-groups.html?pagewanted=all&_r=0

Lending Teams. (2014). Retrieved from http://www.kiva.org/teams

Mainsah, E., Heuer, R., Karla, R., & Zhang, Q. (2004, Spring). Grameen Bank: taking capitalism to the Poor. *Chazen Web Journal of International Business*, 1–28.

Marinescu, A. (2009). Microfinance-a way to confront poverty and stimulate entrepreneurship worldwide. *Young Economists Journal*, 65–71.

McCarthy, C. (2009). Nonprofits next to test Facebook payment platform. *CNet*. Retrieved from, http://news.cnet.com/8301-17939_109-10320315-2.html

Microcredit Summit Campaign. (2005, January 24). *Global microcredit summit coming to Halifax, Canada*. Message posted to Microcredit Summit mailing list. Retrieved from http://www.microcreditsummit.org

O'Reilly, T. (2005). *Web 2.0: compact definition?* Retrieved from http://radar.oreilly.com/archives/2005/10/Web-20-compact-definition.html

Osman, T. (2010). *Egypt on the brink: from the rise of Nasser to the fall of Mubarak*. Retrieved from http://books.google.com/books?hl=en&lr=&id=MWReGwoyjcsC&oi=fnd&pg=PP2&dq=political+scene+in+Egypt+after+2011&ots=lXb93gfxfc&sig=OUPXAJF8szmv6ZNzn-FQW5UMt4c#v=onepage&q=political%20scene%20in%20Egypt%20after%202011&f=false

PlanetFinance. (2008). *National impact survey of microfinance*. Retrieved from http://www.sfdegypt.org/c/document_library/get_file?uuid=d11593f0-46b6-49d3-b64d-2eab2c0e6f12&groupId=10136

Population. (2013). *Central Agency for Public Mobilization and Statistics*. Retrieved from http://www.capmas.gov.eg/pdf/Electronic%20Static%20Book2013/english/population/untitled1/pop.aspx

Prive, T. (2012). What is crowdfunding and how does it benefit the economy. *Forbes.com*. Retrieved from http://www.forbes.com/sites/tanyaprive/2012/11/27/what-is-crowdfunding-and-how-does-it-benefit-the-economy/

Rogers, K. (2012). Small businesses weigh in: should crowdfunding be legal? *Fox Business Journal*. Retrieved from http://smallbusiness.foxbusiness.com/technology-web/2012/02/17/small-businesses-weigh-in-should-crowdfunding-be-legal/

Statistics. (2014). Retrieved from http://www.kiva.org/about/stats

Uganda: A Little Goes A Long Way. (2009). *Pbs.org, Front Line World*. Retrieved from http://www.pbs.org/frontlineworld/stories/uganda601/video_index.html

Waly, G. (n.d.). Microfinance in Egypt: An Overview. *The Egyptian Financial Supervisory Authority*. Retrieved from https://www.efsa.gov.eg/content/efsa2_ar/pool_extra_efsa/UG18915UG18916.pdf

Woolcock, M. (1998). *Social theory, Development Policy, and Poverty Alleviation: A Comparative Historical Analysis of Group Based Banking in Developing Economies*. (Unpublished PhD Dissertation). Brown University.

Yunus, M. (1999). The Grameen Bank: a small experiment begun in Bangladesh has turned into a major new concept in eradicating poverty. *Scientific American*. Retrieved from http://www.anderson.ucla.edu/faculty/bhagwan.chowdhry/Yunus1999%20-%20Scientific%20American.pdf

KEY TERMS AND DEFINITIONS

Consultative Group to Assist the Poor: A group that was founded to aid the poor accessing financial services to elevate their poverty.

Credit Delivery Model: It's a model that is based on lending to groups of five either Men or Women and it was initiated by Grameen Bank in Bangladesh to assist the poor.

Crowdfunding: It's a combined effort by an individual or a group of people to collect the support and the fund needed to carry out an idea through explaining it to the public.

Lending Teams: Lending teams are self-organized groups through Kiva.org platform that share similar interests and lending goals.

Microfinance Institutions: Organizations that provide different financial services to the poor for little or no return.

Microfinance: Microfinance is an approach to provide different financial services to the poor.

Socio-Economic Environment: The social and economic environment for a certain country at a certain period of time.

Section 2

Social Government:

Communicating with the Constituency

Chapter 4
Globalisation, the Internet, and the Nation-State:
A Critical Analysis

Shefali Virkar
University of Oxford, UK

ABSTRACT

Our world today is in the midst of an historical change. Globalisation and spectacular advances in technology have given us an unprecedented peek into the future: a glimpse into a highly interconnected world governed by new paradigms, where the cost of transmitting and accessing an infinite amount of information is virtually nothing, where physical boundaries no longer limit human action – in short, a world characterised by the breakdown of conventional political, social, and economic institutions and systems previously considered rock-solid; spearheaded by the rise of the Internet and its associated technologies, platforms, and applications. This book chapter attempts a critical analysis of the relationship between Globalisation, the Internet, and the State. In evaluating the arguments that present the Internet as a threat to nation-state sovereignty, the work attempts to challenge accepted wisdom; purporting instead to demonstrate that, in many cases, the Internet, far from posing any threat to the attenuation of political power, actually strengthens the hand of the nation-state.

INTRODUCTION

..the biggest technological juggernaut to have ever rolled..-George Gilder on the Internet

Technology, twentieth century modernists prophesied, would dramatically alter the landscape of world politics. Advances in telecommunications and air travel would create a "global village", collapsing vast distances of space and time. The nation-state, dependent as it was on geographical boundaries for its very identity, would be eclipsed by non-state actors such as transnational corporations (TNCs), transnational social movements, and international organisations, all of which harnessed new technology to operate efficiently across borders. Today, much of what the modernists predicted has come to pass

DOI: 10.4018/978-1-4666-8556-7.ch004

(Keohane & Nye, 2002). Globalisation – increased global economic integration coupled with increases in cross-border social and cultural exchanges – has collapsed spatial and temporal barriers to economic and social exchange, a process speeded up by rapid advances in science and technology. The fate of the nation-state, however, remains ambiguous and subject to fierce debate.

Though a relatively recent phenomenon, the accelerating pace of globalisation has prompted the emergence of a large amount of scholarly literature, and to-date has generated a number of disparate "islands of theory" that focus on small parts of the larger question of how globalisation affects governance. Three domains of thinking have emerged as the most popular of the theories over the past few years (Legrain, 2002). The first is that of the 'race to the bottom'; where national governments, locked in fierce competition to keep highly mobile capital within their borders, are forced into lowering labour and environmental regulatory standards and reduce spending on social welfare. Another cluster of literature focuses on the growing importance of non-state actors such as TNCs, non-governmental organisations (NGOs), and transnational activist networks; whilst yet other theories focus on the ability of international institutions to provide effective global governance (Drezener, 2002). These different strands of thinking share one basic assumption: *globalisation attenuates state power relative to other forces in world politics*.

At the same time, many commentators believe that the weakening of the nation state has been accentuated by the deepening of local and global human interactions; a direct result of spectacular innovations in technology (Loader, 1997). The development of new Information and Communication Technologies (ICTs) has, in recent decades, revolutionised the organisation and dissemination of information in a way not thought possible before, bringing in its wake a so-called Information Revolution that has not only seen a dramatic fall in the cost of communications hardware (and software), but has also had a profound effect on political, economic, and social interactions globally (Choucri, 2000). The convergence of three trends – globalisation, world-wide electronic connectivity, and emergent knowledge networking practices – has reinforced the importance of the role played by knowledge and information in the global political and economic arenas. Today, we live in information societies poised on the threshold of the Information Age.

THE STATE, GLOBALISATION, AND THE INTERNET: A CRITICAL ANALYSIS

The influence of the state on the trajectory of human lives is more comprehensive and sustained than that of any other organizational construct. -Michael Zürn and Stephan Liebfried

Chief amongst the new technologies stands the Internet, a vast network of computers and satellites, whose effect on the global political economy has perhaps been more significant than any previous technology revolution. Today, almost anything people can do using conventional communications technology can be done using the Internet (Hill & Hughes, 1998). People meet and maintain friendships with others from different countries, shop online, keep abreast of politics and current affairs, post opinions, and have access to almost limitless information at the click of a button. At the same time, they can use the anonymity of the Internet to dupe and defraud, and to propagate unacceptable or profane ideas (Choucri, 2000). Information and Communication Technology networks such as the Internet thus have the potential to facilitate the deconstruction of national, financial, and cultural boundaries.

Current academic work on the Internet focuses primarily on its ability to transcend 'all barriers of territorial distances and borders'; a quality that, many scholars conclude, contributes directly to a decline of state authority (Drezener, 2002). These scholars believe that frontiers are irrelevant to electronic flows and that, as Internet and ICT usage in everyday life and business increases, the production, marketing, and distribution of goods and services (as well as the exchange of information and cultural products) will be increasingly carried out on a world stage and on a scale that undermines geographic boundaries. They call this global stage *cyberspace*, a "…globally networked, computer-sustained, computer-accessed, and computer generated, multidimensional, artificial or "virtual" reality. In this so-called reality, to which every computer is a window, "seen or heard objects are neither physical nor, necessarily, representations of physical objects but are, rather, in form, character, and action, made up of data, of pure information" (Benedikt, 1991). The ability of cyberspace to slip seamlessly across borders has spurred the rise of private regulatory systems; as national governments, limited to geographical jurisdictions, are backing away from its regulation and instead allowing private entities to move in. The decentralised nature of the network makes it easier for websites to be located and hosted anywhere in the world, whilst falling communication costs and rising computing power make it available as a platform for a larger number of non-state actors to coordinate their activities with greater effectiveness and efficiency (Spar, 1999).

Many commentators feel that globalisation, working in tandem with technologies of the Information Society, has challenged previous practices that served to reinforce the notion of national spaces – from border controls to autonomous governments – and has revealed a new global space: the Global Information Society, which is slowly engulfing the entire world and rendering the concept of the nation-state as 'a site of authority and power' a thing of the past (Birnhack & Elkin-Koren, 2003). On the other hand, others believe that whilst predictions regarding the importance of technology have been proved right, the nation-state has defied its critics and has been more resilient than expected. National identity is still as, if not more, important to people, and the state is still able to command the loyalty of a majority of its population. Governments, to a greater or lesser extent, still have a controlling stake in their countries' resources, and still hold monopoly power in the legislation and enforcement of laws (Drucker, 1997).

Which, then, is the correct view? This book chapter aims to explore whether or not the Internet and its associated platforms and technologies constitutes a threat to national sovereignty. It begins with a brief introduction to the Internet, and its gradual metamorphosis from a military tool to a popular, world-wide entertainment phenomenon. It then explores the main arguments that make the Internet out to be a threat to sovereignty and effective global governance. In doing so, this article attempts to challenge emerging wisdom by demonstrating how the Internet is not the threat it is made out to be, and aims instead to show how in many cases, the use of digital technologies actually strengthens the hand of the state.

HYPER-CONNECTIVITY: GLOBALISATION, THE INTERNET, AND ITS THREAT TO SOVEREIGNTY

All that is solid melts into air -Marshall Berman

Information has been feared by those in authority from time immemorial. The growth of the media and the spread of information have been regarded as serious threats to power ever since the time of James I of England, when the growth in popularity of the printing press was watched with great suspicion by both the Crown and the Church. The nascent print media gave rise to the first stirrings of parliamentary democracy in Britain and allowed, for the first time, the old political order to be challenged by new ideas (Perritt, 1998). The subsequent development of new and improved forms of communication over the next few centuries, beginning with the invention of the radio and the telegraph, and moving thence to the telephone and to television, posed new and greater threats to central power and authority (Thussu, 2000). Today, it may be claimed concretely that the global spread of technology and the subsequent growth of the global media have definitely impacted nation-states in ways that would be seen as 'interference' by many in positions of authority. An oft-quoted example of such media power is the *CNN Effect*: the impact that live, round-the-clock, global news coverage by all-news cable television channels has on public opinion and on foreign policymaking[1]. Many International Relations theorists and self-styled *cyberenthusiasts* believe that the Internet, with its seemingly limitless possibilities for information transmission and access, poses the strongest-ever technological challenge to state sovereignty by any medium in living memory.

The Internet was the brainchild of academic J.C.R. Licklider (who envisioned "a globally interconnected set of computers through which everyone could quickly access data and programs from any site") and engineer Paul Baran (who, concerned with the ease with which military communications could be disrupted by hostile states, proposed the creation of a network of computers using packet-switching technology as an alternative to conventional circuit switching technology, focusing on a single line of communication) (Abbate, 2001). In the early 1970s, the Advanced Research Project Agency (ARPA), a division of the U.S. Department of Defence, put the ideas of both men to the test with the launch of ARPANET. ARPANET aimed to make research on military defence-related issues efficient by enabling researchers and their government sponsors to share resources without having to physically deliver them.

The informal collegial, non-hierarchical working relationships that evolved were the chief cause of the early success of the network, and ARPANET soon transformed from a means of sharing data to a medium of communication (Warkentin, 2001). The late 1980s saw a boom in the sale of personal computers (PCs), and the creation of the World Wide Web in the mid-1990s completed its transformation from a military research tool to a popular communication medium with possibilities for "information gathering, social interaction, entertainment and self-expression". The growth of Internet use has since then been phenomenal, with traffic increasing by 100% a year as compared to less than 10% for conventional voice-based technologies (Nye, 2012).

The past two decades have borne witness to a robust debate on how globalisation alters governance and affects state sovereignty. As a practical arrangement, sovereignty is embodied in the unwritten collective rules of international relations and may be determined using two dimensions: the internal dimension (the ability of the state to exercise power within its national territories), and the external dimension (international recognition of the state as a sovereign body and its freedom from outside interference) (Sassen, 1997). Globalisation is seen to be encroaching on both the realms of internal sovereignty, states are no longer able to exercise exclusive power over their territory; and of external sovereignty, external interference in the working of states especially through international governance organisations.

The effects of globalisation and sovereignty can be distilled into two broad, but complementary, theories (Drezener, 2002). Firstly, globalisation, particularly through uncontrollable global market forces, undermines a government's ability to effectively regulate its domestic affairs. The state is forced into what Thomas Friedman terms "The Golden Straightjacket" that "narrows the political and economic choices of those in power to relatively tight parameters…" (Friedman, 2001). The overriding conclusion seems to be clear: 'omnipotent markets mean impotent politicians.'(Wolf, 2004). The second theory predicts the simultaneous waning of state power and the rise in the power of non-state state actors in global regulation. These actors vary from epistemic communities of technical experts to Non-Governmental Organisations (NGOs), from the formation of private governance structures by transnational corporations to the rise of a Global Civil Society (Haas, 1992).

All the effects of globalisation appear to be enhanced by the advance of the Internet. The Net is inexpensive and increasingly popular - current estimates suggest that over 800 million people were online as of September 2013, up from a little over 300 million in May 2000 (NUA, 2013). Like other technologies before it, the Internet is being used by people and groups who, by having their own 'agendas, resources and visions' for its future, ensure that the Internet follows in the tradition of other media to remain a threat to sovereign authority. However, unlike other earlier technologies, the Net possesses certain characteristics that make it a far more serious danger. Firstly, the Internet is not susceptible to the same physical and regulatory controls that conventional communication technologies like the telegraph, the telephone, the radio, and television are. Such conventional technologies depend on physical circuit technology and easily detectable broadcasting infrastructures constrained by geographical borders, and are thus easily subject to state control (Perritt, 1998).

THE INTERNET AND THE EROSION OF THE NATION-STATE

The Internet might be seen as a subversive's paradise – the ideal environment for polarization and extremism in politics. -Henry H. Perritt, Jr.

The Internet, relying as it does on decentralised packet-switching technology, was also designed to be virtually tamper-proof. The way in which the Net is designed to combine global reach with low entry barriers makes it difficult for individual governments to impose controls on content and information flows (Danitz & Strobel, 2001). At the same time, it is relatively easy for the average user to communicate globally as the Internet works through existing communications infrastructure (phone lines, satellites, etc.) and requires the acquisition of only very basic additional hardware (a computer and a modem at worse) and requisite software. The increase in the number of people participating in online activities has given rise to a unique transnational cyber-culture: one that celebrates freedom (particularly of thought and speech) and distrusts traditional political institutions to an extreme. *Cybersceptics*, who see the Internet as a threat to status quo, believe that the Net challenges **four** historic state functions and hampers the state from controlling power, wealth, and morals within its territory (Perritt, 1998). The four areas in which the Internet is believed to have eroded states' power are discussed below:

A. The Provision of National Security

Many commentators see the Internet as a subversive's paradise. The ease with which anyone may set up a website, and the difficulties faced by those officials trying to control site content, allows for anti-social elements and members of anti-establishment groups to make available their message to a global audience and communicate between themselves with ease (Chroust, 2000). In particular, the Internet allows for extreme opponents of the state – such as the Neo-Nazis or the Taliban – to globalise and modernise, to establish a flourishing 'counter-culture' that they then propagate online. Instantaneous anonymous and (often) un-monitored communication via e-mail, online real-time chat rooms, and message boards is an ideal way for terrorist organisations to plan and co-ordinate attacks, as demonstrated by Al-Qaeda's execution of the September 11th hijackings in 2001. The decentralised nature of the Internet also sits well with the loosely linked, cell-type form of network currently favoured by terrorist groups. Whilst many governments (particularly authoritarian states such as China) attempt to enforce measures to control the leaking of state secrets via the Internet, many believe that the Internet is able to work its way around such legal and bureaucratic obstacles. For example, the e-mails that came out of Russia narrating the attempted coup against Gorbachev in 1990 completely bypassed Russian state censorship (Miles, 2000). In short, the Net is thought to provide the perfect environment for polarisation and extremism in politics, and is considered to be the perfect medium through which national security might be compromised (Perritt, 1998).

B. The Regulation of Economic Activity

The nation-state appears to have lost the greatest amount of its power to the Internet in the economic sphere. In particular, there has been a shift in key economic concepts as assets and products where definitions are becoming more *informational*, and territorial space is becoming less important as factors crucial to economic success need no longer be fixed in specific locations (May, 2002). The Internet is revolutionising national and global commerce by enabling corporations to break the confines of national production systems through the formation of international or global networks that stretch across borders, and making it increasingly difficult for states to apply old models of control to electronic commerce (Perritt, 1998). Using information technology, TNCs are able to evade regulation and taxation by the state whilst at the same time establishing a global market for their goods. Increased capital mobility online has further reduced the nation-state's ability to regulate its own economy. In order to solve regulatory problems, such as those related to the jurisdiction of tax collection, the trying of on-line fraudsters from other countries, and the determination of the size and nature of e-commerce in order to be able to formulate proper economic policies, the nation-state is becoming more and more obliged to increase its dependence on private actors and to watch its power being continuously diluted. In the words of Castells (1997), nation-states 'having become part of a network of powers, are powerless by themselves: they are dependent on a broader system of enacting authority and influence from multiple sources'.

C. The Preservation of Moral, Cultural, and Civic Values

The third fear that haunts critics of cyberspace is that the nation-state has lost the power to control society's use of cyberspace, insomuch as the Net may be used by people for nefarious purposes for personal profit and towards individual gain (Jordan, 1999). This is illustrated by the anxiety that young children surfing the Internet unsupervised might be exposed to information ranging from bomb-making to cyber-pornography; including adult pictures, videos, and solicitations from paedophiles. *Cyberporn*, which may be accessed globally by users in countries where it is banned from servers based in countries that don't outlaw it, is a particularly contentious issue; as are the methods used to prevent the abuse of the Internet by pornographers (BBC News Media Reports, 2001). Whilst some governments seek to curb pornography and other so-called immoral content by restricting their citizens' access to the Internet through fierce monitoring, most intellectuals would argue that this does not stop pornographers from continuing their activities on-line, as they could arguably be based anywhere in the world and operate beyond the arm of any national law. In many countries, for example France and Germany, the Internet is seen as a threat to the government's role as the keeper of the national culture, and the inability of the state to regulate the inflow of information and outside 'foreign' influences points again to declining state power (Perritt, 1998). Many commentators believe that the influx of information via the Internet poses serious ideological challenges to autocratic ideologically-driven governments, and will eventually contribute to their political downfall. In China, for example, the 'rebel' cult Falun Gong was able to attract widespread support by publicising its proposal for a more spiritually-inclined alternative to communism on the Internet much to the anger of the Communist Party of China (Ownby, 2008)

D. The Undermining of International Cooperation

The Internet also stands accused of undermining international co-operation. The interpretation of issues related to the global regulation of the Internet such as privacy, control of content, and rights to free speech differ across borders, and there is a distinct lack of international co-operation in their related regulatory affairs. Each state has its own national laws to tackle those problems it considers priority, and the potential deadlock that might arise out of differing laws can feed back into existing controversies, worsening international relations (Virkar, 2013). The concept of *cyberspace* has underlined the feeling that activity – both economic and social – is no longer linked to particular physical locations. Many scholars believe that an increasing amount of economic activity is not confined to territorial spaces and that any national law concerned with goods and information that might be digitised is in danger of being overturned in an online digital environment (Coyle, 1997). Similarly, the ability of the Internet to enhance the extension of social relationships beyond national boundaries has fuelled speculation that national identity will be eclipsed as people forge stronger non-national affiliations. Sklair's (2002) idea of a transnational capitalist class, comprising of people linked together through their participation in the capitalist system, is a case in point. In cyberspace, nation-states appear unable to govern things as they did in the past, and the notion of 'place' has changed from one defined by governments and their agencies to one which has emerged from private communications networks set up to serve global commercial interests. Cyberspace is thus seen by many to be the place in which the combined effects of globalisation and technology are at their strongest: states are at their most enfeebled and non-state actors at their most powerful (Drezener, 2002). This is certainly the view of most scholars and *cyberenthusiasts*, succinctly summed up by *cyberguru* John Perry Barlow (1995:76): "By creating a seamless global economic zone, borderless and unregulatable, the Internet calls into question the very idea of the nation-state."

'AND STILL IT MOVES': TECHNOLOGY, CYBERBOLE, AND REALITY

Despite all its shortcomings, the nation-state has shown amazing resilience -Peter Drucker

Whilst arguments that claim that the Internet coupled with globalisation undermines sovereignty are plausible, it would be wrong to conclude that the nation-state has been rendered obsolete. Pro-globalisers argue that the fundamental basis of such arguments, which claim that globalisation has been imposed on reluctant nation-states who are then forced to passively let their functions be undermined, is false (May, 2002). They point out, instead, that states have actively pushed globalisation as part of their policies for centuries and have consciously taken decisions towards a greater decentralisation of their functions, resulting in both the emergence of new rules and in the improvement of old ones (Legrain, 2001). Some commentators do not subscribe to the view that ICTs and globalisation will result in a complete breakdown of national borders, and consider those who hold such views to be myopic. They argue instead that the Internet, although undeniably revolutionising the way people work, think, and interact, does not have the capacity to render borders and systems shaped over centuries irrelevant overnight (Perritt, 1998).

Scholars who predict and prophesy the demise of the nation-state as a consequence of the advent of the Internet and its associated technologies, often fail to recognise the way in which political systems across the world have negotiated challenges posed by innovations and developments in communications technology during different periods in history (May, 2002). Though this does not mean to say that the current situation is static and that states are *not* threatened in the Information Age, many commentators feel that theories of state decline project a rather oversimplified version of the relationship between the nation-state and the global information society. Recent academic work recognises that most claims, which predict the marginalisation of the state with respect to the control of the national economy, fail to take into account the central role national governments have played and will always play in the propagation of capitalism as well as in the establishment of a regulatory framework within which transborder capital has thrived (Scholte, 1997). Similarly, many scholars contend that the argument that presupposes that ICTs pose a threat to democracy is equally misplaced, and that although the Internet does pose a threat to authoritarian governments, it enhances significantly the functioning of democratic ones.

The Politics of 'Cybernomics': Net Benefits and Net Profits

Given that at its core, globalisation is an expression of global economic transactions, most arguments concerned with the decline of the state fail to recognise the key role it plays as the guarantor of the legislative infrastructure that governs market activity. Most accounts of globalisation ignore interventions of the state, in essence making the emergence of the market a 'natural' occurrence. Whilst this assumption is good enough to build up basic theories of globalisation with relation to the rise of the network society, they are considered incomplete as they generally ignore the foundations on which these two phenomena rest. In the fast emerging 'knowledge economy', the nation-state is crucial for establishing and enacting the legal framework on which all innovation in the capitalist era depends, viz. property law and in particular intellectual property rights (Birnhack & Elkin-Koren, 2003). Intellectual property is essential for the success of today's profit-driven entrepreneurs and is one area in which even Transnational Corporations would not demand less regulation. In addition, from the piracy of music through the MP3 format to other related software to accusations against multinational companies such as the pharmaceutical industry's 'profiteering' from patent-protected drugs, intellectual property is riddled with disputes of legitimacy and social impacts. Laws to deal with such disputes look to state authority for both their existence and their enforcement, and in that governance role the nation-state has no real competition from non-state actors (Gasser & Ernst, 2006). It must also not be forgotten that many of the central technologies of the Information Age, including the Internet itself, have initially been invented and used by the state as part of its defence or economic policies.

Governments are also users of data and technology on a massive scale and utilise the resources available to maintain their regulatory capacity in society and in the economy, even in areas like the regulation of financial markets (Taylor, 1999). The practice of monitoring financial activities, either through licensing or formalised regulation, remains common. This has led some commentators to believe that "…as anti-money laundering initiatives have become more extensive, they have begun to demonstrate that the new technologies that are said to erode state regulatory capacity today may in fact even *enhance* it", thereby reinforcing territorial integrity and nation-state authority. Similarly, studies have shown that regardless of the spectacular developments in ICTs over the past few decades, over 80% of production in developed economies is still earmarked for domestic consumption and 80% of all investment in those economies is made by domestic investors (Wade, 1996). In sum, although there has been a significant growth in cross-border activities, they are not representative of the majority of economic interactions, which are still very much bounded by national borders and geographical territory.

On the other hand, one area in which some analysts believe that digitalisation and a loss of autonomy might be a good thing is that of financial and monetary policy (Drucker, 1997). They point out that tight monetary and fiscal control have been seen to benefit special interest groups to the detriment of society in general, and, that the loss such sovereignty would strengthen rather than weaken the nation-state. Overall, however, it must be argued that "capitalism has not escaped the state, but rather the state has, as always, been a fundamental constitutive element in the very process of extension of capitalism in our time" (Panitch, 1996: 109). Economies are still subject to rules set by the state, and are driven chiefly by state-dictated economic policy. Many scholars feel that rather than a deregulation of the economy, what we are seeing is its *re-regulation*, a modification of existing rules to suit online digital environments rather than their complete removal. Differentially regulated areas within a state's jurisdiction (with particular regard to offshore markets that have lax regulations compared to the mainland) are considered to be more *innovations* in regulation and development strategies rather than measures of state decline (Mansell, 1996). Consequently, ICTs enabling such operations are not doing so at the cost of nation-state power.

Virtual Democracy: Is Information Power?

A diverse group of commentators, whilst acknowledging that a greater amount of information is in circulation today than ever before, are highly sceptical about the role of the Internet in the enhancement of democracy. Two strands of thinking have emerged in academic circles. The first suggests that the democratic process might be undermined due to a lack of useful public information, with public service broadcasts sidelined in favour of commercially profitable entertainment (Webster, 2002). The Internet, with its on-line chat rooms, gaming sites, and commercial operations is seen as a particular source of 'useless information' and 'frivolous pass-times', and is said to contribute to a decline in civic participation as people retreat into their private worlds of 'infotainment'. The other strand of thinking notes the emergence of a new kind of 'networked politics', which, in mobilising people into interest groups across borders, enhances democracy but contributes to the overall decline of state authority in favour of grassroot networks (Virkar, 2014). Success stories of such groups include the Zapatista movement in Mexico and the global Campaign against the Multilateral Agreement on Investment (MAI). Governments that attempt to resist such changes are labelled 'authoritarian', and are seen to be 'anti-democratic' and 'repressive'. States, in effect, face a lose-lose situation in their bid to maintain democracy online in the Information Age.

Despite such a negative outlook, it may be argued that nation-states today still have a significant role to play in the maintenance of democracy. Proponents of this view believe that the nation-state is the clearest legitimate authority in existence, which, through the enactment of legislation covering issues such as privacy and anti-discrimination rights, can ensure that the Information Age remains 'meaningfully' democratic, and thereby protect minority and individual rights (Virkar, 2013). The state must also provide backing to ensure that *cyberspace*, by giving equal voice to all, becomes a *healthy* public space for vigorous debate and discussion. Governments in some developed countries are taking tentative steps towards *e-governance*, a concept underscored by the idea that governance online loses none of its current bite or legitimacy, and are making themselves more accessible through the setting up of official websites that offer citizens and netizens alike a equal range of services. In the long run, therefore, e-governance policies increase both state efficiency and democratic participation by the public (Mechling, 2002). It must be noted, however, that as a consequence of state participation online a number of contentious issues arise, particularly those regarding surveillance and privacy. Whilst there is a furious debate raging amongst *cyberenthusiasts* as to just how much regulation of the Net is too much, it is generally agreed that there is a general need "to protect public safety and prevent abuses of power while simultaneously promoting governmental flexibility and learning" (Mechling, 2002).

Whilst the Internet is thus seen to cause many jurisdictional problems, some scholars feel that instead of robbing states of the ability to rule, the new Information and Communication Technologies, and in particular the Internet, have enhanced the possibility of the expansion of jurisdictional space (Lenk, 1997). For instance, in November 2000, the French government prosecuted Yahoo!, headquartered in USA, for the sale of Nazi memorabilia, illegal under French national law (The Economist, 2001). Although the government only sought to prevent its citizens from viewing the memorabilia on Yahoo!'s American sites, and the company (being based in America) was not formally subject to French law, Yahoo! nonetheless bowed to French pressure and removed the material from its site altogether. Even though it may not be possible for politically weaker states to wield such clout in the near future, this and many other cases go to show that the nation-state hasn't yet lost its legal bite[2].

The final acid test this section puts forward deals with one of the primary obligations of a 'sovereign' state: the effective maintenance of national security. As discussed earlier, the advent of new ICTs has altered the nature of conflict and has spurred the rise of a new type of battle – the Information War – which involves the disruption of information networks through the use of aggressive software, and where loosely organised terrorist networks use the Net to communicate and co-ordinate attacks (Arquilla & Ronfeldt, 2001). Whilst many consider the state to be too large and unwieldy to respond quickly to small-scale sophisticated *cyberattacks*, others think it able to mobilise significant resources to conduct such a war effortlessly, effectively gathering and using relevant information available both on- and offline. Most theories of the Information War regard the state as playing a major role in equipping society to deal with new threats, and it is generally agreed that the state hasn't withdrawn from the maintenance of national security (May, 2002).

W(H)ITHER THE STATE?: A DEATH FREQUENTLY FORETOLD

The nation-state is dead. Long live the nation-state. -The Economist

At the dawn of the 21st Century we are living in nascent Information Societies, still searching for meanings and concepts to describe the new age that we are creating. In this context, the current debate surrounding the nation-state poses important questions regarding the place of effective sovereignty in the Information Age and beyond. Whilst some commentators would have us believe that Information Technology has replaced the state as the remover of market- and other inefficiencies, and that the importance of physical space and national boundaries has been forever conclusively undermined, this chapter argues that this is not the case. This researcher, along with other scholars, believes that the nation-state is still able to mobilise extensive information resources and has always been a major user, producer, and supplier of the information available in the public domain. In addition, the state remains crucial to the emergence and maintenance of the Information Society and the Information Age Economy.

To date, there has been no other institution capable of political integration, and no other defined unit capable of obtaining membership in the world politico-economic community. That said, whilst the claim that state-oriented politics has gone forever is false, there is undoubtedly space for the reconfiguration of the nation-state in order to make best use of new advantages. Cyberspace undermines the nation-state to the extent that it doesn't allow it to exist in isolation, heedlessly pursuing self-interested policies. Transnational flows of information ensure interdependence between the various nation-states and, as with all technological changes, there will be significant economic, social, and political shifts. The key to being able to respond to rapid change and uncertainty lies in gaining a better understanding of the basic laws that govern both economics and politics, and in recognising that neither ICTs nor globalisation, building (as they do) on the past, require new fundamentals.

Despite all its deficiencies, and all the grandiose claims surrounding its premature 'demise', therefore, it would be safe to predict that the nation-state, greatly changed yet easily recognisable, will 'survive' the globalisation of global commerce, society, and the accompanying Information Revolution, and will continue to play a key role in the new digital international political economy.

REFERENCES

Abbate, J. (2001). Government, Business, and the Making of the Internet. *Business History Review*, *75*(01), 147–176. doi:10.2307/3116559

Arquilla, J., & Ronfeldt, D. (2001). The Advent of Netwar (Revisited). In J. Arquilla & D. Ronfeldt (Eds.), Networks and Netwars: The Future of Terror, Crime, and Militancy (pp. 1-25). Santa Monica, CA: RAND Publications.

Barlow, J. P. (1995). Thinking Locally, Acting Globally. *Time*, *147*(3), 76.

BBC News Media Reports. (2001, April 29). China Internet Café Debate Hots Up. *BBC News Online*. Retrieved from: http://news.bbc.co.uk/1/hi/world/monitoring/media_reports/1302309.stm

Benedikt, M. (1991). Cyberspace. First Steps. Cambridge, MA: M.I.T. Press.

Birnhack, M. D., & Elkin-Koren, N. (2003). The Invisible Handshake: The Re-emergence of the State in the Digital Environment. *The Virginia : The Journal of Law and Technology*, (Summer), 2–3.

Castells, M. (1997). The Power of Identity: *The Information Age: Economy, Society and Culture (vol. 2)*. Oxford, UK: Blackwell Press.

Choucri, N. (2000). Introduction: CyberPolitics in International Relations. *International Political Science Review*, *21*(3), 243–263. doi:10.1177/0192512100213001

Chroust, P. (2000). Neo-Nazis and the Taliban On-Line: Anti-Modern Political Movements and the Modern Media. In P. Ferdinand (Ed.), *The Internet, Democracy and Democratisation* (pp. 102–117). London: Frank Cass. doi:10.1080/13510340008403647

Coyle, K. (1997). Coyle's information highway handbook: A practical file on the new information order. Chicago, IL: American Library Association.

Danitz, T., & Strobel, W. P. (2001). Networking Dissent: Cyber Activists Use the Internet to Promote Democracy in Burma. In J. Arquilla & D. Ronfeldt (Eds.), Networks and Netwars: The Future of Crime, Terrorism and Militancy (pp. 129-169). Santa Monica, CA: RAND Publications.

Drezner, D. W. (2002). *Who Rules? The Regulation of Globalisation*. Paper presented at the 2002 Midwestern Political Science Association Annual Meeting. Retrieved from: http://apsaproceedings.cup.org/Site/abstracts/017/017012DreznerDan.htm

Drucker, P. F. (1997). The Global Economy and the Nation State. *Foreign Affairs*, *76*(5), 159–171. doi:10.2307/20048206

Friedman, T. L. (2001). *The Lexus and the Olive Tree: Understanding Globalization*. New York: Anchor Books.

Gasser, U. & Ernst, S. (2006). *From Shakespeare to DJ Danger Mouse: A Quick Look at Copyright and User Creativity in the Digital Age*. The Berkman Centre for Internet and Society Research Publication No. 2006-2005.

Haas, P. (1992). Introduction: Epistemic Communities and International Policy Coordination. *International Organization*, *46*(1), 1–35. doi:10.1017/S0020818300001442

Hill, K. A., & Hughes, J. E. (1998). *Cyberpolitics: Citizen Activism in the Age of the Internet*. Oxford, UK: Rowman and Littlefield Publishers.

NUA Internet Surveys. (2013). *How Many Online?* Retrieved February 1, 2014, from: http://www.nua.ie/surveys/how_many_online/

Jordan, T. (1999). *Cyberpower: The Culture and Politics of Cyberspace and the Internet*. London: Routledge Press. doi:10.4324/9780203448632

Keohane, R. O., & Nye, J. S. Jr. (2002). Power and Interdependence in the Information Age. In E. C. Kamarck & J. S. Nye Jr., (Eds.), *Governance.com: Democracy in the Information Age* (pp. 161–177). Washington, DC: Brookings Institution Press.

Legrain, P. (2002). *Open World: The Truth About Globalisation*. London: Abacus Press.

Lenk, K. (1997). The Challenge of Cyberspatial Forms of Human Interaction to Territorial Governance and Policing. In B. D. Loader (Ed.), *The Governance of Cyberspace: Politics, Technology and Restructuring* (pp. 126–135). London: Routledge Press. doi:10.4324/9780203360408_chapter_8

Loader, B. D. (1997). The Governance of Cyberspace: Politics, Technology and Restructuring. In B. D. Loader (Ed.), *The Governance of Cyberspace: Politics, Technology and Restructuring* (pp. 1–22). London: Routledge Press. doi:10.4324/9780203360408_chapter_1

Mansell, R. (1996). Innovation in Telecommunication Regulation: Realising National Policy Goals in a Global Marketplace. In W. H. Dutton (Ed.), *Information and Communication Technologies: Visions and Reality* (pp. 371–386). Oxford, UK: Oxford University Press.

May, C. (2002). *The Information Society: A Sceptical View*. Cambridge, MA: Polity Press.

Mechling, J. (2002). Information Age Governance: Just the Start of Something Big? In E. C. Kamarck & J. S. Nye Jr., (Eds.), *Governance.com: Democracy in the Information Age* (pp. 141–160). Washington, DC: Brookings Institution Press.

Miles, J. (2000, January 29). Can Governments Control the Internet? *BBC News Asia Pacific*. Retrieved from: http://news.bbc.co.uk/1/hi/world/asia-pacific/623339.stm

Ownby, D. (2008). *Falun Gong and the Future of China*. New York: Oxford University Press. doi:10.1093/acprof:oso/9780195329056.001.0001

Panitch, L. (1996). Rethinking the Role of the State. In J. Mittelman (Ed.), *Globalisation: Critical Reflections. In International Political Economy Yearbook* (Vol. 9, pp. 83–113). London: Lynne Rienner.

Perritt, H. H. (1998). The Internet as a Threat to Sovereignty? Thoughts on the Internet's Role in Strengthening National and Global Governance. *Indiana Journal of Global Legal Studies*, *5*(2), 431–344.

Sassen, S. (1997). *Losing Control: Sovereignty in an Age of Globalisation*. New York: Columbia University Press.

Scholte, J. A. (1997). Global Capitalism and the State. *International Affairs*, *73*(3), 427–452. doi:10.2307/2624266

Sklair, L. (2002). *Globalization: Capitalism and Its Alternatives*. Oxford, UK: Oxford University Press.

Spar, D. L. (1999). Lost in (Cyber) Space: The Private Rules of Online Commerce. In C. A. Cutler, V. Haufler, & T. Porter (Eds.), *Private Authority and International Affairs* (pp. 31–49). Albany, NY: SUNY Press.

Taylor, J. (1999). The Information Polity. In W. H. Dutton (Ed.), *Society on the Line: Information Politics in the Digital Age* (pp. 127–128). Oxford, UK: Oxford University Press.

The Economist. (2001). Special: The Internet and the Law: Stop Signs on the Web. *The Economist*, *358*(8204), 21-23.

Thussu, D. K. (2000). *International Communication: Continuity and Change*. London: Arnold Publishers.

Virkar, S. (2013). Designing and Implementing e-Government Projects: Actors, Influences, and Fields of Play. In S. Saeed & C. G. Reddick (Eds.), Human-Centered Design for Electronic Government (pp. 88-110). Hershey, PA: IGI Global Inc.

Virkar, S. (2014). Consulting the British Public in the Digital Age: Emerging Synergies and Tensions in the Government 2.0 Landscape. In L. G. Anthopoulos & C. G. Reddick (Eds.), *Government e-Strategic Planning and Management: Practices, Patterns, and Roadmaps* (pp. 185–203). New York: Springer Link. doi:10.1007/978-1-4614-8462-2_10

Wade, R. (1996). Globalisation and its Limits: Reports of the Death of the National Economy are Greatly Exaggerated. In S. Berger & R. Dore (Eds.), *National Diversity and Global Capitalism* (pp. 60–88). Ithaca, NY: Cornell University Press.

Wartenkin, C. (2001). *Reshaping World Politics: NGOs, the Internet and Global Civil Society*. Oxford, UK: Rowman and Littlefield Publishers.

Webster, F. (2014). *Theories of the Information Society* (4th ed.). London: Routledge Press.

Wolf, M. (2004). *The Role of the State in an Era of Globalisation. The Sir Ronald Trotter Lecture, 2004*. Wellington, New Zealand: New Zealand Business Roundtable.

ADDITIONAL READING

Cleaver, H. Jr. (1998). The Zapatista Effect: The Internet and the Rise of an Alternative Political Fabric. *Journal of International Affairs*, *51*(2), 621–640.

Cutler, A. C., Haufler, V., & Porter, T. (1999). *Private Authority and International Affairs*. Albany, N.Y.: SUNY Press.

Deibert, R. J. (2002). International Plug 'n Play?: Citizen Activism, the Internet and Global Public Policy. *International Studies Perspectives*, *1*(3), 255–272. doi:10.1111/1528-3577.00026

Drezner, D. W. (2004). Global Governance of the Internet: Bringing the State Back In. *Political Science Quarterly*, *119*(3), 477–498. doi:10.2307/20202392

Dutton, W. H. (1996). *Information and Communication Technologies: Visions and Realities*. Oxford: Oxford University Press.

Falk, R. (1997). State of Siege: Will Globalisation Win Out? *International Affairs*, *73*(1), 123–136. doi:10.2307/2623553

Foster, W. A., Rutkowski, A. M., & Goodman, S. E. (1997). Who Governs the Internet? *Communications of the ACM*, *40*(8), 15–20. doi:10.1145/257874.257877

Frissen, P. (1997). The Virtual State: Postmodernisation, Informatisation, and Public Administration. In B. D. Loader (Ed.), *The Governance of Cyberspace: Politics, Technology and Restructuring* (pp. 111–125). London: Routledge Press. doi:10.4324/9780203360408_chapter_7

Kaase, M. (2000). Political Science and the Internet. *International Political Science Review*, *21*(3), 265–282. doi:10.1177/0192512100213002

Kalathil, S., & Boas, T. C. (2001). The Internet and State Control in Authoritarian Regimes: China, Cuba and the Counterrevolution. *First Monday*, *6*(8). Retrieved from: http://www.firstmonday.dk/issues/issue6_8/kalathil/

Langmore, K. (2002). Globalisation: Challenges of Governance. *United Nations University/Melbourne University Working Paper*, May 2002. Retrieved from: http://www.unu.edu/p&g/gesgp/workingpapers/Langmore2.pdf

Most, B., & Starr, H. (1984). International Relations Theory, Foreign Policy Substitutability, and 'Nice' Laws. *World Politics*, *36*(3), 383–406. doi:10.2307/2010380

Sassen, S. (1998). On the Internet and Sovereignty. *Indiana Journal of Global Legal Studies*, *5*(2), 545–559.

The Economist (1995). 'The nation-state is dead. Long live the nation-state.'. *The Economist*, *337* (7946), 15-18.

Virkar, S. (2011). *The Politics of Implementing e-Government for Development: The Ecology of Games Shaping Property Tax Administration in Bangalore City*, Unpublished Doctoral Thesis, University of Oxford: Oxford.

Wang, C. (1999). Communications Technology and the Retreat of the State. *Department of Information Studies, University of California (Los Angeles), Student Papers Fall 1999*. Retrieved from: http://is.gseis.ucla.edu/impact/f99/Papers/wang.html. Accessed: 13th April 2014.

Wapner, P. (1995). Politics Beyond the State: Environmental Activism and World Civic Politics. *World Politics*, *47*(3), 311–340. doi:10.1017/S0043887100016415

Woolgar, S. (2002). *Virtual Society? Technology, Cyberbole, Reality*. Oxford: Oxford University Press.

Wriston, W. B. (1997). Bits, Bytes, and Diplomacy. *Foreign Affairs*, *76*(5), 172–182. doi:10.2307/20048207

KEY TERMS AND DEFINITIONS

CNN Effect (The): Refers to the impact that live, round-the-clock, global news coverage by dedicated, all-news cable television channels has on public opinion and on foreign policy making. *The CNN Effect* was first noted when the news channel CNN showed heart-breaking footage of starving children in Somalia that pressured U.S. officials to send troops there. Subsequent horrifying footage of Somalis dragging the body of a dead American soldier through the streets followed; turning American public opinion against the intervention, and prompting U.S. officials to withdraw their troops.

Cyberspace: Refers to the global platform or stage comprising of a highly-decentralised, pan-globally networked, human- and computer-sustained, computer-accessed, and computer generated multidimensional, artificial, and/or otherwise "virtual" reality.

e-Democracy: May be defined by the express intent of an authority to increase the participation of citizens in decision- and policy making processes through the use of digital media and by the application of Information and Communication Technologies to political processes. *e-Democracy* as a concept may be sub-divided into *e-Engagement*(or *e-Participation*), *e-Voting*, and *e-Consultation*.

e-Governance: Refers broadly to the use of Information and Communication Technologies by government, civil society, and/or by political institutions to engage citizens and/or the general public at large in political processes and to promote the greater participation of individual citizens in the public sphere.

e-Government: Refers to the specific use of Information and Communication Technologies by government departments, institutions, and agencies to improve internal governmental functioning and public service provision. Broadly speaking, *e-Government* may be divided into 2 distinct areas: *e-Administration* and *e-Services*.

Globalisation: Refers to the pan-global phenomenon defined as the rapidly developing process of complex interconnectedness between societies, cultures, institutions, and individuals worldwide. It is a social process which involves the compression of time and space; the shrinking of geo-spatial distances through a dramatic reduction in the time taken – either physically or representationally – to cross them, so making the world seem smaller, and in a certain sense bringing human beings 'closer' to one another.

Information and Communication Technologies: Or **ICTs** is an umbrella term employed to identify any so-called 'smart' (usually electronic or digital) communications hardware or technological device, networked platform, or software application utilised to enable and/or to facilitate the rapid electronic production, transmission, processing, communication, and consumption of increasingly vast quantities of information and know-how. (i) *ref. Computing*. Broadly understood, the base definition of *Information and Communication Technologies (ICTs)* encompasses both multimedia and interactive technologies that include radio, television, cellular phones, computer and network hardware and software, and satellite systems; together with the various services and applications associated with them, such as video-conferencing, virtual realities, computer-aided design, the information superhighway, and technologies for consumer profiling and surveillance.

Information Revolution (The): Refers to the global economic paradigm prevalent from the late 1990s onward, characterised collectively by both unprecedented advancements in technological innovation and the rapid global proliferation, appropriation, application, and use of new digital *Information and Communication Technologies (ICTs)* in everyday life. The Information Revolution is further described against three constituent evolutionary strands of scholarship - the *Computer Revolution*, the *Internet Revolution,* and the *Digital Revolution.*

Information Society (The Global): Refers to the global socio-economic paradigm prevalent from the late 1990s onward, characterised by the centrality and the importance of *information* and *Information and Communication Technologies (ICTs)* to human interaction and endeavour; conceptualised and evaluated along several distinct axes including the *technological*, the *economic*, the *occupational*, the *spatial,* and the *cultural.*

Internet (The): Refers to a global computer network consisting of interconnected digital technologies, applications, and platforms bundled together using standardized communication protocols; all of which enable the rapid electronic production, transmission, processing, communication, and consumption of increasingly vast quantities of information and know-how.

ENDNOTES

1 The CNN Effect was first noted when the news channel CNN showed heartbreaking footage of starving children in Somalia that pressured U.S. officials to send troops there. Subsequent horrifying footage of Somalis dragging the body of a dead American soldier through the streets followed, turning US public opinion against the intervention and prompting U.S. officials to withdraw their troops.

2 Two important examples of state primacy are: 1. The Chinese government's success at preventing a majority of its citizens from accessing sites containing 'unwanted' content through setting up a "Great Firewall of China" and 2. The possibility that in the future, Euro-zone laws could allow European citizens to sue businesses based beyond the borders of the European Union (Source: The Economist, 13th January 2001)

Chapter 5
Seeking an Online Social Media Radar

James ter Veen
The George Washington University, USA

Shahram Sarkani
The George Washington University, USA

Thomas A. Mazzuchi
The George Washington University, USA

ABSTRACT

In this paper we identify a method, which rapidly analyzes vast amounts of data present in social media in order to forecast crowd sizes. Based upon comparative analysis of related literature, a conceptual model is proposed and research conducted to develop capabilities to forecast mass collective action behavior such as crowd formation using Social Network Analysis (SNA) tools applied to online social media. We demonstrate that a simple model of online social network parameters can produce situation awareness of crowd sizes in much the same way that radar sensors can produce situation awareness of air traffic density. A prototype online social media 'radar' sensor system is developed and tested in a pilot study with a dataset of tweets gathered regarding the Occupy Wall Street movement. Further work is suggested which could provide anticipated crowd location, movement and intent in addition to size.

INTRODUCTION

Studies have shown that effects of mass collective actions such as diffusion of information across a network, network traffic bandwidth consumption, movie ticket purchases stock prices and Academy Awards winners can all be predicted by leveraging social network analysis tools to evaluate online social media (Asur & Huberman, 2010; Bothos, Apostolou, & Mentzas, 2010; De Choudhury, Sundaram, John, & Seligmann, 2010; Gloor, Krauss, Nann, Fischbach, & Schoder, 2009; Vashist, Mau, Poylisher, Chadha, & Gosh, 2011). With threats from non-state, terrorist organizations on the rise, analysis of online social network media may be a factor in providing an accurate early warning system. The Secretary of De-

DOI: 10.4018/978-1-4666-8556-7.ch005

fense (2011) recognizes recent research advancements made in the analysis of sociocultural behavior and recommends work to develop a "social radar" using multiple technologies to sense changing trends and forecast collective perceptions and actions (Costa & Boiney, 2012). Such a tool would require integration of multiple modalities to support prediction of population attitudes, sentiments and ultimately behaviors using multiple sources of information including online social media (Maybury, 2010). For the purposes of this study, we initially define our prototype 'online social media radar' as a software tool, which uses correlation of data gathered from online social media to sense and forecast sizes of crowd formations. Further development and extension of this prototype, combined with other prototypes under development could meet the needs identified by the Department of Defense (Costa & Boiney, 2012). We survey the trade space of applicable theories and techniques, identify alternative solutions and evaluate the alternatives using a pilot study dataset.

APPROACH

In the course of our investigation of the predictive capability of online social network media we elicit and refine high-level mission needs and user requirements for a radar-like sensor system, which scans social media instead of the skies. Next, we explore the trade space of innovative social networking analysis tools using the sensor system requirements as criteria. Using the trade study results, we identify alternatives for use in our prototype. Then we incorporate candidate approaches in a conceptual prototype model. Using empirical data, we test the approaches for validity in a candidate evaluation. We analyze the test data and report the results obtained from the prototype online social media radar. Finally we validate the results of the analysis for conformance to the high-level sensor system requirements.

BACKGROUND: SENSOR SYSTEM REQUIREMENTS

If the sensor to be evaluated were a new radar system, requirements might include a set of statements such as 'the radar beam shall sweep through 360 degrees of airspace. One complete rotation of the beam shall be accomplished in no more than 5 seconds. Positional error of radar plots shall be no greater than 20%.' For an online social media 'radar' system, we derive a high-level set of generic system requirements shown in Table 1 (for the purposes of prototype development, this is not an exhaustive set of all possible system requirements) from published very high-level electronic radar system specifications (Lockheed Martin, 2010; Thales Raytheon Systems, 2011; Wolff, 2009). First we need to define 'online social media radar'. Maybury (2010) defines a 'social radar' as a tool that "needs to sense perceptions, attitudes, beliefs and behaviors" and that it would use signatures "and correlation to sense, if not forecast, a broad spectrum of phenomena" including changing trends in population behaviors. For the purposes of this study, we initially define our prototype 'online social media radar' as a software tool, which uses correlation of data gathered from online social media to sense and forecast sizes of crowd formations.

Our first requirement parallels the basic operation of a radar system, which is to report the position and altitude of detected radar reflections in near-real time (Lockheed Martin, 2010). The positions of radar reflections are raw data points in a radar system. For our online social media radar system, we derive our first requirement: The sensor system shall report raw social network parameter data points from online social media channels in near-real time. This derived requirement is consistent with the

Table 1. Derived Online Social Media Radar Sensor System Requirements

#	Derived Online Social Media Radar System Requirements
1.0	The sensor system shall report raw social network parameter data points from online social media channels in near-real time.
1.1	The sensor system shall operate without operator intervention, once initiated.
2.0	The sensor system shall reject interfering signals in the monitored social media channels.
2.1	The sensor system shall reduce the effect of clutter found in social media channels.
3.0	The online social network media analysis system sensor shall optimize the probability of prediction of crowd sizes with no greater than 20% error.
4.0	The system sensor shall be capable of processing high volumes of data with near-real time performance.
4.1	The system sensor shall have an operational availability (A_o) > 99% of the time (a mean time between maintenance (MTBM) no less than 1250 hours and a mean maintenance downtime (MDT) no greater than 10 hours such that A_o = MTBM/(MTBM + MDT) = 99.2%) (Blanchard 2004).
4.2	The system sensor shall have a mean time between failure (MTBF) in excess of 1250 hours.

Secretary of Defense (2011) requirement to detect social and cultural signals. A solid basis exists in the relevant literature to support the use of applications to characterize and analyze online social networks to tap the 'wisdom of the crowds' (Anklam, 2005; Hansen, Shneiderman, & Smith, 2011; King, Li & Chan, 2009; Klamma, Spaniol, Cao, & Jarke, 2006; Pallis, Zeinalipour-Yazti, & Dikaiakos, 2011; Tang & Huan, 2010; Zeng, Hsinchun, Lusch, & Shu-Hsing, 2010). Marketers and politicians are already using the demographics that can be gleaned from social media to refine target audiences, increase donations and add votes (King et al., 2009; Wills & Reeves, 2009). These examples show that raw social network parameters can be extracted from online social media. In addition to detection of radar reflections, electronic radar sensors are required to operate automatically without manual inputs required (Wolff, 2009). From this we derive a sub-requirement (shown in Table 1) that our social media radar shall operate without operator intervention.

The next major requirement we find for electronic radar systems is to reduce noise by rejection of interfering signals so that required information can be optimally detected (Wolff, 2009). Noise is always present in signal processing systems and it can be easily introduced into the large volumes of dynamic data in social media channels such as blogs due to the fact that anyone can write a blog anonymously without any constraints of ethics or professional standards that apply to most trained professional journalists (Johnson & Kaye, 2004; Santos, Pan, Arendt, & Pittkin, 2006; Zeng et al., 2010). Challenges to Social Network Analysis (SNA) presented by large amounts of unstructured data subject to the vagaries of language, dialects and slang can be overcome (Bermingham, A., Conway, M., McInerney, L., O'Hare, N., & Smeaton, A., 2009). However, simply counting friends and links alone is not enough to analyze a network; user activity, rate of updates and trust are also important factors (Varlamis, Eirinaki, & Louta, 2010; Wills & Reeves, 2009). The literature indicates that analysis of social network meta-data yields useful results but must be filtered and analyzed further, thus we derive our second major requirement in Table 1 which is to optimize rejection of interfering signals in the monitored social media channels. This requirement is also consistent with the Secretary of Defense (2011) requirement to monitor operationally relevant behavior. In addition to rejection of interfering signals, the next requirement found for electronic radar sensors indicates that clutter shall be reduced. In electronic radar systems designed to track airplanes, reflections of radar waves from uninteresting objects such as rain, ocean surface and land terrain are considered clutter (Wolff, 2009). High quality radar systems can distinguish interesting signals from clutter, thus we derive our requirement 2.1 that the sensor system shall reduce the effect of clutter found in social media channels. In other words, our prototype online social media radar sensor system needs to ignore irrelevant data which is found mixed in with relevant data.

Some SNA techniques proposed to handle large volumes of data on large networks incorporate significant infrastructure and might satisfy our noise reduction requirement but may not meet the performance and processing speed requirements for an online social media radar sensor system (Santos et al., 2006). Categorizing posts on Twitter and identification of influential nodes have been shown to reduce noise in the analysis process without a lot of processing overhead (Behrend & Erwee, 2009; Cucchiarelli, Velardi, D'Antonio, & Curzi, 2008; Jafari, Akhavan, Bourouni, & Amiri, 2009; Khonsari, Nayeri, Fathalian, & Fathalian, 2010; Webb, 2008). Given that noise will be in the system, these techniques may provide utility in reduction of both noise and clutter.

Collective actions such as information diffusion in an online network, online topic and content popularity, movie theater box office revenues, network traffic load, emergent synchronous behavior, prices of stock market shares and winners of Academy Awards, have all been shown to be dependent variables of various SNA techniques and/or parameters (Agarwal, 2009; Asur & Huberman, 2010; Bothos et al., 2010; Cucchiarelli et al., 2008; De Choudhury, Sundaram, John, & Seligmann, 2009; De Choudhury et al., 2010; Gloor et al., 2009; Huberman, Romero, & Wu, 2008; Lee, Moon, & Salamatian, 2010; Park, Shin, Kim, & Chung, 2010; Vashist et al., 2011; Žiberna, 2009). Each of these studies demonstrates that behavior of large groups of people may be predicted by analyzing parameters of online social networks conveying relevant content and topics. For example, using correlation and regression analysis, Asur & Huberman (2010) show the mass collective actions of purchasing movie tickets (and in some cases, crowding movie theaters) can be predicted by analyzing social network parameters. Thus we have theoretical support from multiple sources for an online social media radar system to meet our third major requirement listed in Table 1 that crowd size be forecast using SNA.

To operate successfully, a predictive social network sensor system would need to process large volumes of data at high speed and yield results without manual intervention just as an electronic radar system processes radar reflections in a timely manner (Barbieri et al., 2010; Bermingham et al., 2009; Lou, Lin, Chen, & Lei, 2010; Thales Raytheon Systems, 2011). Many techniques encountered in the survey process data from multiple social media channels, require manual input and are complex and computationally intensive (Dawoud, Alhajj, & Rokne, 2010; Hutter et al., 2010; King et al., 2009; Varlamis et al., 2010). Three categories of SNA centrality measures are described by Hutter et al., (2010); local centrality which counts only direct neighbors of a node and are of linear complexity, distance-based centrality which computes the shortest paths between all nodes and may have cubic complexity, and eigenvector-based centrality which allows nodes to 'pass' their rank to their neighbors and have quadratic complexity. Thus, social network analysis, especially of a large network, is a computationally intensive application as described in the current literature. To meet our fourth system requirement for acceptable near-real time performance, local centrality measures of linear complexity may be the near-optimal solution. Another published specification related to performance in electronic radar systems is operational availability (A_o) which must exceed 0.99 for a Thales Raytheon system (2011). We use this specification to drive our 4.1 requirement for our social media radar and we use the Blanchard (2004, p. 73) definition of A_o = Mean Time Between Maintenance (MTBM)/(MTBM + Mean Maintenance Down Time (MDT). Our final specification derives from a Lockheed Martin specification for reliability of an electronic radar system which indicates that the Mean Time Between Failure (MTBF) must exceed 1250 hours.

Having identified requirements for an online social network media analysis sensor system, the next step is to survey the trade space. This is accomplished in the next section by a review of the scholarly literature related to online social network analysis and predictive capability with respect to mass collective actions.

LITERATURE REVIEW: SURVEY OF THE TRADE SPACE

Our trade study examines alternative approaches to solve a problem (INCOSE, 2006; Sage & Rouse, 2009). First a survey of available alternatives is performed. Not finding a proven, reliable "off-the-shelf" technology, which meets all system requirements we evaluate development of a new solution and possible solutions achieved by combination of existing technologies. Ranking and weighting of the various factors in terms of the system requirements while comparing and contrasting the parameters of candidate solutions contributed to the process of 'trading' higher quality attributes of one solution over attributes of other solutions.

Identify Alternatives

In order to identify the alternatives to be evaluated, we summarize the trade space survey results in the following tables. Predictive, independent variables described in the surveyed literature and discussed in the next paragraphs are grouped into two main factors of Social Network Content (Table 2) and Social Network Structure (Table 3).

Our survey finds social media content described in the literature in terms of communication, coordination, cooperation, sentiment (positive and negative), popularity and credibility. Communication, coordination, and cooperation can be categorized as one-way (broadcast from one node to many) or as more intimate two-way (node to node) links (Cha, M., Perez, J., & Haddadi, H., 2009; Cucchiarelli et al., 2008; Huberman et al., 2008; Lou et al., 2010; Žiberna, 2009). Natural language processing tools (although possibly computationally intensive) can be used to determine positive and negative sentiment attributes of content regarding subjects of interest for mass collective actions including media, places, dates, people, elections, and repression (Asur & Huberman, 2010; Gloor, 2009; Khonsari et al., 2010). Analysis of the rate of change of sentiment displayed in social media content can be a factor utilized for predictive capability (Gloor et al., 2009). Popularity of content (measured in terms of posts, re-posts and the rate of change) is another aspect used to predict actions (Agarwal, 2009; Gloor et al., 2009; Lee et al., 2010; Lou et al., 2010; Park et al., 2010). Credibility of social media content and users trust in the veracity of authors' posts can be influenced by factors including homophily (the tendency of 'birds of a feather' to flock together), closure, gate keeper status (brokerage), degree of separation between users and user demographics (Dellarocas, 2003; Evans & Wensley, 2009; Johnson & Kaye, 2004).

Structural characteristics of social networks are also described as having effects upon the ability of online networks to mobilize participant actions. Density and centralization of networks have been examined by Evans & Wensley (2009) and Spencer (2003). Influence of authors represented as network nodes is frequently measured in SNA terms of Degree Centrality, Closeness Centrality, Betweeness Centrality, Hub Quality, Authority, and Eigenvector Centrality (Hutter, 2010; Lou et al., 2010; Lu, Luo, Polgar, & Cao, 2010; Varlamis et al., 2010). Other network attributes examined in the literature include node PageRank score, rate of change of node authority, mean degree of node distribution, variance of degree

Table 2. Social Network Content Comparative Analysis

Sub-Factors	Variables	Agarwal 2009	Asur & Huber-man 2010	Cha et al. 2009	Cucch-iarelli et al. 2008	Dellarocas 2003	De Choud-hury et al. 2009	Evans & Wensley 2009	Gloor et al. 2009	Huber-man et al. 2008	Johnson & Kaye 2004	Khonsari et al. 2010	Lee et al. 2010	Lou et al. 2010	Park et al. 2010	Spencer 2003	Varlamis et al. 2010	Ziberna 2009
Comm-unication	One way			✓	✓					✓				✓	✓		✓	✓
	Two way				✓					✓				✓	✓		✓	✓
Coordination	One way				✓					✓				✓				✓
	Two way				✓					✓				✓				✓
Cooperation	One way				✓					✓				✓				✓
	Two way				✓					✓				✓				✓
Sentiment (+/-)	Media		✓						✓			✓						
	Places											✓						
	Dates											✓						
	People											✓						
	Election											✓						
	Repression											✓						
	Rate of posts		✓						✓						✓		✓	
Popularity	Posts	✓	✓		✓				✓				✓	✓	✓			
	Re-posts	✓							✓				✓	✓	✓			
	Rate of change		✓		✓				✓				✓	✓	✓			
Credibility (trust)	Homophily					✓		✓			✓						✓	
	Closure					✓		✓			✓						✓	
	Brokerage (Gatekeeper)					✓		✓			✓					✓	✓	
	Degree of separation					✓		✓			✓							
	Demographics										✓						✓	

Table 3. Social Network Structure Factor Comparative Analysis

Sub-Factors	Variables	De Choudhury et al. 2009	Evans & Wensley 2009	Gloor et al. 2009	Hutter et al. 2010	Lou et al. 2010	Lu et al. 2010	Park et al. 2010	Spencer 2003	Varlamis et al. 2010
Density	High, low		✓					✓	✓	
Centralization	High, low		✓					✓	✓	
Node Influence	Degree Centrality				✓		✓			✓
	Closeness Centrality				✓		✓			✓
	Betweeness Centrality				✓	✓	✓			✓
	Out-links (Hub)					✓				✓
	In-links (Authority)					✓		✓		✓
	Rate of change of Authority					✓		✓		
	PageRank score									✓
	Eigenvector				✓		✓			
	Mean of degree distribution	✓								
	Variance of degree distribution	✓								
	Clustering Coefficient	✓								
	Concept Centrality			✓						

distribution, clustering coefficient and the consideration of concepts as nodes in addition to authors (De Choudhury et al., 2009; Gloor et al., 2009; Park et al., 2010). In all of the literature reviewed, we do not find a single technique or system that meets all online social media radar system requirements for near-real time reporting performance, low computational overhead and high confidence in predictions.

In order to define the alternatives to be evaluated in our prototype, we build matrices (See Tables 4 & 5) of the alternative factors in relationship to the system requirements, indicate if the alternatives affect the requirement with a Boolean value, and assign both weights and ranks in order to determine a score. For the purposes of this study, ranks are assigned based upon relative frequency of each factors appearance (as shown in Tables 2 & 3) in the trade space survey. Without direct customer contact at this initial phase, weights were assigned based upon Costa & Boiney (2012), Secretary of Defense (2011) and Maybury (2010).

All six social network content factors would affect the system's ability to report actionable information from online social media channels (Requirement 1.0) and therefore a Boolean of 1 is assigned to each for the first requirement. Weights for all requirements sum up to 1. Thus Requirement 1 at a weight of .25 is the most heavily weighted and critical requirement. The ability to operate without operator intervention, Requirement 1.1 is easily achieved for the Communication and Popularity factors as these involve simple counting and those factors are assigned a Boolean of 1 (Cha et al., 2009; Cucchiarelli et al., 2008; Huberman et al., 2008; Lou et al., 2010; Park et al., 2010; Varlamis et al., 2010; Žiberna, 2009). Coordination, Cooperation, Sentiment and Trust require either offline analysis or computationally intensive processing so these factors are assigned Boolean 0 for Requirement 1.1 (Asur & Huberman, 2010; Cucchiarelli et al., 2008; Dellarocas, 2003; Evans & Wensley, 2009; Huberman et al., 2008; Khonsari et al., 2010; Lee et al., 2010; Lou et al., 2010; Park et al., 2010). Similarly, rejection of redundant signals (Requirement 2.0) is also straightforward for the Communication and Popularity factors and much more complex for Coordination, Cooperation, Sentiment and Trust (Asur & Huberman, 2010; Cha et al., 2009; Cucchiarelli et al., 2008; Dellarocas, 2003; Evans & Wensley, 2009; Huberman et al., 2008; Khonsari et al., 2010; Lee et al., 2010; Lou et al., 2010; Park et al., 2010; Varlamis et al., 2010; Žiberna, 2009).

Requirement 1.1 is not as critical to the system operation as are Requirements 1.0 and 2.0, therefore 1.1 is ranked at 0.05 and 2.0 is assigned a rank of 0.2. This pattern of analysis is repeated for all of the Social Network Content and Social Network Structure factors shown in Tables 2 & 3 resulting in the scoring shown in Tables 4 & 5. The Social Network Content factors of Communication and Popularity score the highest in Table 4 and the Social Network Structure factors of Node Influence and Density score the highest in Table 5. The highest scoring alternatives are evaluated and utilized to develop related hypotheses in the next section.

CONCEPTUAL PROTOTYPE MODEL

Given the Sensor System Requirements, the Trade Space Survey and the alternatives identified above, this study sets forth the conceptual model for a prototype Online Social Network Media Sensor System presented in Figure 1. The review of scholarly literature conducted in the Trade Space Survey demonstrates that the factors of Social Network Structure and Social Network Content may have sufficiently significant relationships to Mass Collective actions such as crowd size that could provide the basis for a 'social media radar' sensor.

Table 4. Social Network Content Trade Study Scores

	Requirements		Communication	Coordination	Cooperation	Sentiment	Popularity	Trust
1.0	The sensor system shall report actionable information from online social media channels	Y/N	1	1	1	1	1	1
		Weight	0.25	0.25	0.25	0.25	0.25	0.25
		Rank	7	4	4	5	7	5
		Score	1.75	1	1	1.25	1.75	1.25
1.1	The sensor system shall operate without operator intervention, once initiated.	Y/N	1	0	0	0	1	0
		Weight	0.05	0.05	0.05	0.05	0.05	0.05
		Rank	7	4	4	5	7	5
		Score	0.35	0	0	0	0.35	0
2.0	The sensor system shall reject irrelevant signals in the monitored social media channels.	Y/N	1	0	0	0	1	0
		Weight	0.2	0.2	0.2	0.2	0.2	0.2
		Rank	7	4	4	5	7	5
		Score	1.4	0	0	0	1.4	0
2.1	The sensor system shall reduce the effect of clutter found in social media channels.	Y/N	1	0	0	0	1	1
		Weight	0.05	0.05	0.05	0.05	0.05	0.05
		Rank	7	4	4	5	7	5
		Score	0.35	0	0	0	0.35	0.25
3.0	The sensor system shall optimize the probability of prediction of crowd sizes with no greater than 20% error	Y/N	1	1	1	1	1	1
		Weight	0.2	0.2	0.2	0.2	0.2	0.2
		Rank	7	4	4	5	7	5
		Score	1.4	0.8	0.8	1	1.4	1
4.0	The system sensor shall be capable of processing high volumes of data with near-real time performance.	Y/N	1	0	0	0	1	0
		Weight	0.05	0.05	0.05	0.05	0.05	0.05
		Rank	7	4	4	5	7	5
		Score	0.35	0	0	0	0.35	0
4.1	The system sensor shall have an operational availability (A_O) > 99% of the time.	Y/N	1	0	0	0	1	1
		Weight	0.1	0.1	0.1	0.1	0.1	0.1
		Rank	7	4	4	5	7	5
		Score	0.7	0	0	0	0.7	0.5
4.2	The system sensor shall have a mean time between failure in excess of 1250 hours	Y/N	1	0	0	0	1	1
		Weight	0.1	0.1	0.1	0.1	0.1	0.1
		Rank	7	4	4	5	7	5
		Score	0.7	0	0	0	0.7	0.5
	TOTAL SCORE:		7	1.8	1.8	2.25	7	3.5

Table 5. Social Network Structure Trade Study Scores

	Requirements		Density	Centralization	Node Influence
1.0	The sensor system shall report actionable information from online social media channels	Y/N	1	1	1
		Weight	0.25	0.25	0.25
		Rank	3	3	7
		Score	0.75	0.75	1.75
1.1	The sensor system shall operate without operator intervention, once initiated.	Y/N	1	0	0
		Weight	0.05	0.05	0.05
		Rank	3	3	7
		Score	0.15	0	0
2.0	The sensor system shall reject irrelevant signals in the monitored social media channels.	Y/N	1	0	0
		Weight	0.2	0.2	0.2
		Rank	3	3	7
		Score	0.6	0	0
2.1	The sensor system shall reduce the effect of clutter found in social media channels.	Y/N	1	1	1
		Weight	0.1	0.1	0.1
		Rank	3	3	7
		Score	0.3	0.3	0.7
3.0	The sensor system shall optimize the probability of prediction of crowd sizes with no greater than 20% error	Y/N	1	1	1
		Weight	0.2	0.2	0.2
		Rank	3	3	7
		Score	0.6	0.6	1.4
4.0	The system sensor shall be capable of processing high volumes of data with near-real time performance.	Y/N	1	1	1
		Weight	0.05	0.05	0.05
		Rank	3	3	7
		Score	0.15	0.15	0.35
4.1	The system sensor shall have an operational availability (A_o) > 99% of the time.	Y/N	0	0	1
		Weight	0.1	0.1	0.1
		Rank	3	3	7
		Score	0	0	0.7
4.2	The system sensor shall have a mean time between failure in excess of 1250 hours	Y/N	1	0	1
		Weight	0.05	0.05	0.05
		Rank	3	3	7
		Score	0.15	0	0.35
TOTAL SCORE:			2.7	1.8	5.25

Figure 1. Conceptual Model of Social Networking and Collective Action

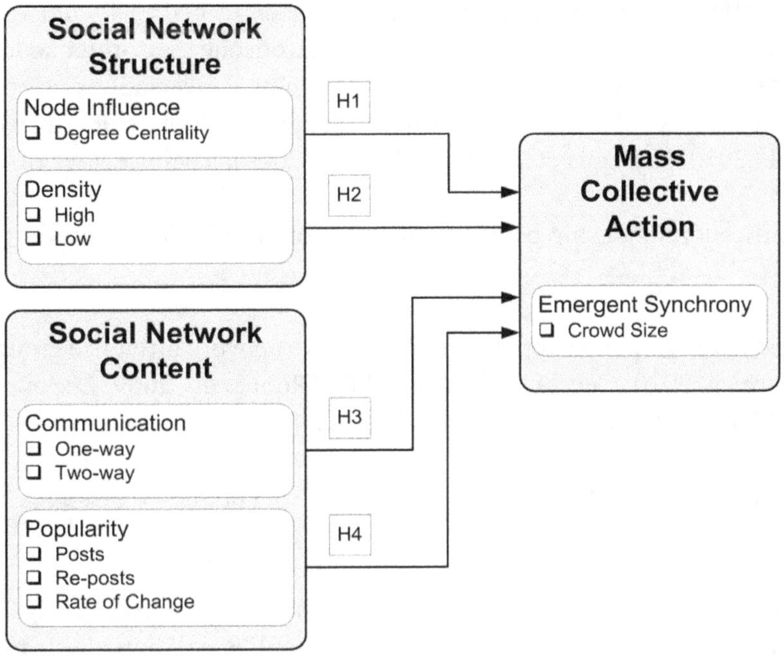

Based upon the identified alternatives and trade study scores, four hypotheses indicated in Figure 1 are presented for the conduct of evaluation of this conceptual model. We base the hypotheses for the effect of Social Network Structure on Mass Collective Action shown in Figure 1 (H1 & H2) upon the high Trade Study scores of 5.25 and 2.7 for the sub-factors Node Influence (measured in terms of Degree Centrality) and Density, respectively.

Feasibility of predicting mass collective actions (such as information diffusion, movie ticket purchases, stock prices, and emergent synchrony) by analysis of Node Influence variables (including multiple measures of Centrality), are demonstrated in De Choudhury et al. (2009), Gloor et al. (2009), Hutter et al. (2010), Lou et al. (2010), Lu et al. (2010), Park et al. (2010) and Varlamis et al. (2010). Dellarocas (2003) theorizes a negative correlation of Degree Centrality to cooperative behavior. Therefore, we offer hypothesis H1:

H1: There is a significant relationship between network Node Influence variables (especially Degree Centrality) and reported crowd sizes.

Network density factors are shown to be predictors of strength of trusting relationships within communities of practice, discovering social media streams with topical relevance, and identification of emergent dominant industrial designs (Evans & Wensley, 2009; Park et al., 2010; Spencer, 2003). Therefore, we offer hypothesis H2:

H2: There is a significant relationship between network density measures and reported crowd sizes.

Social network content factors of Communication and Popularity outscored the other factors in the trade study, thus we offer hypotheses H3 & H4 in the model described by Figure 1.

The effects of unidirectional communications broadcast from one transmitter node to a large audience and predictive factors of bidirectional communications forming conversations on collective actions are described in multiple studies (Cha et al., 2009; Cucchiarelli et al., 2008; Huberman et al., 2008; Lou et al., 2010; Park et al., 2010; Varlamis et al., 2010; Žiberna, 2009). Therefore, we offer hypothesis H3:

H3: There is a significant relationship between network Communication measures and reported crowd sizes.

Many researchers have shown that popularity of topics in a network influence group actions (Agarwal, 2009; Asur & Huberman, 2010; Cucchiarelli et al., 2008; Gloor et al., 2009; Lee et al., 2010; Lou et al., 2010; Park et al., 2010). Therefore, we offer hypothesis H4:

H4: There is a significant relationship between measures of Popularity of topics in the network and reported crowd sizes.

For each hypothesis proposed, we will test the null hypothesis, i.e. that no significant relationship exists between the independent and dependent variables. Should a null hypothesis be rejected, a regression analysis will be performed to determine if the relationship is strong enough to qualify as a suitable candidate for an online social media radar sensor.

RESEARCH METHOD: EVALUATION OF ALTERNATIVES

To evaluate the alternatives identified in the trade study and data collection process for testing of the conceptual model and hypotheses set forth above, a study was conducted of the movement initially known as the 'March on Wall Street' and which became the 'Occupy' movement when it spread out from New York to over 200 cities across the U.S. and to 2000 cities around the world (Barton, 2011). A dataset of approximately 150,000 tweets, associated meta-data and crowd size estimates was collected over a 100-day period from 10/26/11 to 02/15/12 from web crawls of millions of tweets in the online social network channel Twitter, using scripts written in the Python programming language, the Twitter Application Programming Interface (API) and keywords related to the Occupy movement. Key Occupy movement events organized, communicated or reported on via online social media channels, which occurred during the study period are highlighted in the timelines presented in Figures 2 & 3.

In addition, the social network analysis tool NetMiner was used to analyze the discovered networks' characteristics including social network parameters such as Density, Degree Centrality, Mean Degree Centrality, Closeness, Betweeness and Authority (Cyram, 2012). This combination of tools formed our prototype online social media 'radar'. An iterative process of design parameterization was followed to develop the prototype system configuration and to establish system parameters (Sage & Rouse, 2009). Although the prototype system was not a fully integrated system, we tracked two Technical Performance Measurements (TPMs) of throughout the process in order to measure compliance with requirements 1.0, 4.0 and 4.1; percentage of average CPU usage and average physical memory usage. All processes were performed using an Intel ® Core ™ i5 CPU @ 2.3 GHz with 4 GB of RAM running the 64-bit

Figure 2. Occupy movement events 10/26/11 to 12/31/11

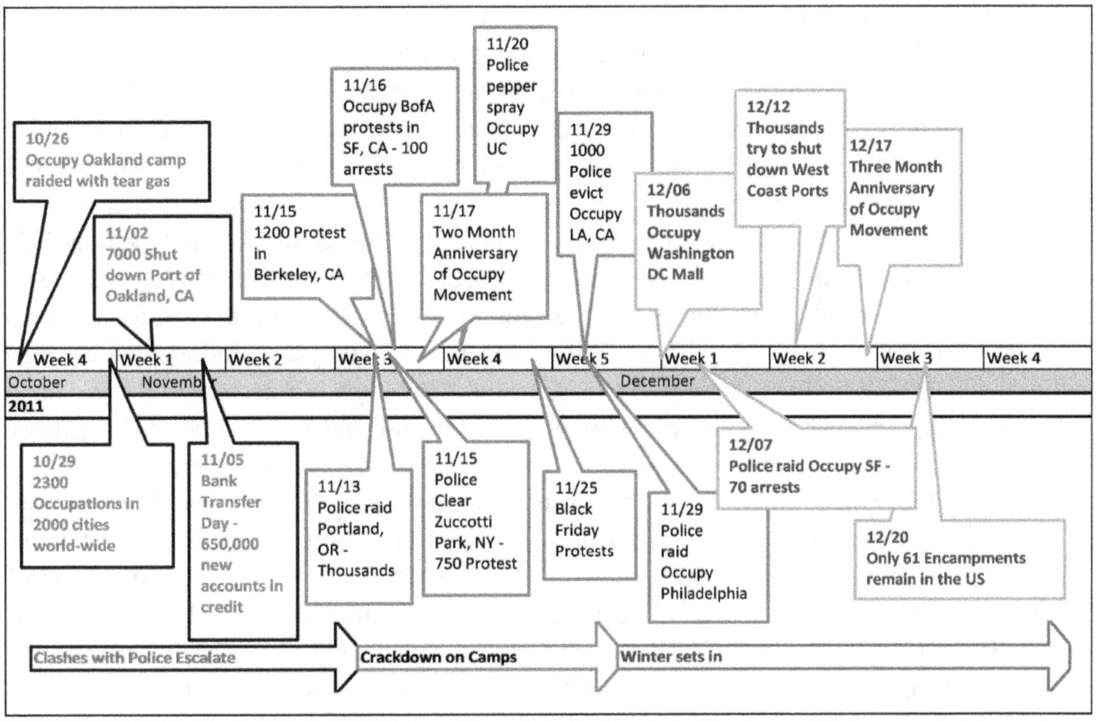

Figure 3. Occupy movement events 1/01/12 to 2/15/12

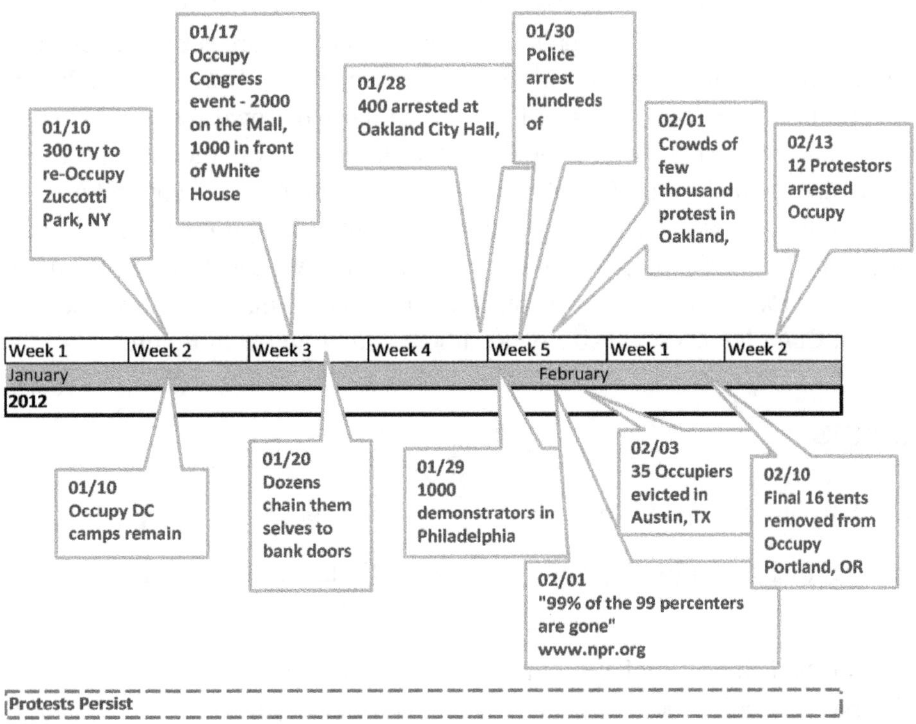

Microsoft Windows 7© operating system. During collection of tweets, our system CPU usage peaked twice at 67% and averaged 33% well below our design parameter upper limit of 50%. System memory usage averaged 1.7 GB, below our design parameter limit of 2.5 GB.

After gathering tweet data using the Twitter API, additional Python scripts were written to build two matrices: one showing the 'hidden' author-to-author network (those authors who communicated directly to specific other authors as opposed to those who simply broadcast to all of their followers) described in Huberman et al. (2008); the second matrix traces authors to the thread subjects to which they posted tweets. TPMs gathered during the matrix processing demonstrated usage of CPU and memory below our prototype design limits. Using the Cyram, Inc. NetMiner tool (2012), Network Density, Degree Centrality, the Mean Degree Centrality, the number of Transmitter nodes and the number of Receiver nodes of both matrices were measured. TPMs gathered during the network analysis demonstrated usage of CPU and memory, which were higher than in previous steps but still below our design limits. We focused on Twitter data for the purposes of this study and assumed that the normal distribution of Centrality among Twitter authors found in Khonsari et al. (2010) also applies to the dataset we used.

Two hundred Occupy movement locations sprouted up in the US and over two thousand total sites around the world. For this study, crowd size data from over 40 locations in the US was tabulated from a combination of news media reports, blog posts, YouTube videos and web sites including ABC News (2012), Barton (2011), Businessweek (2012), Bernton (2011), Finextra (2011), FireDogLake.com (2011), Los Angeles Times (2012), NBC News (2012), News-record.com (2011), NPR (2012), Slate.com (2011), Telegraph (2012), and the Wall Street Journal (2011). Additional crowd size data was obtained from detailed records posted on Occupy movement web sites and Facebook pages of meeting attendance, vote tallies, and demonstration event crowd sizes.

If no specific numbers were available, Police estimates and estimates reported in the established news media were used when available. When multiple sources reported conflicting numbers the most conservative estimates were used unless a majority of the sources consistently reported a larger number. Crowd size estimates of bystanders, journalists, police and activists are subject to error due to perspective, location, bias, crowd density variations and experience (Kong, Gray, & Tao, 2006; Lin, Chen, & Chao, 2001; Ma, Li, Huang, & Tian, 2004). Many of the reported crowd sizes cited in this study simply indicated that a crowd was 'small' or numbered in the 'hundreds' or 'thousands'. Therefore, in order to quantify crowd sizes, which were described only in relative terms, we applied a seven-point Likert scale: No event = 0, very small = (1 – 49), small = (50 – 199), medium = (200 – 499), large = (500 – 1999), very large (2000 – 4999), massive (5000 and up). For these events we utilized the mid-point of the assigned range and totaled up the crowds reported from the 40 major locations studied for each day of the study period. Cumulative crowd size data gathered are graphed in Figure 4.

DATA ANALYSIS

First, we examine the null hypothesis $H1_0$:

$H1_0$: There is no significant relationship between network Node Influence variables and reported crowd sizes.

Using the Cyram NetMiner™ tool, we collected Node Influence variables Degree Centrality and the mean of the Degree Centrality measures on both the Author-to-Subject network matrix and the 'hid-

Figure 4. Cumulative Crowd Size Data

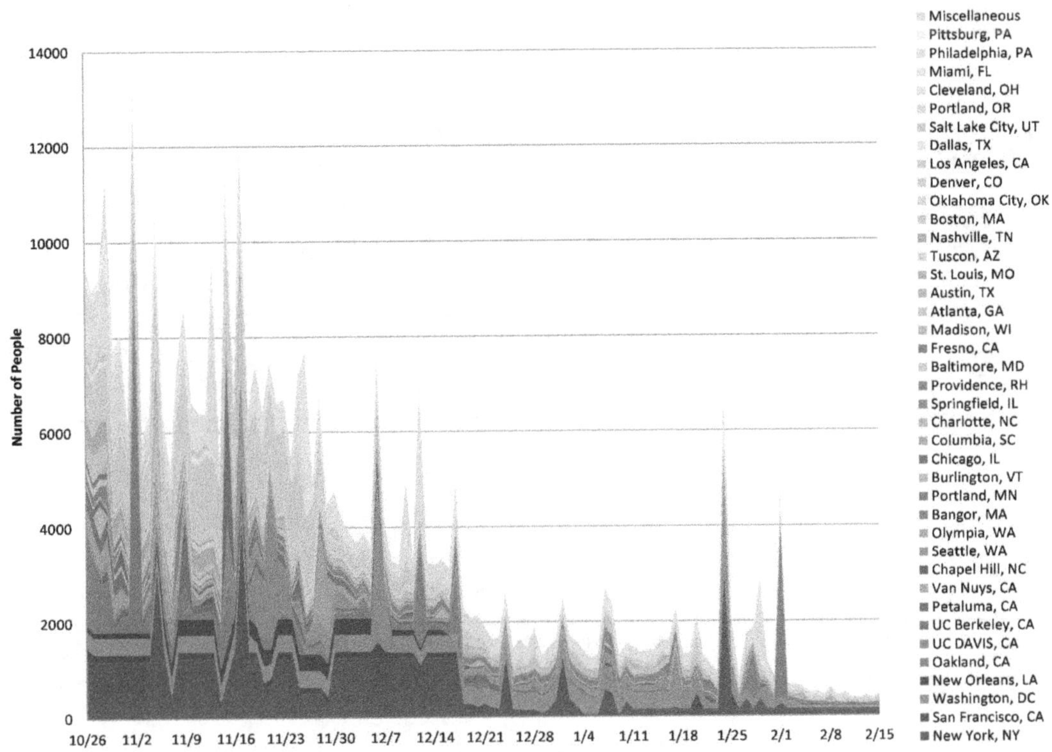

den' Author-to-Author matrix extracted from the collection of tweets regarding the Occupy Wall Street movement. Graphing the linear trends of Degree Centrality and Crowd size in the Author-to-Subject network showed an inverse relationship, as expected (Dellarocas, 2003).

However, the crowd size trend is better described with an exponential curve. Therefore, we applied a natural log transformation to the crowd size data. In Figure 5, we see a distinct inverse relationship between the Mean Degree Centrality and Crowd Size (Ln) in the Author-to-Subject network.

When we look at the Author-to-Author network Node Influence variables of Degree Centrality and Mean Degree Centrality as compared to the natural log of the crowd size, we again see inverse relationships. Figure 6 demonstrates the expected inverse relationship of Degree Centrality in the Author-to-Author network to the natural log of the crowd size.

Given these trends, we expect to see a correlation between these node influence variables quantified in terms of Degree Centrality and crowd sizes.

Graphical representation of the data related to the null formulation of our second hypothesis suggests that no correlation will be found, and that we are not able to reject $H2_0$: There is no significant relationship between Network Density measures and reported crowd sizes.

Graphical representation of the Author-to-Subject transmitter and receiver nodes shows less of a relationship between crowd size and transmitter nodes than between crowd size and receiver nodes. The Author-to-Author trends for transmitter and receiver nodes suggest relationships that may support rejection of our third null hypothesis:

Figure 5. Crowd Size (Ln) vs. Mean Degree Centrality in Author-to-Subject Network

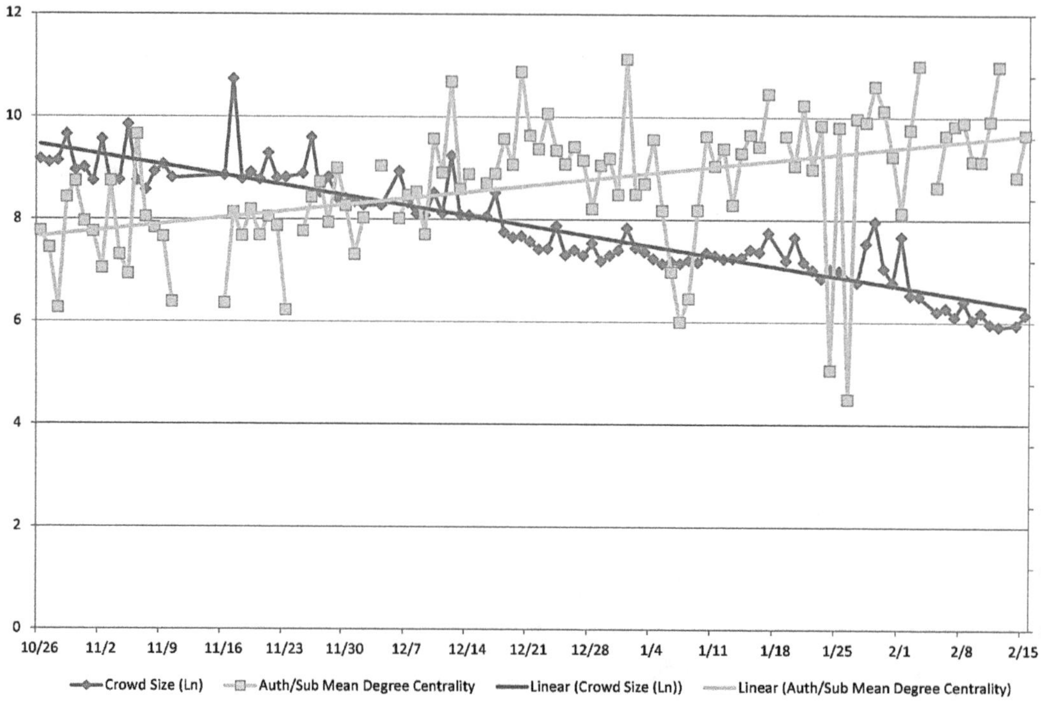

Figure 6. Crowd Size (Ln) vs. Degree Centrality in Author-to-Author Network

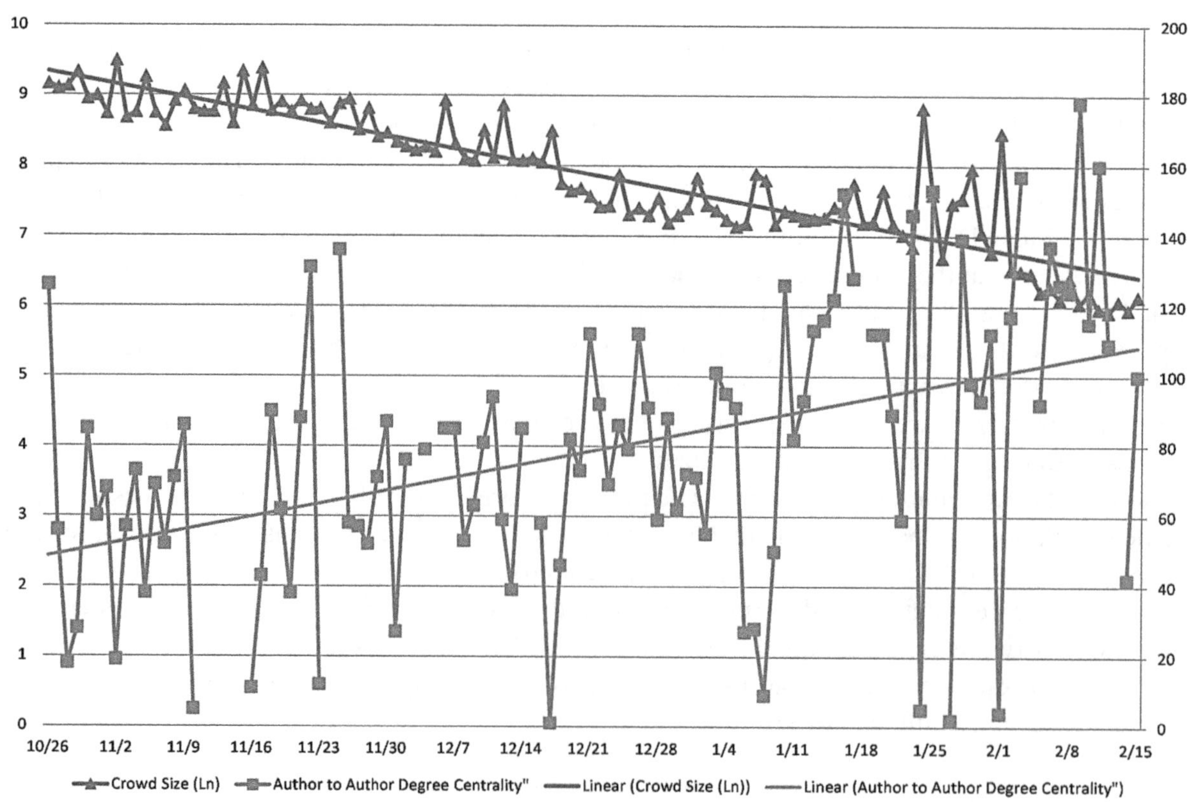

$H3_0$: There is no significant relationship between network Communication measures and reported crowd size.

Many advanced measures of topic popularity have been shown to have predictive capability (Agarwal, 2009; Asur & Huberman, 2010; Cucchiarelli et al., 2008; Gloor et al., 2009; Lee et al., 2010; Lou et al., 2010; Park et al., 2010). In addition, several parameters can be dynamically calculated with low computational overhead. For the purposes of this study to evaluate predictive factors suitable for inclusion in an online social media radar system, we chose to analyze the straightforward temporal order of Tweet intensity as a measure of popularity related to crowd events. The fourth null hypothesis is presented as:

$H4_0$: There is no significant relationship between measures of Popularity of topics in the network and reported crowd sizes.

Graphical representation of the temporal order of tweet intensity vs. crowd size leads us to expect to see some relationship demonstrated in a correlation analysis.

Author-to-Author Network Data Analysis

The linear trend line of the Crowd Size data shown in the graphs presented above does not offer the best fit to the data; an exponential trend was found to provide a better bit. Therefore, Crowd Size data were transformed to natural logs. A Pearson correlation study was performed on the Author-to-Author network parameters of Network Density, Degree Centrality, Mean Degree Centrality, number of Transmitter nodes, number of Receiver nodes and Tweet Quantity with the natural log of the daily total crowd sizes summed from the forty tabulated locations. The parameters of Network Density and Tweet Quantity demonstrated no linear relationship at all. The study was repeated with Degree Centrality, Mean Degree Centrality, the number of Transmitter nodes, and the number of Receiver nodes. Results of the correlation are presented in Table 6.

Author-to-Subject Network Data Analysis

Pearson correlation of the Author-to-Subject network parameters shows the strongest correlations to crowd size with the parameters Mean Degree Centrality, Degree Centrality, Receiver Nodes and Tweet Quantity. Transmitter Nodes showed a very small relationship and Network Density demonstrated almost no linear relationship at all to the natural log of the crowd size. Results of the correlation are presented in Table 8.

RELIABILITY: VERIFICATION AND VALIDATION

The final step is to verify if we engineered the product right and to validate that we engineered the right product. First we look at conformance to our system requirements. Did the prototype social media radar meet the requirements for the online social media radar sensor system? Yes. The results are shown in Table 10.

Table 6. Pearson Correlation of Author-to-Author Network Parameters to Natural Log of Crowd Size

	LN(Crowd Size)	Degree Centrality	Mean Degree Centrality	Transmitter Nodes
Degree Centrality	-0.438345171			
Mean Degree Centrality	-0.235867772	0.695986809		
Transmitter Nodes	-0.447023433	0.936847391	0.442326299	
Receiver Nodes	-0.387713298	0.984939098	0.652877707	0.919491264

We observe the parameters Transmitter, Degree Centrality, and Receiver are ones most strongly correlated to Crowd size, in that order. Multiple linear regression analysis reveals the equation:

Ln (Crowd Size) = – 0.108 Degree Centrality + 16.383 Mean Degree Centrality + 0.036 Transmitter Nodes + 0.070 Receiver Nodes

S = 0.8569 R-Sq = 98.8% R-Sq(adj) = 97.8%

Analysis of the variance is presented in Table 7.

We see the system developed meets the requirements, which can be examined with the prototype (development of an integrated system that could demonstrate operational capability and MTBF is beyond the scope of this study; however the TPMs collected during the study indicate acceptable performance on a low-end computer system). Next, we examine the validity of the hypotheses given the data analysis performed. For the Social Network Structure factor, the results of testing the following null hypotheses are presented:

$H1_0$: There is no significant relationship between network Node Influence variables and reported crowd sizes.

Significant negative correlations were demonstrated as expected for Degree Centrality and Mean Degree Centrality and in both the Author-to-Author and Author-to-Subject networks, and both passed the goodness of fit test as part of a multiple linear regression analysis, therefore we can reject $H1_0$ and accept the alternate hypothesis: *H1: There is a significant relationship between network Node Influence variables (especially Degree Centrality) and reported crowd sizes.*

Testing of our second null hypothesis, $H2_0$*: There is no significant relationship between Network Density measures and reported crowd size,* did not support rejection. No significant correlation was demonstrated with either the Author-to-Author or the Author-to-Subject network in order to support rejection of this hypothesis. Thus we cannot reject $H2_0$.

For the Social Network Content factor, the results of testing the following hypotheses are presented:

$H3_0$: There is no significant relationship between Network Communication measures and reported crowd size.

Negative correlations were demonstrated for Transmitter and Receiver nodes in the Author-to-Author network matrix and for Receiver nodes in the Author-to-Subject network matrix. Both passed the goodness of fit test as part of a multiple linear regression analysis. Thus we can reject $H3_0$ using the selected parameters evaluated in this study and accept the alternate hypothesis: *H3: There is a significant relationship between network Communication measures and reported crowd sizes.*

The Tweet quantity parameter did not significantly contribute to the Author-to-Author matrix correlation but it did demonstrate a small negative correlation in the Author-to-Subject matrix and it did

Table 7. Analysis of Variance: Author-to-Author Network Parameter Regression

	df	SS	MS	F	Significance F
Regression	4	6181.046681	1545.2617	2104.43191	8.80187E-92
Residual	96	70.49176528	0.7342892		
Total	100	6251.538446			

An inverted beta (F)-distribution test with alpha of 0.01 shows $F_{TV} = 2104 > F_{(1, 96)}$; therefore we have a significant result with a 99% confidence level. The goodness of fit shows that 98.8% of the crowd size variability is accounted for and this meets our online social media radar system requirements.

Table 8. Correlation of Author-to-Subject Network Parameters to Crowd Size

	LN(Crowd Size)	Degree Centrality	Mean Degree Centrality	Receiver Nodes
Degree Centrality	-0.330119815			
Mean Degree Centrality	-0.464254455	0.744836615		
Receiver Nodes	-0.280242876	0.943153445	0.597892849	
Tweet Qty	-0.258824699	0.566196932	0.32892944	0.506830508

A multiple linear regression analysis reveals the equation:

Ln (Crowd Size) = -0.006 Degree Centrality + 6.075 Mean Degree Centrality + 0.015 Receiver Nodes + 0.001 Tweet Quantity

S = 1.4167 R-Sq = 96.9% R-Sq(adj) = 95.8%

Analysis of the variance is presented in Table 9.

Table 9. Analysis of Variance: Author-to-Subject Network Parameter Regression

	df	SS	MS	F	Significance F
Regression	4	6058.83578	1514.70895	754.59287	4.85251E-71
Residual	96	192.7026661	2.00731944		
Total	100	6251.538446			

An inverted beta (F)-distribution test with alpha of 0.01 shows $F = 754.83 > F_{(1, 96)}$; therefore we have a significant result with a 99% confidence level. The goodness of fit shows that 96.9% of the crowd size variability is accounted for and this meets our online social media radar system requirements.

Table 10. System Requirements and Results

#	Online Social Media Radar Sensor System Requirements	Results
1.0	The sensor system shall report actionable information from online social media channels in near-real time	Pass
1.1	The sensor system shall operate without operator intervention, once initiated.	Pass
2.0	The sensor system shall reject irrelevant signals in the monitored social media channels.	Pass
2.1	The sensor system shall reduce the effect of clutter found in social media channels.	Pass
3.0	The online social network media analysis system sensor shall optimize the probability of prediction of crowd sizes with no greater than 20% error	Pass
4.0	The system sensor shall be capable of processing high volumes of data with near-real time performance.	Pass
4.1	The system sensor shall have an operational availability (A_o) > 99% of the time.	N/A
4.2	The system sensor shall have a mean time between failure in excess of 1250 hours	N/A

contribute to the multiple linear regression analysis, which shows that 96.9% of the crowd size variability can be accounted. Therefore, we can reject $H4_0$: *There is no significant relationship between measures of Popularity of topics in the network and reported crowd sizes.*

Tweet-rate per hour regarding new movies and the count of movie theaters has been shown to predict box office revenues with an adjusted R^2 value of 0.973 (Asur & Huberman, 2010). Box office revenues are easier to quantify than crowd size however, further analysis of the rate of tweets in combination with other popularity parameters in this dataset and others along with a count of event locations may prove to be valuable avenues for future research.

FUTURE RESEARCH DIRECTIONS

Given the conceptual model defined above and the study results obtained so far, further research using other online social networks such as Facebook, YouTube and Instagram as potential orthogonal sensors may also show significance to provide additional support for the hypotheses presented in this study. Additionally, further analysis of tweet metadata may be shown to provide predictions of crowd formation time and location. Analysis of tweet (and other media) content using tools such as natural language processing should be examined for capabilities to predict crowd attitude, goals and direction of movement.

CONCLUSION

In summary, our Systems Engineering process shows that an online social media radar system may be developed to predict crowd sizes from social network variables with a high degree of confidence, low processing overhead and in near-real time in a manner analogous to the use of radar information to predict the movement of air traffic. Our trade study of alternative approaches and social network factors led to our model and study design, which demonstrates that significant relationships between crowd sizes and social media network Node Influence variables, network Communication Measures, and network Popularity of Topics, can be used as predictive factors. Using the open Twitter API we wrote simple code to gather tweets referring to the Occupy movement (for the purposes of this study the tweets were gathered from static data however, this code can be run in real-time). We have shown that scripts combined with commercially available tools can perform the social media network analysis required to reveal the Author-to-Author and the Author-to-Subject networks within the set of tweets. Theoretically, network parameters obtained from the tools could be used in the regression equations to predict crowd sizes from several blind data sets. This testing should be included in further analysis to demonstrate widespread applicability and to estimate the fraction of crowds, which may not be correlated with online social media network parameters.

We have seen that studies have shown diffusion of information across a network, network traffic bandwidth consumption, movie ticket purchases, stock prices and Academy Awards winners can all be predicted by leveraging social network analysis tools with online social media (Asur & Huberman, 2010; Bothos et al., 2010; De Choudhury et al., 2010; Gloor et al., 2009; Vashist et al., 2011). This study has demonstrated the application of a Systems Engineering approach to the analysis of online social media can yield a new sensor system. With threats from non-state, terrorist organizations on the rise, analysis of online social network media may be a factor in providing an accurate early warning system, which can forecast, if not predict, mass behaviors as desired by the Department of Defense.

REFERENCES

Agarwal, S. (2009). *Social networks as Internet barometers for optimizing content delivery networks.* Paper presented at the 3rd International Symposium on Advanced Networks and Telecommunication Systems (ANTS), New Dehli, India. doi:10.1109/ANTS.2009.5409895

Anklam, P. (2005). Social Network Analysis in the KM Toolkit. In M. Rao (Ed.), *Knowledge Management Tools and Techniques* (pp. 329–346). Burlington, MA: Elsevier. doi:10.1016/B978-0-7506-7818-6.50029-3

Asur, S., & Huberman, B. A. (2010). *Predicting the Future with Social Media.* Paper presented at the International Conference on Web Intelligence and Intelligent Agent Technology (WI-IAT), Toronto, Canada. doi:10.1109/WI-IAT.2010.63

Barbieri, D., Braga, D., Ceri, S., Valle, E. D., Yi, H., Tresp, V., & Wermser, H. et al. (2010). Deductive and Inductive Stream Reasoning for Semantic Social Media Analytics. *IEEE Intelligent Systems, 25*(6), 32–41. doi:10.1109/MIS.2010.142

Barton, C. (2011, October 29). Occupy Auckland protest speaks with many voices. *New Zealand Herald.* Retrieved August 12, 2012 from http://www.nzherald.co.nz/nz/news/article.cfm?c_id=1&objectid=10762353

Behrend, F. D., & Erwee, R. (2009). Mapping knowledge flows in virtual teams with SNA. *Journal of Knowledge Management, 13*(4), 99-114. doi: 10.1108/13673270910971860

Bermingham, A., Conway, M., McInerney, L., O'Hare, N., & Smeaton, A. (2009). *Combining Social Network Analysis and Sentiment Analysis to Explore the Potential for Online Radicalisation.* Paper presented at the 2009 Advances in Social Network Analysis and Mining. doi:10.1109/ASONAM.2009.31

Bernton, H. (2011). Portland police arrest 50 protesters. *Seattle Times.* Retrieved September 1, 2012 from http://www.seattletimes.com/seattle-news/portland-police-arrest-50-protesters

Blanchard, B. (2004). *Logistics and Engineering Management* (6th ed.). Upper Saddle River, NJ: Pearson Prentice Hall.

Bothos, E., Apostolou, D., & Mentzas, G. (2010). Using Social Media to Predict Future Events with Agent-Based Markets. *IEEE Intelligent Systems, 25*(6), 50–58. doi:10.1109/MIS.2010.152

Businessweek. (2012). Occupy DC camps persist as rally planned next week. *Businessweek.* Retrieved September 5, 2012 from http://www.businessweek.com/occupydc

Cha, M., Perez, J., & Haddadi, H. (2009). *Flash floods and ripples: the spread of media content through the blogospere.* Association for the Advancement of Artificial Intelligence.

Costa, B., & Boiney, J. (2012). *Social Radar.* McLean, VA: The MITRE Corporation. Retrieved December 5, 2012 from https://www.mitre.org/sites/default/files/pdf/12_0581.pdf

Cucchiarelli, A., Velardi, P., D'Antonio, F., & Curzi, M. (2008). *Modeling collaborations content in Social Network Analysis*. Paper presented at the IEEE/WIC/ACM International Conference on Web Intelligence and Intelligent Agent Technology, Sydney, Australia. doi:10.1109/WIIAT.2008.119

Cyram. (2012). *Cyram NetMiner*. Retrieved January 15, 2012 from www.netminer.com

Dawoud, K., Alhajj, R., & Rokne, J. (2010). *A Global measure for estimating the degree of organization of terrorist networks*. Paper presented at the 2010 International Conference on Advances in Social Networks Analysis and Mining, Odense, Denmark. doi:10.1109/ASONAM.2010.84

De Choudhury, M., Sundaram, H., John, A., & Seligmann, D. (2009). Social Synchrony: Predicting mimicry of user actions in online social media. In *Proceedings of the 2009 International Conference on Computational Science and Engineering*. Vancouver, Canada: IEEE. doi:10.1109/CSE.2009.439

De Choudhury, M., Sundaram, H., John, A., & Seligmann, D. (2010). Analyzing the dynamics of communication in online social networks. In Handbook of Social Network Technologies and Applications (pp. 59 – 94). Academic Press.

Dellarocas, C. (2003). The Digitization of Word of Mouth: Promise and Challenges of Online Feedback Mechanisms. *Management Science*, *49*(10), 1407–1424. doi:10.1287/mnsc.49.10.1407.17308

Evans, M. M., & Wensley, A. K. P. (2009). Predicting the influence of network structure on trust in knowledge communities: Addressing the interconnectedness of four network principles and trust. *Electronic Journal of Knowledge Management*, *7*(1), 41–54. Retrieved from www.ejkm.com/issue/download.html?idArticle=158

Finextra. (2011). *Social media spreads Bank Transfer Day message*. Retrieved August 15, 2012 from www.finextra.com/news/fullstory.aspx?newsitemid=23133

FireDogLake.com. (2011). *List of active Occupy encampments*. Retrieved September 21, 2012 from http://fdlaction.firedoglake.com/2011/12/20/list-of-active-occupy-encampments-across-the-country-now-at-61

Gloor, P., Krauss, J., Nann, S., Fischbach, K., & Schoder, D. (2009). *Web Science 2.0: Identifying trends through semantic Social Network Analysis*. Paper presented at the 2009 International Conference on Computational Science and Engineering, Vancouver, Canada. doi:10.1109/CSE.2009.186

Hansen, D. L., Shneiderman, B., & Smith, M. A. (2011). *Analyzing social media networks with NodeXL: insights from a connected world*. Burlington, MA: Morgan Kaufmann.

Huberman, B., Romero, D. M., & Wu, F. (2008). Social networks that matter: Twitter under the microscope. *First Monday*. Retrieved September 21, 2012 from http://firstmonday.org/htbin/cgiwrap/bin/ojs/index.php/fm/article/view/2317/2063

Hutter, C., Hartmann, B., Bohm, K., Heistermann, T., Kohlmeyer, K., Reckling, R., . . . Parra, D. S. (2010). *SONAR: Towards User-Centric Social Network Analysis and Visualization*. Paper presented at the ACM International Conference on Web Intelligence and Intelligent Agent Technology, Los Alamitos, CA. doi:10.1109/WI-IAT.2010.67

INCOSE. (2006). *Systems Engineering handbook, version 3* (C. Haskins, Ed.). Seattle, WA: INCOSE.

Jafari, M., Akhavan, P., Bourouni, A., & Amiri, R. H. (2009). A framework for the selection of knowledge mapping techniques. *Journal of Knowledge Management Practice*, *10*(1), 40–53.

Johnson, T., & Kaye, B. (2004). Wag the blog: How reliance on traditional media and the internet influence credibility perceptions of weblogs among blog users. *Journalism & Mass Communication Quarterly*, *81*(3), 622–642. doi:10.1177/107769900408100310

Khonsari, K. K., Nayeri, Z. A., Fathalian, A., & Fathalian, L. (2010, 9-11 Aug. 2010). *Social Network Analysis of Iran's Green Movement opposition groups using Twitter.* Paper presented at the 2010 International Conference on Advances in Social Networks Analysis and Mining (ASONAM), Odense, Denmark. doi:10.1109/ASONAM.2010.75

King, I., Li, J., & Chan, K. T. (2009). *A brief survey of computational approaches in social computing.* Paper presented at the International Joint Conference on Neural Networks, Atlanta, GA. doi:10.1109/IJCNN.2009.5178967

Klamma, R., Spaniol, M., Cao, Y., & Jarke, M. (2006). Pattern-Based cross media Social Network Analysis for technology enhanced learning in Europe. In W. Nejdl & K. Tochtermann (Eds.), *EC-TEL 2006* (pp. 242–256). Berlin, Germany: Springer-Verlag. doi:10.1007/11876663_20

Kong, D., Gray, D., & Tao, H. (2006). *A viewpoint invariant approach for crowd counting.* Paper presented at the 18th International Conference on Pattern Recognition, Hong Kong, China. doi:10.1109/ICPR.2006.197

Lee, J. G., Moon, S., & Salamatian, K. (2010). *An approach to model and predict the popularity of online contents with explanatory factors.* Paper presented at the 2010 IEEE/WIC/ACM International Conference on Web Intelligence and Intelligent Agent Technology, Los Alamitos, CA. doi:10.1109/WI-IAT.2010.209

Lin, S. F., Chen, J. Y., & Chao, H. X. (2001). Estimation of number of people in crowded scenes using perspective transformation. *IEEE Transactions on Systems, Man, and Cybernetics. Part A, Systems and Humans*, *31*(6), 645–654. doi:10.1109/3468.983420

Lockheed Martin. (2010). *TPS-79 Tactical Military Air Traffic Control Radar.* Lockheed Martin Corporation. Retrieved September 1, 2012 from http://www.radartutorial.eu/19.kartei/pubs/TPS-79_productcard.pdf

Los Angeles Times. (2012). Timeline Occupy Wall Street. *LA Times.* Retrieved January 2, 2012 from http://timelines.latimes.com/occupy-wall-street-movement

Lou, J., Lin, S., Chen, K., & Lei, C. (2010). *What can the temporal social behavior tell us? An estimation of vertex-betweenness using dynamic social information.* Paper presented at the 2010 International Conference on Advances in Social Networks Analysis and Mining, Odense, Denmark. doi:10.1109/ASONAM.2010.46

Lu, Y., Luo, X., Polgar, M., & Cao, Y. (2010). Social network analysis of a criminal hacker community. *The Journal of Computer Information Systems, 51*(2), 31-31-41.

Ma, R., Li, L., Huang, W., & Tian, Q. (2004). *On pixel count based crowd density estimation for visual surveillance.* Paper presented at the 2004 IEEE Conference on Cybernetics and Intelligent Systems, Singapore.

Maybury, M. (2010). *Social Radar for smart power.* McLean, VA: The MITRE Corporation. Retrieved September 15, 2012 from http://www.mitre.org/publications/technical-papers/social-radar-for-smart-power

ABC News. (2012). *Occupy Philly shows national solidarity.* Retrieved September 21, 2012 from http://www.nbcphiladelphia.com/news/local/Occupy-Philly-Shows-National-Solidarity-138304044.html

News, N. B. C. (2012). Occupy Oakland: 400 arrested near Oakland city hall. *NBC News.* Retrieved September 15, 2012 from http://usnews.nbcnews.com/_news/2012/01/30/10268080-occupy-oakland-400-arrested-after-violent-protest

News-record.com. (2011). Black Friday occupations. *News & Record.* Retrieved September 11, 2012 from www.news-record.com/2011/black-friday-occupations

NPR. (2012). Occupy movement has captured the nation's attention, if not its cities. *National Public Radio.* Retrieved October 1, 2012 from http://www.npr.org/2012/02/01/146205425/occupying-the-nations-attention-if-not-its-cities

Pallis, G., Zeinalipour-Yazti, D., & Dikaiakos, M. (2011). Online social networks: Status and trends. In New Directions in Web Data Management (pp. 213 - 234). Berlin: Springer-Verlag.

Park, J., Shin, Y., Kim, K., & Chung, B.-S. (2010). Searching Social Media Streams on the Web. *IEEE Intelligent Systems, 25*(6), 24–31. doi:10.1109/MIS.2010.150

Sage, A., & Rouse, W. (2009). *Handbook of Systems Engineering and Management* (2nd ed.). Hoboken, NJ: Wiley.

Santos, E., Pan, L., Arendt, D., & Pittkin, M. (2006). *An effective anytime anywhere parallel approach for centrality measurements in Social Network Analysis.* Paper presented at the 2006 IEEE International Conference on Systems, Man, and Cybernetics, Taipei, Taiwan. doi:10.1109/ICSMC.2006.385045

Secretary of Defense. (2011). *Sociocultural behavior research and engineering in the Department of Defense context.* Retrieved August 15, 2013 from http://www.dtic.mil/dtic/tr/fulltext/u2/a549230.pdf

Slate.com. (2011). *Portland police arrest 50.* Retrieved September 10, 2012 from www.slate.com/blogs/2011/portland_police_arrest_50.html

Spencer, J. W. (2003). Global gatekeeping, representation, and network structure: A longitudinal analysis of regional and global knowledge-diffusion networks. *Journal of International Business Studies, 34*(5), 428–442. doi:10.1057/palgrave.jibs.8400039

Tang, L., & Huan, L. (2010). Understanding group structures and properties in social media. In P. S. Yu (Ed.), Link Mining: Models, Algortithms, and Applications (pp. 163 - 185). Berlin, Germany: Springer Science+Business Media, LLC.

Telegraph. (2012). Police Clash with Occupy Wall Street Protesters. *Telegraph*. Retrieved September 12, 2012 from http://www.telegraph.co.uk/finance/8811627/Police-clash-with-Occupy-Wall-Street-protesters.html

Thales Raytheon Systems. (2011). *Thales awarded tactical control radar contract*. Retrieved September 1, 2012 from http://www.thalesgroup.com/Press_Releases/Countries/Canada/2011/Tactical_Control_Radar_Contract

Varlamis, I., Eirinaki, M., & Louta, M. (2010). *A study on social network metrics and their application in trust networks*. Paper presented at the 2010 International Conference on Advances in Social Networks Analysis and Mining (ASONAM), Odense, Denmark. doi:10.1109/ASONAM.2010.40

Vashist, A., Mau, S., Poylisher, A., Chadha, R., & Gosh, A. (2011). *Leveraging social network for predicting demand and estimating available resources for communication network management*. Paper presented at the IEEE IM 2011: Mini Conference, Dublin, Ireland. doi:10.1109/INM.2011.5990558

Wall Street Journal. (2011). *Thousands occupy mall*. Retrieved August 5, 2012 from http://blogs.wsj.com/washwire/2011/11/15/occupy-dc-allowed-to-keep-on-occupying-d-c-park

Webb, C. (2008). Measuring social capital and knowledge networks. *Journal of Knowledge Management, 12*(5), 65-78. doi: 10.1108/13673270810902948

Wills, D., & Reeves, S. (2009). Facebook as a political weapon: Information in social networks. *British Politics, 4*(2), 265-281. doi: 10.1057/bp.2009.3

Wolff, C. (2009). *Requirements for Radar Receivers*. Retrieved September 1, 2012 from http://www.radartutorial.eu/09.receivers/rx04.en.html

Zeng, D., Hsinchun, C., Lusch, R., & Shu-Hsing, L. (2010). Social media analytics and intelligence. *IEEE Intelligent Systems, 25*(6), 13–16. doi:10.1109/MIS.2010.151

Žiberna, A., & Vehovar, V. (2009). *Using social network to predict the behavior of active members of online communities*. Paper presented at the 2009 Advances in Social Network Analysis and Mining, Athens, Greece. doi:10.1109/ASONAM.2009.24

ADDITIONAL READING

Akinwale, A., Adekoya, A., & Onashoga, A. (2010). Egocentric Database Operations for Social and Economic Network Analysis. *Interdisciplinary Journal of Information, Knowledge, and Management, 5*, 293–303.

Hassanpour, N. (2011, 8/31/11). *Media disruption exacerbates revolutionary unrest: Evidence from Mubarak's natural experiment*. Paper presented at the APSA 2011 Annual Meeting Paper.

Jones, Q., Ravid, G., & Rafaeli, S. (2004). Information overload and the message dynamics of online interaction spaces: A theoretical model and empirical exploration. *Information Systems Research, 15*(2), 92–101. doi:10.1287/isre.1040.0023

Knoke, D., & Yang, S. (2008). *Social Network Analysis*. Thousand Oaks, CA: SAGE Publications, Inc.

Lakshmanan, G., & Oberhofer, M. (2010). Knowledge discovery in the Blogosphere. *IEEE Internet Computing*, *14*(2), 24–32. doi:10.1109/MIC.2010.26

Liebowitz, J. (2005). Linking social network analysis with the analytic hierarchy process for knowledge mapping in organizations. *Journal of Knowledge Management, 9*(1), 76-76-86.

Shin, D. (2010). Analysis of online social networks: A cross-national study. *Online Information Review*, *34*(3), 473–495. doi:10.1108/14684521011054080

Steinberg, A. (2005). *An Approach to Threat Assessment*. Paper presented at the 2005 7th International Conference on Information Fusion.

Wassermann, S. F. K. (1994). Social Network Analysis: Methods and Applications. New York, NY: Cambridge University Press.

Yuan-Chu, H. (2009, 20-22 July 2009). *Mining Proximal Social Network Intelligence for Quality Decision Support*. Paper presented at the Social Network Analysis and Mining, 2009. ASONAM '09. International Conference on Advances.

KEY TERMS AND DEFINITIONS

Authority: The number of directed edges in a graph that point into a node.

Betweeness: A measure of the quantity of links from one user to two or more distinct groups of nodes and who therefore may be able to control the flow of information between groups.

Crowd Formation: Chaotic assembly of a large number of people.

Degree Centrality: A measure of the quantity of links from one user to other users; a highly central user has a greater number of links to other users.

FaceBook: An online social media network which enables users to post text, photos and videos for sharing with "friend" users and the public in addition to other features.

Mass Collective Action: An action taken together by a large group of people with a common objective.

Node: A representation of a user in a mathematical model of a communications network.

Occupy Wall Street: A protest movement against economic inequality that started with the occupation of the Wall Street financial district in New York City in September of 2011 and then spread to over 200 cities across the U.S. and 2000 cities around the world.

Situation Awareness: Knowing what is going on around an environment.

Social Network Analysis (SNA): Mathematical and statistical methods of describing patterns of relationships among users involved in social networks. Users are typically modeled as nodes in a mathematical representation of a communications network as a graph with links between the nodes indicating a parameter such as the distance between the nodes.

Systems Engineering: Design and management of complex engineering systems throughout their lifecycles.

Twitter: An online social media network which enables users to send and receive 140-character messages called "tweets".

Chapter 6
Large-Scale Disaster Response Management:
Social Media and Homeland Security

Kimberly Young-McLear
The George Washington University, USA

Thomas A. Mazzuchi
The George Washington University, USA

Shahram Sarkani
The George Washington University, USA

ABSTRACT

This chapter provides readers with an overview of how social media has enhanced large-scale natural disaster response at the Department of Homeland Security and its partners. The authors of this chapter present the history of the Federal Emergency Management Agency and how its successes and failures have shaped how the Department of Homeland Security has managed trends in increased community participation and information technology. Concepts from Systems Engineering frame the discussion around resilience engineering, network analysis, information systems, and human systems integration as they pertain to how social media can be integrated more effectively in large-scale disaster response. Examples of social media in disaster response are presented including a more in-depth case study on the use of social media during the 2012 Hurricane Sandy response. The chapter concludes with a proposed framework of a decision support system which integrates the benefits of social media while mitigating its risks.

INTRODUCTION

The devastation of Hurricane Katrina and the lack of federal response has been the subject of scholarly research for more than a decade. This tragedy, among others since 2005, has brought immediate attention to the need for an improvement to large-scale disaster response and management. Hurricane

DOI: 10.4018/978-1-4666-8556-7.ch006

Katrina claimed the lives of more than 1,800 people (Zimmermann, 2012) as the nation and the world monitored federal response on various media sources, from television to print. Public outrage deepening further after it was apparent that the disaster management resources were grossly underestimated and not timely. Various local and national media sources were reporting that there were hundreds of people in need of rescue and aid, but there was a gap between what was being reported and where disaster response resources were actually deployed. This chapter will provide an in-depth background on large-scale natural disaster response and propose a framework for using social media to bridge the gap between media reporting and adequate resource deployment.

Large-scale disasters undeniably make response very complex. "Every year, disasters put millions of Americans in danger and destroy billions of dollars worth of property" (Department of Homeland Security [DHS], 2013b). Disasters of this scale can be labeled as a catastrophic incident. The National Response Framework (DHS, 2008) defines a catastrophic incident as "any natural or manmade incident, including terrorism that results in extraordinary levels of mass casualties, damage, or disruption." The sheer magnitude of these disasters typically occurs in multiple jurisdictions and has greater risk to catastrophic loss of life and property. Additionally, with advances in technology and a culture of immediate social media reporting, it is often difficult to coordinate resources at a federal level ahead of public perception of slow response.

To further understand disaster response management, it is also important to understand other commonly used terms, such as, emergency management and crisis management. An emergency can also be considered as an event which has made it unlikely or "impossible for an organization to conduct business as usual" (Turoff, Chumer, Hiltz, & Klashner, 2004b). Typically in crisis situations, it is expected that there will be an element that could not be predicted. The main purpose of an emergency management system is assuring that there are minimum loss of life and property given the investment of resources towards this effort (Nikolic, Savic, & Stankovic, 2007). Similarly, Perry (2003) argues that "emergency management refers to the implementation of plans, and the use of personnel and equipment to achieve the tactical and task requirements of response to address a given threat." The Federal Emergency Management Agency defines crisis management as the "measures to identify, acquire, and plan the use of resources needed to anticipate, prevent, and/or resolve a threat or act of terrorism" (Federal Emergency Management Agency [FEMA], 2003). Although disasters and crises can be different events, they are certainly highly related. "The disaster is similar to the crisis in terms of the developing phases (pre-event phase, event phase, and recovery phase)" (Shaluf, Ahmadun, & Said, 2003) Therefore, this chapter will apply research from disasters, crises, and emergencies.

Effective disaster response management requires the right resources to be deployed to the right location at the right time. This triangulation is the foundation of precise decision-making. Decision making is a very complex behavior in large-scale disaster management because of the "half-structured or a non-structured problem [and] … the value of timely information is immense" (Nikolic et al., 2007). One such location of disaster-related information is that found in social networks and media. Bhanot (2012) states that "social media are media for social interaction, using highly accessible and scalable communication techniques… [and] is the use of web-based and mobile technologies to turn communication into interactive dialogue."

From Web 1.0 in the 1990s to current uses of Web 2.0, social media has evolved over the years. Platforms, such as, Facebook, Twitter, and YouTube have millions of users that are connected to each other over a broad network where their feelings and observations are posted in only a short few moments. Nearly instantly, those that are connected to the same network, linked by common interest, are updated on a new post from a user in their network. There are tremendous advantages to social media. For example, social media is easy to use, quickly connects a large number of people, capable of gathering real-time data, and is inexpensive. Some argue that social media may create isolating behavior due to lack of personal physical interaction, however, (Billington & Billington, 2012) assert that "social media actually aids the formation of communities." Parallel to the evolution of social media platforms, is the evolution of its applications by corporations and agencies. Corporations and agencies have used social media to "push" information out to the public, substituting information from traditional public affairs media to that of on social media. Agencies, such as FEMA, have a history since the 1980s of adopting new technology. FEMA's use of social media, in particular for disaster response management, has evolved significantly in the past decade. In addition to pushing information, they are also pulling information. There are some disadvantages, however, that must be considered when an organization is considering leveraging social media in a Web 2.0 reality. Some of the limiting areas include privacy concerns, monitoring data could be time consuming, data may include slang and/or abbreviations, and the accuracy may be challenged.

This chapter emphasizes the need for a systematic approach to integrating social media into the disaster response phase for large-scale incidents. Systems engineering topics, such as resilience engineering, network analysis, information systems, and human systems integration, will also be explored and applied to analyze specific examples of social media in emergency management. These examples include the Shadow Lake Fire, Harris County Fire, Haiti Earthquake, and Hurricane Irene. Systems engineering is important in the social media discussion in this chapter because it provides a context for further understanding the National Incident Management System (NIMS) and the Incident Command System (ICS). Additionally, (Crowe, 2010) outlines clear differences between NIMS and social media; "NIMS was built on the foundation of formally organized command, control, and approval of all emergency actions, while social media was built on open, organic, and informal response." Systems engineering concepts will illustrate why these seemingly two opposing perspectives exist and how they can be better aligned and integrated.

Lastly, an in-depth case study on the 2012 Hurricane Sandy response is presented in this chapter to identify key perspectives from various levels of government and public sector, including FEMA, New York City, and the Red Cross. The objective of this chapter is to present a case study to analyze the strengths, weaknesses, challenges, and opportunities of social media in incident response. Since Hurricane Katrina, DHS has campaigned heavily to the public, "If You See Something, Say Something" (DHS, n.d.). This chapter and case study will specifically address how effectively the Department has leveraged what everyday citizens are saying real-time on social media during incidents and proposes a framework to integrate social media into NIMS and ICS, as well as appropriate measures of effectiveness.

Background

Conceived after the terrorist attacks on September 11th 2001, the Department of Homeland Security (DHS) was formed to protect the United States from threats. "The vision of homeland security is to ensure a homeland that is safe, secure, and resilient against terrorism and other hazards" ("Our mission," n.d.). The Department brought together 22 Federal agencies, each with its own history and culture. Totaling more than 280,000 personnel, these agencies were brought together to form the third largest Federal department. DHS, from its onset, made tremendous effort to create a "unified and integrated Department" which was built upon the principles of resiliency and transparency ("Who joined DHS," n.d.). The jurisdictions and borders of America are complex; the Department does not have enough resources to be effective without partnerships. The political and geographical make-up of the United States created the need for a unique partnership with "hundreds of thousands of people from across the federal government, state, local, tribal, and territorial governments, the private sector, and other nongovernmental organizations" ("Our mission," n.d.). These partnerships are critical to the effectiveness of the Department's five broad missions: preventing terrorism, securing the borders, enforcing immigration laws, securing cyberspace and ensuring resilience to disasters.

The fifth mission, ensuring resilience to disasters, is discussed in this chapter. From the 2010 Quadrennial Homeland Security Report, DHS defines resiliency as the ability to "foster individual, community, and system robustness, adaptability, and capacity for rapid recovery" (DHS, 2010). In terms of risks, including natural disasters, it is not possible to eliminate all threats, therefore, the robust capacity to respond "must be effective, efficient, and grounded in the basic elements of incident management" for emergencies and crises (DHS, 2012). From a strategic perspective, to accomplish this mission, the Department's incident management is based on four phases of an emergency management cycle. These elements are "hazard mitigation, enhanced preparedness, effective emergency response, and rapid recovery" (DHS, 2010). These phases form the resilience to emergencies for a given incident or disaster. This chapter will focus on the third phase of the cycle – effective emergency (disaster) response.

The Department has adopted a systems approach to disaster management. Understanding that despite the Department's best efforts towards hazard mitigation and disaster preparedness, disasters will still happen. DHS has communicated the value of open government and transparency "also require[s] a shift from a reliance on top-down emergency management to a process that engages all stakeholders" (DHS, 2010). "Individuals, families, and communities—and the systems that sustain them—must be informed, trained, and materially and psychologically prepared to withstand disruption, absorb or tolerate disturbance, know their role in a crisis, adapt to changing conditions, and grow stronger over time" (DHS, 2010). To achieve the goal of ensuring effective disaster response, the Department established three objectives. The first objective is to provide accurate information to the public in a timely manner. The second objective describes broadly that response operations should be unified and effective. And lastly, the Department in working with its partners should provide timely disaster relief assistance (DHS, 2010). Another key component to ensuring resiliency is transparency with it stakeholders. FEMA's Administrator Craig Fugate, in a 2011 testimony stated that his agency fully recognizes "that a government-centric approach to disaster management will not be enough to meet the challenges posed by a catastrophic incident." (Improving the Nation's Response: How to minimize, 2011) This is important because these partnerships are a part of the disaster response system. The stakeholders across the federal government, state, local, tribal, and territorial governments, the private sector, and other nongovernmental organizations "regularly interact with the public, who are responsible for public safety and security, who own and operate our nation's critical infrastructures and services" ("Our Mission." n.d.). Regardless of where disaster strikes in the Unites States, it always occurs at the local jurisdiction level (FEMA, 2010). The local government is the first to assess and respond to the disaster. As a result, the local community is the first to cope with the loss of life and property destruction. The local government "maintains control of all assets used in the response and recovery efforts, regardless of the source of those assets" (FEMA, 2010). If the resources needed to manage the disaster are exceeded by the local government, there is a legal and sequential process to request more resources at the State level. As the scale of the disaster increases, the sequence of requesting resources expands to accommodate. There may be multiple local governments requesting resources from the state government at the same time frame. The same sequence is managed at the Federal level if the State's resources are exhausted.

The government plays a major role in large-scale disaster response management. Specifically, the Department of Homeland Security's component charged with overseeing its disaster resilience mission is the Federal Emergency Management Agency (FEMA). From the Department of Homeland Security's website, "FEMA coordinates federal operational and logistical disaster response capability needed to save and sustain lives, minimize suffering, and protect property in a timely and effective manner in communities that become overwhelmed by disasters" (DHS, 2013). The agency works with State and local governments to achieve system resiliency. Ward describes large-scale disaster management as the destruction of resources such that the "entire system of societal infrastructure exists in a state of chaos" (Ward, Wamsley, Schroeder, & Robins, 2000). From this perspective, the system's equilibrium is restored when the resources located beyond the affected area are deployed to the areas in need. FEMA received high praise for their management of Hurricane Sandy, the large-scale disaster that devastated the northeastern United States in 2012.

FEMA's disaster management history has not always been successful. FEMA in the 1980s was plagued with organizational, leadership, and information technology challenges. The Agency "was created as a centralized body, but it has been restricted largely to a reactive role with respect to federal disaster funds management and no significant mission ability for threat assessment, response planning on an integrated government wide basis, and most importantly, overall command and control in an emergency situation" (Turoff, Chumer, Van de Walle, & Yao, 2004a). Figure 1 shows the early organizational structure of FEMA.

When FEMA was established under the Carter Administration, it had two missions. Carter envisioned that its primary mission was to deal with natural disasters and then also have the legislative authority for

Figure 1. 1980s FEMA Structure (Ward, Wamsley, & Robins 2000)

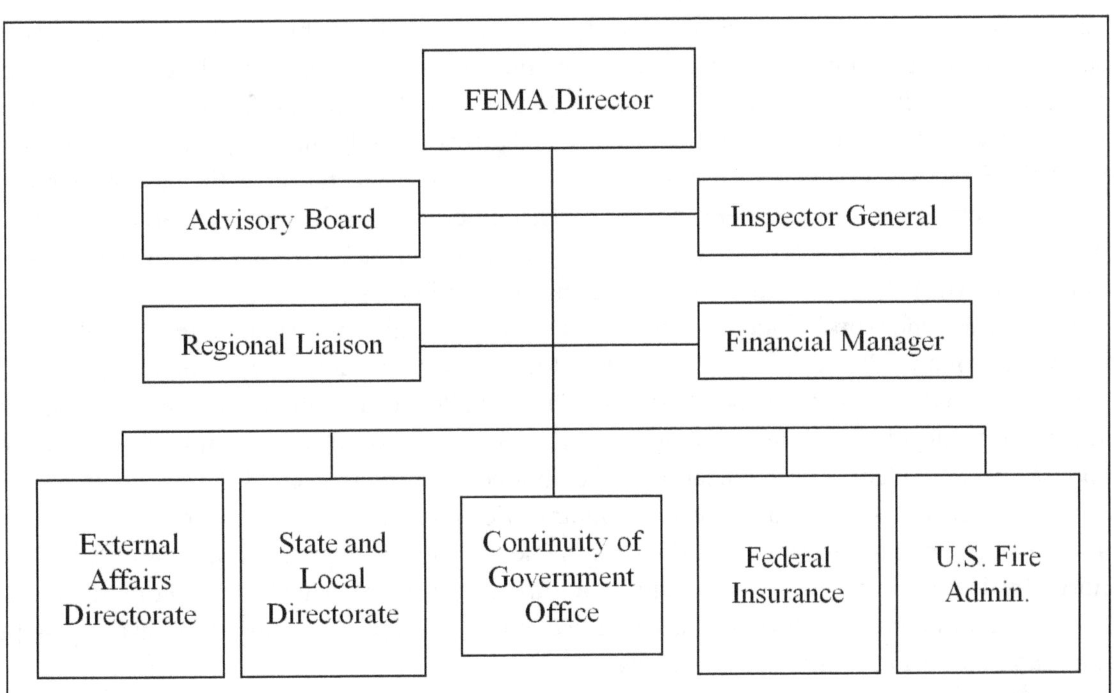

total civil defense to protect the nation from other than natural threats, such as nuclear attacks. FEMA developed a disaster response system that was capable of managing disasters of any type or scale. FEMA however refused to deploy this system in support of civilian emergency management. FEMA's leadership feared that this would put more emphasis on their secondary mission, thereby potentially losing control of that mission.

Similar leadership issues continued into the 1990s under the George H.W. Bush Administration. On August 24, 1992, Hurricane Andrew affected more than 250,000 people in several counties in Southern Florida. FEMA's Information Technology support was poor in the wake of Hurricane Andrew. The Office of Emergency Management had to purchase hundreds of personal computers and train local operators filling Disaster Assistance Centers (Ward et al., 2000). After significant negative attention from the public, Congress, and even President Bush about FEMA's lack of full participation in Hurricane Andrew, this motivated a new phase of leadership based on collaboration, information sharing, and an emphasis on IT. Only one year later FEMA, in cooperation with the Army Corps of Engineers developed a fairly robust Disaster Management Information System (DMIS). This system consisted on linking portable laptops with satellite data connections. These computers were installed in relief worker vehicles. It provided workers full access to grid maps and assessments of the affected disaster areas.

In August 2005, Hurricane Katrina revealed yet again serious failures in FEMA. Despite the agencies strides since the early 1990s, the American public witnessed first-hand FEMAs lack of preparedness and response. The Government Accountability Office (GAO) reported that "the government's performance was aimed at the highly bureaucratic structure of the emergency management system and the ability of such an inflexible and unresponsive system to deal with similar disasters in the future" (Tsai & Chi, 2012). Ultimately, this led to the passing of legislation, called the Post-Katrina Emergency Management Reform Act of 2006. This "new FEMA is forward leaning, able, agile and reliable, and has inspired public trust...helping families and communities overcome all hazards and helping America build a culture of preparedness" ("Disasters Overview," n.d.).

In 2003, the Department of Homeland Security and FEMA outlined three goals to achieve resilience to disasters, in which communication and information management are critical. To ensure effective disaster response, DHS must systematically "provide timely and accurate information to the public … conduct effective and unified response operations … and provide timely and adequate disaster assistance" (DHS, 2010). Effective disaster response management requires "cooperation on the part of a large group of autonomous organizations…suddenly placed into a temporary organizational hierarchy" (Ward et al., 2000). Nikolic et al. (2007) asserts that disaster response should be based on systems which "systems are also characteri[z]ed by a specific working environment of decision makers…which often causes a high level of stress and working load as well as a higher probability of oversight and misinterpretation of available information." In examining FEMAs history, there are clear themes to challenges associated with large-scale disaster response and resilience. These challenges include managing decision support systems through information technology and sharing; deployment of resources; and public expectations of response to the incident. Communication is the common ground to each of these challenges. Turoff et al. (2004) asserts that some characteristic communication challenges which occur in disaster response are driven by information overload and the heterogeneous nature of team coordination of individuals who normally do not work together. In Horsley and Barker's (2002) model of crisis communication, they propose that communications during a disaster are cyclical; it emphasizes the need for ongoing public relations and monitoring of new crisis updates

To overcome the known challenges and risks associated with managing large-scale disasters, national guidelines was developed that outlines specific systems which are required by law, through Presidential Directives, to use during disasters. The National Response Framework (DHS, 2008) is a guide to how the country responds to disasters and emergencies. The framework was directed by The George W. Bush Administration to be developed and implemented for all types of emergencies, including natural disasters. This unclassified document was first released in 2008 as one of five comprehensive national planning frameworks. Each incident management framework covers each phase of the emergency management cycle. In 2003, the Bush Administration authorized Homeland Security Presidential Directive-5, Management of Domestic Incidents. This was created "to enhance the ability of the United States to manage domestic incidents by establishing a single, comprehensive National Incident Management System (NIMS)" (HSPD-5, 2003). Within NIMS, the government's objective was to align crisis management and consequence management into a single, integrated concept. It is by this authority that the Secretary of Homeland Security is the Federal official in charge of national incidents and emergencies. NIMS framework has five core components: preparedness; communications and information management; resource management; command and management; and ongoing management and maintenance (DHS, 2008). "NIMS includes a core set of concepts, principles, terminology, and technologies covering the Incident Command System (ICS); multi-agency coordination systems; unified command; training; identification and management of resources (including systems for classifying types of resources); qualifications and certification; and the collection, tracking, and reporting of incident information and incident resources" (HSPD-5, 2003).

The operational component within NIMS is the Incident Command System (ICS). "The ICS was developed in the early 1970s to help coordinate the response to forest fires, and it has since been used in a wide variety of crises" (Moynihan, 2008). ICS is a unified command structure that manages an incident and is responsible for deploying resources to the right location at the right time. Figure 2 shows that structure of the Incident Command System. Typically, personnel are assigned to standard positions in the three main management functions: logistics, planning, and operations sections. The planning section is the heart of the system. It is the section that allows the logistics section to stage resources and the operations section to execute decisions. The incident goes through iterations, called planning cycles. In these cycles, strategic and operational action plans are discussed with leadership and developed. ICS is a system that reacts to the known threats and works towards responding to set objectives.

ICS leadership, called the unified command, ensures that these three sections are coordinating effectively through pre-planned meeting cycles. As seen in the Figure 2, the structure of ICS is from

Figure 2. Incident Command System Structure, Source: (FEMA, n.d.)

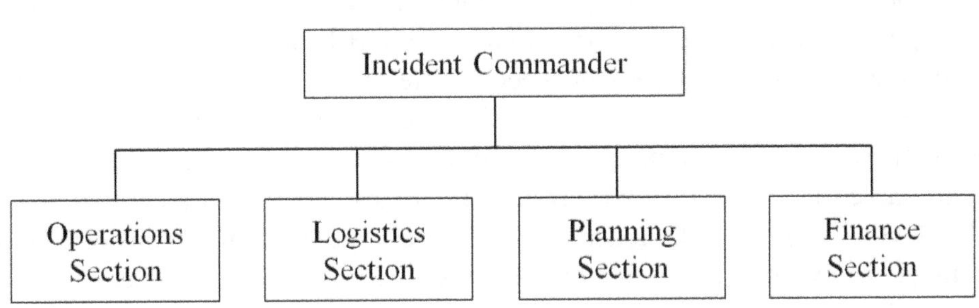

spans from the field level to the incident commander. At each level in between, each individual reports to only one individual. ICS is "based on the scalar principle, with its size and complexity depending on the size and complexity of the disaster or emergency incident to which it is applied" (Buck, Trainor, & Aguirre, 2006). There is organizational learning that occurs through the required use NIMS and ICS during an incident because of the standardization of the common terminology, practices, and principles. ICS, if implemented successfully, it will reduce the stress and risks of uncertainty (Kapucu, Arslan, & Demiroz, 2010).

ICS is also a sociotechnical system with the integration of technology and people. Tsai and Chi (2012) acknowledges that ICS is a temporary and synthetic where the system is joined by the technical and the social components. This system, comprised of ICS personnel from State, local, private, Federal, etc interact with the general public affected by the disaster to deliver aid and respond appropriately. Although there are common terminology and practices, both technology and humans must work together effectively for the system to reach an optimal state. Because the system relies heavily upon the massive numbers of humans from the public, nonprofit, and government playing a particular assigned role in the response, there is an opportunity for inherent stress and errors associated with human behavior in crises. As the scale of the disaster increases, the "social aspect of the system, [which are] the connections between different organizations and individuals, becomes increasingly complicated" (Tsai & Chi, 2012). Therefore, the culture of the organization is also critical for ICS to be successful.

Critics of ICS "suggest that the emphasis on formal organizations and rationality fails to recognize a series of key factors in disaster response, particularly the presence of unorganized volunteers and emergent groups as well as the transformations of the structure and function of established organizations during the response phase of disasters" (Buck et al., 2006). This horizontal and vertical communication and coordination must be interoperable; however, "technology itself cannot solve the interoperability problem because the core of the problem lies with how people in different organizations choose to work together rather than how they choose to talk to each other" (Kapucu et al., 2010).

Finally, Homeland Security Presidential Directive 8, National Preparedness supports the National Response Framework; it is aimed at strengthening the resilience of the United States. It accomplishes this "through systematic preparation for the threats that pose the greatest risk to the security of the Nation, including acts of terrorism, cyber-attacks, pandemics, and catastrophic natural disasters" (HSPD-8, 2003). This systematic preparation includes using the entire community as problem solvers. This holistic approach to disaster management is achieved both philosophically and in practice. In 2009, the Obama Administration emphasized transparency and open government initiatives. This philosophy continued down to the Department of Homeland Security and inspired complementary disaster response policies. In 2013, the National Response Framework (DHS, 2013d) was expanded to include the whole community philosophy and how it would conceptually work together with response and preparedness efforts.

The Whole Community concept is a "philosophical approach on how to think about conducting emergency management" (FEMA, 2011). Specifically, the DHS and FEMA disaster response management system not only includes employees assigned to roles within NIMS and ICS, but it also includes everyday citizens in the general public. The "system" is more inclusive than ever before in FEMA history. With this perspective, the whole community can "collectively understand and assess the needs of their respective communities and determine the best ways to organize and strengthen their assets, capacities, and interests" (FEMA, 2011). This approach to crisis and consequence management does not mean that FEMA has relinquished its role as the federal response coordinator. Richard Serino, FEMA's Deputy Administrator, testified that his agency recognizes that a large part of its mission is to support "our citizens and first responders to ensure resilience to all hazards" (Accountability at FEMA: Is Quality, 2011). Evidence and experience has shown that when disaster strikes in the local community, regardless of scale, the first line of "defense rests with individuals, families, and communities, who must take responsibility for their own safety and must be prepared to respond to and possibly endure a crisis" (DHS, 2010). The whole community approach benefits state, local, and federal partners, as well as empowers everyday citizens to contribute.

INTEGRATING SOCIAL MEDIA AND DISASTER RESPONSE

Challenges and Opportunities with Social Media

People of all ages and factions of society have incorporated social media into their daily routines. Social media has been intertwined in the lives of millions around the world. These applications communicate information exchanged electronically over a social network (Auer, 2011). Electronic information can be accessed on a suite of devices including, personal computers, laptops and on hand-held mobile devices. Commonly used social media applications include Facebook, Twitter, Instagram, YouTube, and LinkedIn. "In recent years, social media [has] expanded access to resources and increased the speed at which information can be distributed…. [causing an] exponential amounts of socially-generated data [which] are publicly available" (Hughes & Palen, 2012). Because users of social media post and exchange a wide range of information, it continues to be a source of public scrutiny. Critics of social media argue that Facebook, along with other platforms, are riddled with content that is too sensitive or private to share online. Advocates of social media argue that it is a powerful platform for the community to share constructive information and perhaps solve problems. Facebook, for example, has more than 500 million users in its network with more than 50% of its users logging in daily into their accounts. On average, the user's individual network consists of approximately 80 different users, groups, or event pages (Billington & Billington, 2012). Social media advantages and cautions, as shown in Figure 3, should be addressed as factors for Web 2.0 based policies.

Figure 3. Social Media Advantages and Risks (Billington and Billington, 2012)

Advantages	Risks
• Connects people in networks efficiently • Tool to monitor the environment • Aggregation of data reveals insights • Simple syntax • Adopted by younger generation • Possibly faster method of communication	• Privacy concerns • Consumes time • Depersonalizes communications • Unclear or ambiguous syntax • Questionable content quality • Learning curve

In 2009, President Barack Obama signed a memorandum that promoted transparency and open government. His Administration made it clear that government officials at the state, local, and federal levels could benefit from having access to the dispersed knowledge in society. To achieve this, the use of Web 2.0 technologies by federal agencies was highly encouraged. "Emerging technologies open new forms of communication between a government and the people.... It is important that policies evolve to realize the potential of technology for open government" (White House, 2009). Additionally, the memorandum encouraged agencies to create easily accessible websites which to publish non-sensitive information for the public. Government websites were also enhanced with features which allowed the option of the public posting feedback and contributing ideas. This validated the concept that in American society, exchanging information online was the norm. In response to President Obama's open government initiative, the Office of Management and Budget (OMB) authored and released several documents in 2010 which offered guidance for increasing transparency and information sharing, and reducing paperwork. Guidance for Agency Use of Third-party Websites and Applications (White House, 2010a); Guidance for Online Use of Web Measurement and Customization Technologies (White House, 2010b), and Social Media; Web-Based Interactive Technologies, and the Paperwork Reduction Act (White House, 2010c), were three relevant documents released by OMB. The guidance from OMB also states that there is tremendous opportunity for collaboration with the public. It also recognizes that the open government initiative will inevitably rely on social media use. OMB emphasized that information should only be collected if it was necessary and provided a practical function. The guidance stated that federal agencies must be especially vigilant about privacy issues when using Web 2.0 technology.

At DHS, the Office of the Chief Information Officer (OCIO) is responsible for the department's social media policy and guidance "including those that are related to accessibility, privacy, and security" (DHS, 2013a). Generally speaking, DHS employees are restricted from accessing social media on official computer terminals. There are a few exceptions, as long as it in compliance with the Federal Information Security Management Act of 2002. For example, DHS components, such as FEMA, Fusion Centers, Coast Guard, and Citizenship and Immigration, may request special permission to bypass standard policies to access third-party sites if it is necessary and will likely result in improved situational awareness or decision-making. The 2009 National Institute of Technology (NIST) issued guidance on the use of a risk-based method to establish security controls when accessing third-party social media sites. The OCIO is also responsible for guidance on the use of official social media sites created by DHS and its components. All of the department's seven operational components from ICE to CBP to TSA have standardized their websites and created social media sites to enhance traditional public affairs and situational awareness methods.

Since Hurricane Katrina, the Department of Homeland Security and FEMA have been scrutinized for not actively engaging with the public. Lessons learned from Deepwater Horizon also highlighted the lack of engagement with the public, both for problem-solving and also for public affairs. Horseley and Barker (2002) claim that, "in any type of crisis situation, an organization can benefit from a proactive strategy to work with the press rather than wait for the press to bombard it with tough questions and speculation." Perhaps using previous failures as motivation, DHS launched several official social media "communications to reach a wider audience, support operational activities such as investigations, and maintain situational awareness" (DHS, 2013a). By 2012, all of the major Federal agencies had Web 2.0 and a social media presence. Since launching its Twitter account in 2010, DHS "had 118,977 Twitter followers and had issued 2,796 tweets" (DHS, 2013a). For comparison, NASA had more than three million followers on Twitter. FEMA Administrator, Craig Fugate testified to Congress that "social media provides the tools needed to minimize the communication gap and participate effectively in an active, ongoing dialogue" (Evolution of Emergency Management, 2011). DHS has attempted to address if these policies and guidelines for social media are adequate given the fast pace of technology and increased public interest in government performance during disasters. Administrator Fugate boasted at Congressional hearings in 2011 that his agency must "engage the public as a critical resource by facilitating two-way communication that allows [FEMA] to communicate with the public in a disaster environment rather than talking at the public" (Evolution of Emergency Management, 2011). He made it clear that in step with President Obama's open government initiative, FEMA must adapt the way they communicate, instead of expecting the public to fit their traditional system. He also believes that social media, as it pertains to disaster-related information, allows FEMA and other partners "to incorporate critical updates from the individuals who experience the on-the-ground reality of a disaster" (Evolution of Emergency Management, 2011). Specific examples of benefits and use of social media by FEMA and its partners from a systems engineering perspective will be covered in next section. The major benefits of a system-wide application of social media include (DHS, 2013b):

- Centralization and Aggregation of Information and Technology Support;
- Multi-Way Information Sharing;
- Rumor Control and Misinformation;
- Alternative Utilization of Private Resources and Partnerships; and
- Maps and Crowd-sourced Information

From Social Media in the Emergency Management Field (2012), data collected by Su et al provides further insight on viability of the benefits of an ICS system-wide application of social media. Respondents from various levels of government answered specific questions about social media usage and goals. The study revealed that at the state, county, and local levels of government, at least 60% of the respondents rated that there was some coordination among the ICS areas of PIO, Incident Commander, Planning Section Chief, and Operations Section Chief. With respect to government goals for social media operations, Figure 4 depicts that although the state level has active goals to be engaged with the public on social media, there are opportunities to better construct mapping products and leverage the public to aid in disaster response tasks.

Figure 4. Goals for Social Media Operations (Adapted from Su et al., 2012)

Emergency Manager Goals for Social Media Use	State	County	Local
No goal	2%	36%	21%
Maintain a social media account or website	95%	50%	69%
Monitor real-time information	88%	47%	60%
Relationship building with the community	61%	24%	47%
Aggregate public information	46%	11%	19%
Crisis Mapping	24%	15%	21%
Crowdsourcing and Public participation	17%	11%	17%

Systems Engineering Perspective to Disaster Response

To effectively leverage social media in large-scale disaster response, disaster response managers can benefit immensely from a practical understanding of key systems engineering concepts. Because "emergency management requires management systems with specific characteristics, which make them quite different from traditional management systems," (Nikolic et al., 2007) a deeper understanding of systems engineering facilitates a better understanding of these systems. Disaster management is a system, comprised of four components: mitigation, preparedness, response, and recovery. NIMS and ICS, which are required to be implemented for any scale of response, are systems. In addition to state, local, and federal resources, the newer concept of "whole community" extends the disaster management system to include participation of individuals to contribute to disaster response. To further understand disaster management as a system, it is important to outline how a systems engineering perspective adds depth to the disaster management discussions. Using topics in resilience, network analysis, information systems, and human systems integration, this section will discuss DHS examples of when media, both traditional and social, has been effective or ineffective for disaster response. There will also be examples on how social media and emerging technologies have started to transform DHS disaster response capabilities.

Systems engineering is an interdisciplinary, highly-structured discipline that enables the development of successful systems. "It focuses on defining customer needs and required functionality early in the development cycle, documenting requirements, then proceeding with design synthesis and system validation while considering the complete problem" (INCOSE, 2004) from concept to production to operation. Sub-topics in systems engineering can also facilitate how the Department of Homeland Security manages large-scale disasters. These foundational topics include resilience engineering, network analysis, information systems, and human systems integration.

"Resilience to disasters" is how the Department of Homeland Security describes their mission to manage threats. The deliberate goal of "resilience" is examined further from a systems engineering perspective. Resilience "engineering is an approach which emphasizes how people manage complexity under stress to achieve success" (Williams & Smart, 2010). Similarly, resilience can be defined as "a system's capacity to maintain or restore an acceptable level of functioning despite perturbations or failures" (Hémond & Benoît, 2012). The term resilience has also been traced to the discipline of civil engineering. From this perspective, the absence of resilience engineering "would have a debilitating effect on the public's safety and security" (Anonymous, 2013). Williams & Smart (2010) argues that resilience is a proactive approach to safety which allows the system to function within an "operating envelope."

From a resilience perspective, there are three system states, pre-disruption, survival, and recovery. Prior to the actual disaster, or disruption, the system should take measures to anticipate or avoid the threat. This is analogous to the mitigation and preparedness phases of the DHS disaster management system. In the survival phase, this is equivalent to the DHS response phase, the system should have the capacity to absorb the disruption such that it is still in a functioning state after the initial threat or impact. In the recovery phase, the system attempts to restore what has been affected by the disruption. The "system resumes some degree of its original goals, including the survival of the humans in it" (Pellissier, 2011). For a system to operate successfully within this "envelope," and achieve resiliency, the literature contends that there must be specific system attributes. Pellissier (2011) states the primary four attributes are "capacity, flexibility, tolerance, and inter-element collaboration." Capacity is defined as the requirement for the system size to be able to "handle the maximum and most likely events, such as terrorist attacks and natural disasters" (Pellissier, 2011) and have functional redundancy. Functional redundancy includes physical and information-based system redundancy. If a particular sub-system failed, the redundant sub-system would take over. For example, response officials can update disaster information to the public through traditional media channels, but if communications systems failed, they may also release information through social media outlets. Flexibility describes how a system can transform or adapt to accommodate new information. For example, if there is a major disruption to disaster response, such as a new threat, an agile system would allow ICS leadership to adapt quickly and shift resources. "The adaptability of an organization or group of organizations and the way in which it should be evaluated constitute the first major challenges for the emerging field of resilience engineering applied to organizations" (Hémond & Benoît, 2012). Tolerance describes the system's ability to withstand disruptions or threats. In practical terms, a tolerant system has resources that are not rendered "inoperative" after the first stage of disruption. Disaster response systems must be able to withstand multiple disruptions. The initial onset of the disaster itself could be view as the first threat to resources. Throughout the disaster response phase, there could be several small or large threats, which could include weather updates, infrastructure damage, and communication system failures (Pellissier, 2011). Inter-element collaboration pushes effectiveness of resources beyond tolerance. It "allows all elements of the system to interact and cooperate with each other as in collaborative innovation systems (Pellissier, 2011). To achieve an optimal disaster system state which leverages social media from the public, this collaborative attribute is especially critical. "The majority of recommendations [from Hurricane Katrina] concern coordination, integration and planning. The focus mainly involves interagency interactions. The need for a "shared common operating picture and situational awareness" during an extreme event is stressed" (Jefferson, 2006). From this list of traits below, we can also see the common themes of capacity, flexibility and collaboration (Pellissier, 2011).

1. Actively monitor internal and external indicators of change
2. Prepare for disruptions
3. Build in flexibility
4. Cultivate a culture with clearly shared purpose and values
5. Foster a culture of continuous innovation and ingenuity
6. Strengthen and extend their communications networks
7. Social networks are built on trust and are visible

An example of a disaster response related social media case with resilience principles is evident in Los Ranchos, New Mexico. In 2009, the Los Ranchos de Albuquerque Emergency Management Office set up a social media site. One of the observations that coordinators witnessed, was that they needed to recruit a team that could be dedicated solely social media related responsibilities during a crisis. They found that was difficult to sustain traditional ICS duties and monitor social media. As a result, the office developed a team of trusted agents whose aim was to provide operational support using social media during incidents" (Katims, 2013). The Virtual Operations Support Team, or VOST, had the flexibility to actively monitor social media for indicators of change over the duration of the response period. The trusted agents were recruited from the very network that they would ultimately monitor. In March of 2011, the Los Ranchos coordinator solicited for volunteers on Twitter. This resulted in a total of 16 volunteers that were trained on the ICS structure and terminology. The team was officially incorporated into the ICS structure for several large-scale disasters, such as Hurricane Irene, of which they had specific social media responsibilities and assignments (DHSLLIS 2013). Some of these responsibilities included: following social media and traditional media trends and report back; communicating issues and concerns being expressed by the public; identifying misinformation; pushing out key message each day on official Twitter accounts; and representing the citizen's perspective (St. Denis, Palen, & Hughes, 2012). Prior to Hurricane Irene making landfall, VOST members used Skype to communicate and collaborate with leadership at the ICS. As the storm progressed, the team generated maps from information gathered from social media. They also posted warnings and communicated directly with the public. VOST efforts made the response system more resilient; it allowed the threats to be addressed more efficiently.

Disaster response systems can also be framed from a network perspective. There are both interorganizational and intraorganizational networks (De Wever, Martens, & Vandenbempt, 2005). The ICS, which would be best described as an interorganizational network, is "a large organization made up of a network of smaller organizations" (Betts & Stouder, 2004) including social media networks, and other partnering response organizations. In some literature, prefers "the term community over inter-organizational, because it refers not only to a legal entity, but also to the entire community involved" (Bharosa, Lee, & Janssen, 2010). Information is shared across the interorganizational or community network through the social connection of individuals and groups. "The ICS essentially overlays a hierarchical structure on a network, using a central command to manage conflict, coordinate action, and reduce classic network characteristics, such as a reliance on consensus" (Moynihan, 2008). Network analysis is a set of methods that can be used to analyze systems which is most appropriate for "relational type data which deals with the linkages between agents" (Betts & Stouder, 2004). Network analysis describes the patterns and relationships among the social players, or actors, in the network (Betts & Stouder, 2004).

Social network analysis is a specific subset of network analysis. "Social network analysis should not be outside the capability of a systems engineer, but rather its benefits would fold nicely into mission engineering—a core component of systems engineering" (Selke, 2007). Social network analysis "involves actors (seeing how actors are located or 'embedded' in the overall network) and relations (seeing how the whole pattern of individual choices gives rise to more holistic patterns)" (Liebowitz, 2005). "Centrality, distance, and clusters" (Betts & Stouder, 2004) are the basic concepts which can be measured in networks. Systems engineers could benefit from conducting a centrality analysis, social structure analysis, blockmodeling, and a clique analysis. Centrality analysis determines the extent to which the network is centered on various individuals or groups (Selke, 2007). Social structure analysis and blockmodeling aims to uncover group connections, or lack thereof, not necessarily observable to initial inspection. Clique, or cluster, analysis is helpful in studying disaster response systems because it will help identify how many and cliques exist, which may be contrary to the assumed ICS organization chart or structure (Selke, 2007).

"One of the major lessons we learned from the January 2010 earthquake in Haiti was that even if the physical infrastructure of an area is completely destroyed, the cellular infrastructure may be able to bounce back quickly, allowing emergency managers to relay important disaster-related information and enabling the public to request help from local first responders" (Evolution of Emergency Management, 2011). During the Haiti earthquake crisis, there was difficulty with delivering aid to the ground by aircrafts. The U.S. Air Force, who were in command of the airport responded to an urgent request on Twitter that relayed information of an aircraft in need. The Air Force responded back to the message on Twitter, communicating that the aircraft was cleared to land (Auer, 2011).

Another element in a network is the degree of uncertainty revealed between the social actors in the network. "The highest priority tasks necessary to save and sustain lives and stabilize following a catastrophic incident during the crucial, first 72 hours" (Evolution of Emergency Management, 2011). It is not possible to predict every behavior of the network, including its actors. One element that intensifies uncertainty in a network is turnover beyond the first 72 hours. The Incident Command System is a prime example of uncertainty in turnover of expertise. The ICS is staffed primarily with temporary actors who have been taken from their primary occupations, sometimes even specialty, to serve in a new disaster response capacity. ICS responders are expected to have the appropriate training when first deployed, however, it takes time to learn their roles and adapt to the latest information about the disaster. Just as the actor mastered their role, they were rotated out of the network. "Even those serving a second or third term in the network might find themselves working in a different role than they had before, adapting to unfamiliar supervisors, with different preferences and ways of operating" (Moynihan, 2008).

Quantum thinking and resilient organizations and systems are better aligned to manage network uncertainty. Pellissier (2011) highlights that quantum thinking is enabled in part by a shift toward systems thinking and viewing the system as a relationship of networks. The integration of this dynamic system is through three elements of identity, relationships, and information. Here, identity describes the organization's value or reason for existence. Relationships, in quantum organizations, extend beyond the traditional boundaries, and are formed ultimately anywhere in the organization (Pellissier, 2011). These three elements validate the overarching theme that systems are networks of people and information that exist for a specific purpose. FEMA and the Department of Homeland Security have steadily evolved much closer to a quantum organization since its Hurricane Katrina. "We work not just with governmental entities and private sector organizations, but with the individuals, families and communities who are our nation's 'first' first responders" (Evolution of Emergency Management, 2011). The Department has emphasized the value of whole community relationships and empowered individuals to share information beyond traditional communication boundaries.

Information systems are critical for disaster response. Information can be viewed as the link between data and knowledge. One of the most well-known models to understand the chain of information is the DIKW (data-information-knowledge-wisdom) chain. In this model, specific activities bridge the links together. Author describes four activities in the DIWK model to link the chain. Observation activity translates reality in the form of data. Structuration activity translates data into information. Training activity transforms information into knowledge (Saulais & Jean-Louis Ermine, 2012). Information itself is critical, but we can also conceptualize from the DIKW model that observation, structuration, and training activities can play a large role in disaster response in developing a shared operating picture that managers rely on to make decisions. "Social media are electronic communication platforms that convey content generated and exchanged by networks of users" (Auer, 2011). Information extracted from social networking and media in the DIWK model can thus be expressed as the result of observation and structuration activities.

Within NIMS and ICS, communicating "correct, up-to-date information concerning the status of emergency incidents to the media, members of the public, and other directly or indirectly affected stakeholders" is the primary responsibility of the Public Information Officer (PIO) (Hughes & Palen, 2012). The PIO coordinates with the ICS leadership to ensure that when a crisis occurs, decision makers are reacting to new information to assess how to prioritize resources to fill a specific need. The effectiveness of this iterative process is dependent upon accurate, up-to-date information. Mistakes can certainly occur in the decision making process and mistakes made "during the later stages of an emergency can lead to truly catastrophic results" (Belardo, Karwan, & Wallace, 1984). "In order for crisis and disaster management efforts to succeed it is necessary for data and information collection and analysis be inherently linked to decision making and action" (Jefferson, 2006). (Yang, Prasanna, & King, 2009) stated that there are four categories of information which require collection, analyzing, and sharing. Information can be categorized into status on available resources, casualties, environmental conditions, response participants.

Decision Support Systems (DSS) are mitigation tools to aid the Incident Commander during a crisis. Given the increase of information received from the public, it is increasingly more critical that "such a DSS must be made with care so as not to "overload" decision makers with unnecessary or confusing information" (Belardo et al., 1984). Decisions in crisis situations are typically made with the best data available. "Data in an emergency contains a great deal of subjective input and requires dynamic updating as the estimates and the quality of the data undergo change" (Turoff et al., 2004b). Therefore, real-time data is not always reliable even if obtained to official channels. Disaster response leaders must manage the tradeoffs between the reliability and the timeliness of the data. "The people and the intelligent agents designed to conduct analysis on both quality of the data and on the anticipated data changes must be sensitive to the constraints and temporal nature of decisions that have to be made" (Turoff et al., 2004b). Zheng et al. (2013) argues that "what is not very well known is how to effectively discover, collect, organize, search, and disseminate real-time disaster information." Turoff et al. (2004a) "developed a set of general and supporting design principles and specifications for a dynamic emergency response management information system (DERMIS)." Turoff et al. (2004a) developed a total of eight design principles which are summarized in the list below.

1. System Directory
2. Information Source and Timeliness
3. Open Multi Directional Communication
4. Content as Address
5. Up-to-Date Information and Data
6. Link Relevant Information and Data
7. Authority, Responsibility, and Accountability
8. Psychological and sociological factors

In 2006, response efforts during Hurricane Katrina demonstrated the early difficulty with decision support systems that captured data from social media and agency websites. There was a general lack of coordination across communications and information was dispersed. Following landfall of the hurricane, there were multiple official websites to share disaster related information with responders and the community. However, "so many websites sprang up that it became difficult to find the specific website for the information, resources, or reconnection one needed" (DHS, 2013c). Response officials must be organized because "it is not likely to accommodate a multidimensional crisis situation as the verbal mediation necessary for improvisation and flexibility will fall apart from information overload and the heterogeneous nature of all the individual agencies involved" (Turoff et al., 2004a).

In 2010, to overcome some of the past information and communication challenges, FEMA developed a Facebook page, blog, mobile-site for smartphones, YouTube page, and 16 Twitter accounts. On FEMA's official Facebook page, there were "more than 33,000 followers [that could] receive updates on current situations and get preparedness tips through text, photos and videos" (Understanding the Power, 2011). Administrator Fugate stated in his official testimony that "a blog which provides information before, during and after a disaster strikes, and highlights the best practices, innovative ideas and insights that are being used across the emergency management community" (Understanding the Power, 2011). "FEMA launched its first-ever mobile website, which allows the public to view web pages easily loaded on their smartphones." The site featured "information on what to do before, during and after a disaster, along with the ability to apply for federal disaster assistance" (Understanding the Power, 2011). On FEMA's YouTube page, "users can watch videos detailing FEMA's response and recovery efforts" (Understanding the Power, 2011) on their official YouTube page. FEMA's Twitter account[s] offers brief updates to those looking for disaster preparedness or situational updates. These advances, however, were not robust enough to solve all of their disaster information systems challenges.

By 2011, social media and mobile technology use were becoming increasingly more popular with the government. For example at the city level in Harris County, Texas, social media resources were leveraged. After a massive wildfire in George Bush Park threatened Harris County, Texas, the local Homeland Security Emergency Management office activated its emergency operations center. Harris County's EOC also included its own social media staff. The Harris County Office of Homeland Security and Emergency Management utilized a software application which allowed them to identify common Twitter hashtags related to the incident (DHSLLIS, 2011). "Twitter users can also follow topics of conversation that are of interest to them by following a 'hashtag,' which is the name given to a common topic of conversation on Twitter" (Understanding the Power, 2011). The staff observed that adding "#hounews" and "#txfires," increased public awareness (DHSLLIS, 2011).

It is evident that the public is engaged on all types of media, including social media, during disasters. FEMA's whole community approach to disaster management hinges on public participation; however, information security cannot be ignored. Information security is also the responsibility of the Public Information Officer, in addition to managing the public's expectations. "Information security is a discipline that seeks to promote the proper and robust use of information in all forms and in all media" (Van de Walle & Turoff, 2008). "Misinformation can place persons in danger, cause public panic, and disrupt the critical flow of proper information (FEMA, 2008b)." FEMA has made strides to make "non-sensitive disaster-related information like evacuation routes and shelter locations available and accessible to the public .... [and] view this information geospatially by plotting it onto a map, in order to improve our situational awareness during and after a catastrophic event" (Understanding the Power, 2011). In late 2011, a few agencies responding to Hurricane Irene were mapping real-time data from social media. They "began including tweets from response agencies within the impact zone as an additional data layer" (DHS, 2013c). One of the immediate observations was that "the information collected from social media (or crowd-sourced data), however, imposed a specific concern; the information was not vetted or verified and therefore was not used for more than informational purpose (DHS, 2013c). As social media usage and public participation increases, the PIO has seemingly more challenging role of managing and securing disaster related information.

In systems engineering, human systems integration, abbreviated HSI, "focuses on ensuring all human elements are properly accounted for in research and design initiatives when developing large configurations of people-operated equipment and systems" (Krueger, 2013). The military was the first entity to fully integrate human systems into their systems engineering design process. The U.S. Army developed MANPRINT, which is a multi-domain comprehensive approach to incorporating HSI into their systems engineering projects. The domains of MANPRINT are "manpower, personnel capability, training, human factors engineering (HFE), system safety, and health hazards" (Krueger, 2013). Human systems integration enhances the discussion on disaster response management. In analyzing just a few of the domains, it is evident that the absence of effective human systems integration will result in a less capable disaster response system. Disaster response systems rely heavily on people, not just technology, to function optimally. The systems that support the community, including individuals, "must be informed, trained, and materially and psychologically prepared to withstand disruption, absorb or tolerate disturbance, know their role in a crisis, adapt to changing conditions, and grow stronger over time" (DHS, 2010).

The first domain, manpower, is defined as the act of identifying "the number of people needed to operate and maintain new systems" (Krueger, 2013). "ICS is based on the scalar principle, with its size and complexity depending on the size and complexity of the disaster or emergency incident to which it is applied" (Buck et al., 2006). Personnel capability identifies the skills and competencies needed to perform various tasks within ICS (Krueger, 2013). Although sometimes there are political decisions behind the assignment selection process, "task assignments [are] made rationally to the person most qualified for the assignment regardless of rank in the organization" (Buck et al., 2006). The four primary positions in ICS are the Incident Commander, Operations Section Chief, Planning Section Chief, and Logistics Section Chief. Each of these positions has the following competencies in common: assuming position responsibilities, leading assigned personnel, communicating effectively, and ensuring completion of assigned tasks to meet objects (FEMA, 2007a). Each of the competencies has specific, albeit generic, behaviors for the specific position. Due to the rapid pace and nature of social media, compared to traditional media, a designated staff to with the expertise to effectively monitor and respond to social media traffic would serve the system well. Furthermore, since individual employees in DHS cannot access social media on their work computers, it would not be expected that once deployed to a crisis under ICS they would somehow know the formal competencies necessary to monitor social media effectively.

Manpower and capability are not sustainable without training. Training for "both new equipment and sustainment" (Krueger, 2013), can often be overlooked in system design. Personnel are required to complete basic courses in ICS as well as receive training specific to their position, if it is highly technical or involved with leadership. FEMA launched "Social Media in Emergency Management," which is an optional course to help train government employees involved with social media during a response (Emergency Management Institute, n.d.). There is also some evidence of training social media volunteers from the public. In 2011, the Los Ranchos VOST coordinator provided training on ICS structure and procedures for those who were unfamiliar (Katims, 2013). While this is a great example, it is not being done systematically.

Human factors engineering (HFE) ensures both physical and cognitive elements are incorporated into system design. Some areas of HFE includes information overload, situational awareness, trust, and interface design. Research in the advances of cognitive-level theory and improvisation may benefit emergency management because "improvisation has often been depicted as a type of problem solving" (Mendonca & Wallace, 2007). Disaster response systems "are also characteri[z]ed by a specific working environment of decision makers (which is dynamic and loaded with information), which often causes a high level of stress and working load as well as a higher probability of oversight and misinterpretation of available information" (Nikolic et al., 2007). Studies in collective cognitive accuracy assess the perception of actors in the system; "when most participants perceive that network structures are comparable, [they] are likely to share an accurate operational picture" (Choi & Kim, 2007). Technologies that present the operating picture interface between the information gathered and the operator. This should factor in research from interface design to better mitigate information overload, mistrust, and over trust, by the actor.

The last domains, system safety and health hazards are aimed at reducing "exposure to operators and maintainers of the systems" (Krueger, 2013). FEMA makes this domain a top priority to "[help] emergency officials prepare for all hazards, and helping people get back on their feet after their lives

Figure 5. Shadow Lake Fire Documents and Online Materials (St. Denis et al, 2012)

Resource	Description
InciWeb Incident Information System	The online incident information system used for official communication between the incident management team and the public
ORfireinfo Blog	Blog maintained by the VOST team containing latest information, photos, maps and videos during fire (orfireinfo.net)
ORFireInfo Facebook Page	Postings for the Shadow Lake Fire
ORfireInfo Twitter Account	ORfireInfo Twitter activity during the Shadow Lake Fire
ORFire Business Card	Used by PIOs with printed information and QR code with links to online sources
VOST Team Document	The Google document containing team information, operational objectives and daily activity log
VOST Group 'Skype Chat' Archive	An archive of the Skype Chat communication between members of the VOST team during the Shadow Lake Fire
Facebook Analytics File	Facebook Analytics File for ORFireInfo
Keepstream Archive	File archive of media monitoring activities and communication

are disrupted by a disaster" (DHS, 2013b). In 2011, the Shadow Lake Fire grew to 10,000 acres which forced the evacuation of residents and business owners. For 19 days, recently recruited VOST volunteers, in coordination with the Portland National Incident Management Organization, communicated safety & hazard information to the public via social media (Katims, 2013). "The ORFireInfo blog served as the primary communication interface used by the VOST to communicate with the public" (St. Denis et al., 2012). Social media was instrumental in this response because it helped communicate fire hazard warnings to citizens returning to the evacuated area. Other resources used during the response are summarized in the Figure 5.

Since 2011, the federal government has made huge strides at leveraging social media and involving the public in disaster management. There are substantial trends of emerging technology and how social media was utilized to support various large-scale incidents. It wasn't all perfect, however. Social media channels are prevalent with rumors, there are public privacy concerns with government monitoring, VOST efforts were not always seamlessly integrated into the official response (St. Denis et al., 2012) and some may find receiving information on social media impersonal. "The most serious emergency situations involve the most critical decisions and are subject to considerable public scrutiny" (Belardo et al., 1984). The Sandy response marked a major shift in how the federal government and its homeland security partners could use social media in a large-scale natural disaster.

Hurricane Sandy Case Study

Using a theme-based approach, the Sandy Case Study will discuss and analyze social media use in the response efforts. Within each theme, the Sandy Analysis Team indicated areas of where FEMA performed well and areas which could be improved. The case study presented in this chapter will highlight functional areas where social media can offer better approaches to disaster response. Four major themes were identified (FEMA, 2013):

1. Coordinating Across the Federal Government;
2. Shifting Leadership Focus in Disaster Support;
3. Enhancing Coordination Between Whole Community Partners; and
4. Fostering a Qualified, Adaptive Workforce

On Monday, October 22, 2012, most Americans were unaware that a record mega-storm would soon be approaching the East Coast. The NOAA National Weather Service issued advisories that Tropical Depression #18 had become a tropical storm. FEMA and dozens of its federal partners tracked the storm. The storm passed through Cuba as a Category 2 hurricane and weakened to a Category 1 as it moved over the Atlantic Ocean. However, as the storm continued to move north, it gained extreme sustained winds, of up to 105 miles per hour, after merging with an arctic cold front. Prior to Sandy making landfall, President Obama issued emergency disaster declarations for 11 states and the District of Columbia (American Red Cross, 2012). DHS, FEMA, and some it's operational components made preparations for the large-scale disaster response operations. The National Response Coordination Center (NRCC) was activated to coordinate multi-agency Federal support. For example, DHS deployed its all-volunteer Surge Capacity Force, made up of DHS employees, for the first time during Hurricane Sandy. FEMA deployed more than 1,500 personnel, some of which were Incident Management Teams, along the East Coast. DHS also coordinated its Office of the Infrastructure and Protection Security Advisors with state Emergency Operations Centers (EOCs) to identify potential impacts to critical infrastructure. For logistics with the anticipated need for emergency communications, FEMA deployed its Mobile Emergency Response Support (MERS) teams throughout several eastern states. The U.S. Coast Guard warned mariners and boaters to evacuate as necessary. The issued rip current advisories and closings for various ports up the East Coast. FEMA regional leadership engaged with their state counterparts to ensure effective coordination of disaster response.

Despite the preparations by state, local, and federal agencies, Hurricane Sandy tragically claimed the lives of more than 115 individuals in the United States alone (CDC, 2013). This tragedy affected millions of people, including 8.5 million without power and more than 23,000 people, primarily in New York and New Jersey, were directly displaced by the storm that ravished the East Coast of the United States on October 29, 2012. Hurricane Sandy produced severe flooding and in some states, multiple feet of snow. Storm damaged areas included flooded roads, blocked streets, and destroyed infrastructure. Ultimately, the 18th named storm of the season, resulted in billions of dollars to the American taxpayers. \$1.4 billion was issued in Individual Assistance to 182,000 survivors and \$2.4 billion for small businesses. \$7.9 billion was approved by the Flood Insurance Program and emergency response and infrastructure repair was funded with another \$3.2 billion (FEMA, 2014).

The first theme, coordinating across the Federal Government, identified several strengths and areas for improvement. Of particular interest is the WebEOC system. This system was deployed for one of the first times during Hurricane Sandy. WebEOC is an online crisis management designed to facilitate the coordination for Federal response to a disaster. This system that was very similar to a traditional Emergency Operations Center, but it expanded the concept to a semi-virtual environment. Operations at different physical locations could "see" what was happening at FEMA headquarters and other major regional centers. From a survey taken, more than 60% of deployed workers in the National Response Coordination Center (NRCC) rated the system as "effective" or "very effective" (FEMA, 2013). They did, however, identify several areas where the WebEOC could be improved, such as; better incorporating real-time feeds to enhance situational awareness and the common operating picture. The common operating picture is really at the foundation of decision-making under ICS. Although the findings of the report indicated that use of an online crisis management system was a strength during Sandy, leveraging social media as an input for real-time feeds could add further value to the common operating picture. "Displaying data on maps can be an enormous help to emergency response professionals, providing situational awareness with regard to weather, key resources and personnel, their location, and more" (DHS, 2013). These maps, including services such as Google Crisis Map, can display layers of real-time information from traffic conditions to crowd assessments to disaster declared areas. A few short days after Sandy made landfall, the White House also notified the group responsible for another real-time mapping service, called CrisisCommons, requesting their assistance to determine how the community could contribute to their efforts (DHS, 2013).

Emergency Support Functions (ESFs) were constructed on the basis that they provide the best possible access to federal resources regardless to which agency owns the actual resource. There are 15 ESFs, each with a different function (FEMA, 2008a). Operations during Sandy highlighted the need to consider a community-based approach to compliment the traditional government-centric approach. Millions of people and businesses were without power, which in part triggered massive fuel shortages in New York and New Jersey. Emergency Support Function #12, Energy, was led by the Department of Energy (DOE). They were tasked with coordinating efforts with the necessary parties to resolve the fuel shortage. It was evident that the Department of Energy had difficulty engaging with the full spectrum of key partners in government and private sector. To further address these challenges, the President directed FEMA and DOE to achieve this objective, thereby establishing an Energy Task Force. Amidst these federal challenges, the public turned to social media to solve parts of the fuel shortage problem. Students from Franklin High School, New Jersey leveraged Mappler, an online mapping service (DHS, 2013). They successfully posted information gathered from social media and other sources on which gas stations were open, had electricity and power, and most importantly had fuel. These efforts greatly assisted FEMA and private-sector partners to identify fuel and power resources. Individuals also greatly benefited because it minimized wait times and streamlined information on where to obtain fuel.

FEMA developed a mission assignment process to direct Federal partners, such as the U.S. Coast Guard, Department of Defense, Department of Transportation, and others, to conduct response activities. One day prior to Sandy making landfall to nearly three weeks into its response, approximately 40% of the assignments took longer than 24 hours to process (FEMA, 2013). Furthermore, Federal partners had no visibility on which assignments were already issued or requested to another responding entity. The first day the Hurricane Sandy made landfall, approximately $6.3 million had been issued in mission assignments (FEMA, 2013). The mission assignment process not only directs specific tasking and reviews proposed assignment requests, but it obligates the government financially. A responding homeland security partner that completes something other than the mission assignment will not be reimbursed. Similar to other financial processes, the mission assignment process is complex, which resulted in time-sensitive tasking to be delayed. Building upon the WebEOC platform, FEMA could benefit from "posting" mission assignments in a virtual, centralized location. There are areas where there could be synergy between those individuals are reporting on social media and how mission assignments are requested and directed. The WebEOC system could perhaps be configured to track and centralize mission assignments.

Because Sandy was a multi-state response, there was incident management structure challenges associated with coordinating with different states. ICS does allow incident managers to select the best structure, given the variables of the response. In the case of Hurricane Sandy, FEMA selected a combined organizational structure. This included geographical branches and specifically for the most affected states, New York and New Jersey, an area coordination group. Their goal to this organizational structure was to centralize decision-making while making sure that there was geographic coverage. It was clear that at times, there were conflicts between the branches and the area regional leadership. There were reports of unclear lines of authority, disagreements about decisions and mission assignments. There were also numerous cases of workforce misalignment with the combined structure. Of the 70 people designated as division supervisors, only 39 were fully qualified from the FEMA Qualification System (FQS) (FEMA, 2013). This presented an ongoing dynamic of challenges with turnover in these positions. FEMA was struggled to implement the combined structure; there were a large number of vacancies which accelerated turnover and instability within the ICS network. Leveraging analysis from social media and mapping may have aided in resolving the disputes between the field and the regional leadership. Social media analysis could have been used as a tool for leadership to rationalize which structure was optimal and offer transparency in its decision-making.

Sandy revealed an alarming trend that outside of the NRCC, 64% of the planners surveyed acknowledged they did not use pre-incident planning products, which are critical for decision making (FEMA, 2013). Planners complained that the products did not consistently meet the needs of operational decision-makers. Planning products derived from social media could be helpful to operational decision-makers. For example, when individuals post publicly on social media, there are tools that can aggregate the information to determine specific trends. Operational leaders may benefit from the insight of the trends identified ahead of or in contrast to information passed through official channels of communication. An organization called, Disaster Tech Lab, developed their Hurricane Sandy Communications Map, indicating where the public could access or not access various types of emergency communications. The map was driven by the Ushahidi platform, which was the same free platform used to crowd-source survivor information on Twitter in the 2010 Haiti 7.0 magnitude earthquake. During the first three weeks of the Sandy response, The American Red Cross, for example with their three-person national social engagement team, coordinated volunteers to track more than two million social media posts to help in decision-making relevant to their humanitarian missions (DHS, 2013c). They flagged 10,447 posts for further review. Of those posts selected, the team of 31 digital volunteers responded directly to nearly 2,500 posts. This was a significant increase from the approximately 500 posts reviewed a few short months earlier following Hurricane Isaac (DHS, 2013c). Several posts by the public resulted in a change of action for ground operations, thereby placing the public at the forefront of decision-making.

FEMA valued continuous improvement of disaster doctrine, policies, and plans, however, recognized areas where it could be improved. FEMA lacked formal and informal processes for documenting lessons learned and best practices. After Sandy, FEMA developed their Lessons Learned Continuous Improvement Program (LLCIP), which was established to address this challenge. FEMA and other state and federal agencies are required to conduct regular training and exercises. Leveraging social media posts can certainly aid in continuous improvement efforts. For example, historical posts and tweets can be analyzed retroactively to see if current policies and disaster response practices are adequate. Additionally, trends from social media can be directly simulated in training or exercises. Resilience to future disasters is heavily dependent on continuous improvement approaches.

Sandy leadership responded very well to survivor needs through innovation. FEMA established several programs and initiatives to meet the broad requirements of survivors, thereby improving survivor outcomes. The city of New York saturated various social media channels including third-party websites, NYC.gov, Twitter, Facebook, and official mayor media press releases on YouTube. The city pushed out and reinforced critical messages for survivors using traditional media and social media. The New York Fire Department initially directed the public not to use Twitter for emergencies during Sandy, however, as the storm surged, phone lines severely affected people in some areas which prevented the use of the 9-1-1 system. Individuals reported emergencies such as physically being trapped, severe flooding, etc. on the New York Fire Department's social media pages. FDNY began responding to these emergencies and even communicating directly with individuals using social media (DHS, 2013c). FEMA created a "Check Your Home" mobile app to provide information to survivors. This tool, praised by the New Jersey Office of Emergency Management, permitted home owners and renters to input their home address to determine the status of their home, including aerial photos and information on hazards. In another innovation, New York City worked with Airbnb, which is an online housing service that connected more than 4,000 people displaced by the storm with temporary housing (FEMA, 2013).

There are still opportunities to better meet survivors' needs during initial interactions. Several organizations, such as the Red Cross, FEMA, and the U.S. Department of Veterans Affairs (VA) developed smartphone apps for use in disaster response. Often times the psychological impact to survivors is often overlooked during a crises. The VA developed a psychological first aid app, which was downloaded more than 1,800 times by the public. The Red Cross Hurricane app was downloaded more than 100,000 times, which was used to track the storm and mapped locations for aid and shelter. And finally, FEMA's Emergency Preparedness app was downloaded by more than 50,000 users, which provided the public with an interactive checklist to prepare for storms (FEMA, 2013). FEMA stood up the Innovation Team, which is an office that examines innovation opportunities to support FEMA's missions. The team partnered with dozens of groups during Sandy. The team worked closely with a group called OccupySandy, where the primary source of information was social media. Some of their accomplishments include recruiting approximately 15,000 volunteers through social media, which resulted in serving 10,000 meals a day and coordinating transportation for medical teams to survivors in remote locations (FEMA, 2013). During the response and recovery phase of the hurricane, OccupySandy raised more than $1 million in four months (DHS, 2013c). Each of these examples represents how social media can enhance survivors' needs during a major disaster.

The Sandy response demonstrated the strength of innovation and collaboration of non-governmental partners. OccupySandy was not the only non-governmental partner that collaborated with FEMA. To FEMA's credit, an abundance of groups were integrated to achieve a higher level of response for Hurricane Sandy. Furthermore, volunteers from all over the country participated in assisting survivors through an online presence. Some of the key partners leveraged crowd-sourced data to collaborate with FEMA. Organizations, such as, Humanity Road, National Voluntary Organizations Active in Disaster (NVOAD), and Geeks Without Bounds, actively monitored traditional media, but concentrated on social media channels (DHS, 2013c). Many of these volunteer organizations directly prevented further loss of life and property. Many of the groups used free of open source software tools. Sahara Eden, used by OccupySandy, managed high volume information sharing. Humanity Road deliberately sought out disaster status information from the public on social media. They solicited information from the public to use in coordination with state volunteers from the Maryland Emergency Management Agency. Response efforts improved with the integration of social media inputs from the community. Social media improved maps for situational awareness and facilitated decision making during the incident.

The scale of Sandy revealed challenges in coordination among various levels of jurisdiction. There were numerous cases where the state and city did not prioritize or consolidate resource requests. It was evident that this level of coordination could be improved for future disaster responses. Social media network monitoring may offer a means to improve some of the challenges associated with large-scale coordination of resource requests. Monitoring social media channels and leveraging the full capabilities of the WebEOC system can be used to evaluate and potentially jointly develop resource needs from the state and local communities. Although FEMA, did however, at the NRCC, activate a tribal affairs liaison for overseeing resource requests in tribal territories affected by Sandy, monitoring posts on social media may enhance the clearer common operating picture, especially in underrepresented areas.

Recognized as an area for improvement, the personnel deployed from all across the nation could be better supported. The deployed groups of individuals faced many challenges due to the large-scale nature of the incident. For the first time during Sandy, DHS launched its Surge Capacity Force, which was a team of DHS federal employees who could be activated during a response. Many non-FEMA employees from DHS lacked the information technology equipment and smartphones to be effective at their duties in the field (FEMA, 2013). This caused a delay, and in some cases a gap, between information from the Surge Capacity Force and information other DHS incident responders. Many of these responders presumably already had personal smartphones that could have been leveraged in the absence of DHS' lack of IT investment.

Figure 6. summarizes the areas where social media-based technologies would likely improve disaster response. Reinforced by the many successes presented in this chapter, it is evident that emerging technology from social media can add tremendous value to large-scale disaster response at the federal, state, and local levels. Although social media will not improve every aspect of disaster response, it will improve the information paths and coordination of the people affected by the disaster, including survivors and responders. The case study demonstrates that social media technologies are a phenomenal resource for fostering information sharing, detecting rumors, and aggregating crowd-sourced information on maps or other products.

SOLUTIONS AND RECOMMENDATIONS

A framework is needed to integrate social media effectively into disaster response management (IG Report 2013). There is data and information that decision makers can utilize to create knowledge and deploy the appropriate resources. Given the importance of timely and accurate information during disasters, in recent cases, social media has been recognized as a powerful source of data and information because of its rapid nature. DHS's third Secretary, Janet Napolitano launched the nationwide "If You See Something, Say Something" campaign (DHS, n.d.) to elevate public awareness and promote a culture of citizens raising their observations with government officials. She recognized the power of the public and empowered FEMA to leverage the public for a citizen-centric approach to resiliency.

Typically the government relies on official communication channels for information used to make critical decisions. However, since social media channels are considered unofficial channels of commu-

Figure 6. Social Media Value-Added Areas from Sandy Response (Adapted from FEMA, 2013)

Strengths and Weaknesses during Sandy Response	Can Social Media Add Value?
Strength: The President expedited Federal disaster declarations	No
Strength: Using an online crisis management system to coordinate Federal response operations	Yes
Area for Improvement: Integrating Federal senior leader coordination and communications into response and recovery operations	No
Area for Improvement: Coordinating Emergency Support Functions (ESFs) and Recovery Support Functions (RSFs) to support disaster response and recovery	Yes
Area for Improvement: Refining the mission assignment process	Yes
Area for Improvement: Implementing incident management structures	Yes
Area for Improvement: Using planning and analysis to drive operational decision-making	Yes
Area for Improvement: Ensuring continuous improvement of disaster doctrine, policies, and plans	Yes
Strength: Meeting survivor needs through innovation	Yes
Area for Improvement: Meeting survivors' needs during initial interactions	Yes
Area for Improvement: Ensuring survivors have equal access to services	No
Area for Improvement: Reducing the complexity of the Public Assistance program	No
Strength: Integrating response and recovery efforts with nongovernmental partners	Yes
Area for Improvement: Coordinating among states, localities, and tribes	Yes
Strength: Completing one of the largest and most diverse personnel deployments in FEMA history	Yes
Area for Improvement: Ensuring a qualified disaster workforce	No
Area for Improvement: Mobilizing the FEMA workforce for disaster response	No
Area for Improvement: Supporting deployed personnel	Yes
Area for Improvement: Ensuring continuity of operations	No

nication, there may be hesitation on the part of decision makers to use these channels in times of emergency. FEMA has not always had a positive track record for adopting new technology and processes, but it's Administrator, Craig Fugate in his June 2011 testimony stated that "Our state and local emergency management experience has taught us that, in the event of a disaster, individuals and communities are not liabilities; rather, they are our greatest resources and the key to our success" (Evolution of Emergency Management, 2011). FEMA has embraced the department's campaign of information sharing and has encouraged citizens to continue reporting disaster related information through any appropriate channel.

Of the five core NIMS framework components, two will be explored in greater detail in this chapter. Communications & information management and command & management, are likely components where social media could be infused during a response (FEMA, 2008b). Communications and Information Management "provide a common operating picture to all command and coordination sites .... based on the concepts of interoperability, reliability, scalability, and portability, as well as the resiliency and redundancy of communications and information systems" (FEMA, 2008b). The Command and Management component provides "is based on three key organizational constructs: the Incident Command System, Multiagency Coordination Systems, and Public Information" (FEMA, 2008b) Within the Public Information construct, the "Joint Information System (JIS) provides the mechanism to organize, integrate, and coordinate information to ensure timely, accurate, accessible, and consistent messaging across multiple jurisdictions and/or disciplines with NGOs and the private sector" (FEMA, 2008b). For Public Information Officers, the Joint Information Center is the "central location that facilitates operation of the JIS, where personnel with public information responsibilities perform critical emergency information functions, crisis communications, and public affairs functions" (FEMA, 2008b). The NIMS framework does discuss the possibility that the JIC may be virtual JIC. There isn't any extensive guidance, however, on how it would be best set up and integrated into the ICS response (FEMA, 2008b).

Other gaps in the NIMS framework, with respect to joint information systems, include the overall lack of language to support utilizing information from the public to improve disaster response. The framework outlines an ongoing cycle of four steps which pushes information to the public. Although it specifies that "internet and other web-based tools can be resources for emergency management/response personnel" (FEMA, 2008b), it states that information should be primarily pushed to offer situational awareness. The four steps are outlined in Figure 7, including areas where social media is lacking:

Using the research from systems engineering topics and social media-based technologies, a virtual "whole community" disaster response framework for NIMS and ICS is proposed in Figure 8. The main components of the framework are the command center and the community, which are connected virtually. The virtual command, which is scalable to include an area command, EOC, and multiagency coordination center, consists at a minimum of the incident commander, planning section chief, operations sections chief, and the public information officer. The virtual community consists of community actors and ICS actors, in particular those that are on-scene. Social media-based technologies proposed would be a multimedia platform integrated between the virtual command and the virtual whole community. "Multimedia platform makes it possible to integrate different media (data, text, picture and sound), and multimedia systems are not primarily defined by their data structures but by the nature of their communication" (Nikolic 2007).

As shown in the framework, the Public Information Officer is vital to integrating the physical actors into the virtual community. We can see that the virtual command would be inclusive of the Command and Management component of the NIMS framework and the virtual community would fall under Communications and Information Management. The NIMS and ICS framework are enhanced by the virtual whole community concept because it leverages the natural social channels of communication to effectively respond to disasters. It reduces the burden of slower information, which is plagued by official channels. It accelerates the rate of information coming into the command, which has an impact on the type and timing of decision made. The framework is iterative and the social media information can flow in multiple directions.

Overview of the Virtual Whole Community DSS Framework

The Virtual Whole Community Decision Support System Framework, which is supported by systems engineering principles, formalizes the methodology for integrating social media in the response phase

Figure 7. Social Media Gaps Identified in the Joint Information System (JIS), Adapted from (FEMA, 2008b)

Four Steps of Getting Information to the Public	Gaps in Incorporating Social Media
1. Gather Information a. On-Scene Responders b. On-Scene PIOs c. Media Monitoring, (GAP 1) d. News Media e. Public and Elected/Appointed Officials f. (GAP 2)	1. Media Monitoring does not elaborate on the use of social media. 2. The public, including individuals, should be included as potential contributors of decision-support information.
2. Verifying Information a. Other PIOs in the JIC b. EOC Liaison Resources c. On-Scene PIOs d. (GAP 3) e. (GAP 4)	3. The public, including individuals, should be included as potential contributors to verify disaster related information. 4. On-scene ICS responders public should be included as potential contributors to verify disaster related information.
3. Coordinating Information a. Unified Messaging b. Obtaining Authority/Approval to Release	
4. Disseminating Information a. Multiple Methods, phones, fliers, etc. (GAP 5) b. Monitoring the Media	5. Extend method to virtual channels like social media

of disaster management. This framework empowers the Public Information Officer in new ways; they should be perceived as the common thread which keeps the social network intact and resilient. They are the synthesizers of information which directly impact the decision-makers in the system.

The Community Actors are any groups or individuals that have the ability to send and receive information transmitted on social media channels. These are the initial first responders. This includes individual citizens and emergency offices at the state, local, federal levels. Other partners, such as universities, humanitarian organizations, and faith-based groups, also make up the community. ICS Response Actors are a subset of this community, however, are individuals specifically assigned to respond to the incident. It is proposed that they communicate through the Public Information Officer to the Incident Command using online or mobile-based technology in addition to traditional methods. For example, on-scene re-

Figure 8. Proposed Virtual "Whole Community" DSS Framework

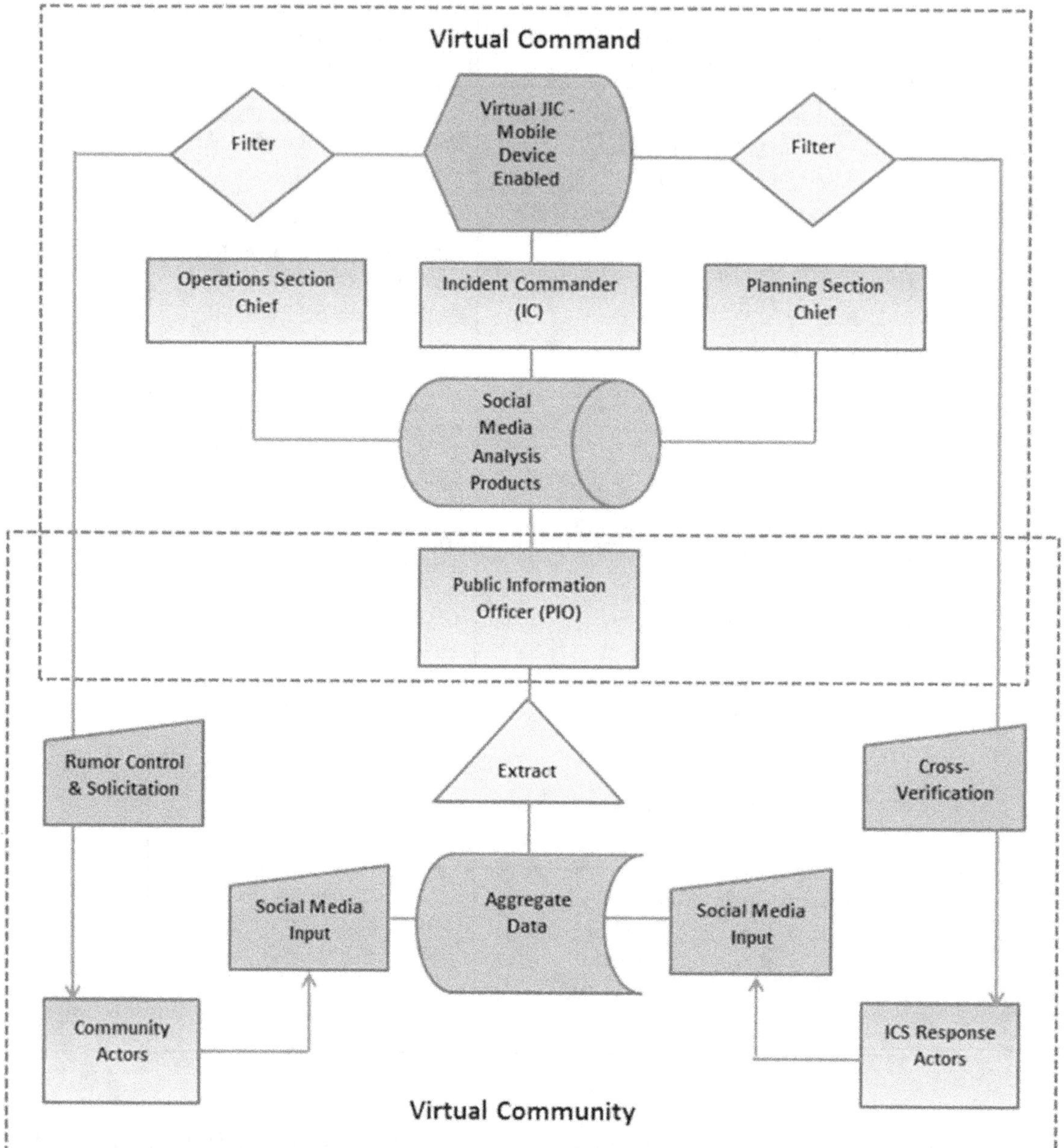

sponders can upload pictures and post comments on the status of a mission objective. Their geographic location, experience and competencies can also be linked to their profile as a disaster responder.

Publically available multimedia data and information from both the community and ICS responders should be aggregated and stored in one directory, keeping in mind privacy and information security requirements. This "pulling" requires the Public Information Officer to monitor social media channels from sources like Facebook, Twitter, Instagram, YouTube, and LinkedIn. After this listening and interpreting, the data should be aggregated and extracted using data mining tools. Zheng (2013) states "to quickly reassemble or create information flows for multiparty coordination activities during disaster situations, technologies that are capable of extracting information from recent updates, delivering that information without conflict or irrelevance, and representing preferential information are needed." Social Media Analysis Products After the data has been mined, the PIO should use robust applications to convert the information into planning and decision-making products, such as maps. Other products may include crowd-sourced trends identified from the public or ICS responders.

ICS leadership leadership positions make up the basic construct of the incident command. The Incident Commander, Operations Section Chief, and Planning Section Chief should utilize any products derived from social media analysis. These planning and decision-making products will have implications for the traditional 12 hour planning and operational cycles. As data and information is retrieved and processed closer to real-time than official, non-social media channels, ICS leadership will need to determine the appropriate rhythm of providing new direction or assignments based on new information. The Public Information Officer, along with digital volunteers, such as the VOST, is also a required assigned position in the ICS response. The PIO is responsible for synthesizing all of the information captured from social media channels. The PIO ensures that information is stored in compliance with privacy and security policies.

An expansion of the WebEOC and InciWeb Incident Information System concept, the virtual joint information center with mobile-device compatibility is the heart of the DSS; it would enable members of the community, responders, and the command to receive and transmit disaster related information virtually. To avoid information overload, the platform would have filters or layers of information that is only visible if it is pertinent to the individual. Information would be pushed to ICS responders and community members. The four types of NIMS (FEMA, 2008) framework communications – strategic, tactical, support, and public address communications would be fully supported. Furthermore, each responder would have mission assignments, access to leadership decisions, be able to post pictures, view web/mobile situation maps, and view other relevant activities happening near them in their network such trends on survivor's needs and hazards. As new information comes into the online platform virtual command system, ICS responders, community members, and VOST members, can help verify the accuracy of the information collected over social media through a cross-verification process. With the online and mobile interface, responders will be able to upload pictures to the online platform for comparison and cross-verification. As new information is vetted through the DSS, ICS leaders will have an opportunity to push messages and recruiting opportunities to community members and organizations. Rumor control is vital; it helps to align public expectations of the response with actual performance and results.

The framework lays the foundation for key DSS design principles. The aggregation methods allow information to be stored in a system directory. Communication is facilitated in multiple directions in an open, transparent manner. "A decision support system that incorporates and adjusts the decision is often necessary.... because information often changes as the understanding of the situation improves." (Chen et al., 2007) The Incident Commander, Operations Section Chief, Planning Section Chief, and other actors in the virtual community are linked to relevant disaster response information. The layers filter information, which addresses the psychological impact of potential information overload. The framework expands on the existing authority and responsibilities of the Public Information Officer. The PIO is not just responsible for traditional press releases in this model. The PIO has additional responsibilities that require a closer coordination with not only the responders, but the incident leadership. Therefore, the PIO will need additional resources to be formally assigned to the incident command system. These should be trusted digital volunteers (St. Denis et al., 2012) from the community and from employees at the state, local, or federal level, as appropriate. Volunteers are needed for monitoring; coordinating with the incident commander to manage the filters of the online DSS platform, safeguarding content, enforcing privacy policies, ensuring quality of data, and validating the content. They will also have the unique responsibility to ensure the cultural climate of the incident command is supportive of information garnered from and disseminated to social media networks.

Implementation of this proposed framework could have assisted in the 2012 Sandy response. From the case study, there were several areas identified where social media added value. Because the framework compels the PIO to actively monitor social media channels and verify information, this iterative process allows managers the opportunity to examine the effectiveness of disaster doctrine, policies, and plans. This monitoring and extracting resilience process will support better coordination of the Emergency Support Functions; federal agencies and other partners can exchange information more quickly. Given the profiles of the ICS responders in the network, the Incident Commander can use the online platform to tailor mission assignments, which may help reduce turnover from lack of expertise and better support a geographically-dispersed incident structure. Additionally, the velocity of information exchanged in the DSS framework facilitates better integration of response efforts of community and nongovernmental partners in the network. The PIO in this framework must select the appropriate suite of analytical tools to deliver planning and situational awareness products to the virtual command. Human-systems integration is present throughout the framework. The innovative DSS framework ensures that human elements of the responders are addressed, such as trust, situational awareness, and information overload. Furthermore, the DSS framework is survivor-centered. For example, information about gas stations and humanitarian services can be pushed to the community based on feedback from the community. The advances in social media technology will meet the needs of survivors and support deployed personnel.

The benefits of the Virtual Whole Community DSS are far reaching; it includes a more sophisticated understanding of community needs and shared risks. The community is empowered to contribute directly to disaster response more efficiently by offering volunteer services, posts of disaster related information, and feedback to ongoing operations. Information is collected more efficiently through mining techniques, which allows leaders to make decisions timelier. Involving the community in this virtual network increases the resources available to the incident command. They can be used as problem-solvers and present information faster that otherwise may not have been known or discovered. The cross-verification component of the framework allows the system to be more resilient than traditional disaster response. The system can monitor and pull a substantially larger data set to verify clusters of information to determine its accuracy. This framework increases the capability, flexibility, and tolerance, and inter-organization collaboration (Pellissier, 2011) of the incident response and fills in the gaps of the NIMS and ICS frameworks.

Proposed Measures of Effectiveness

Published in 2010, the Quadrennial Homeland Security Review names three objectives to achieve the DHS goal of effective disaster response. The desired goal is to "strengthen [the] response capacity across the nation" (DHS, 2010) by concentrating on the following objectives:

1. "Provide timely and accurate information to the public: establish and strengthen pathways for clear, reliable, and current emergency information, including effective use of new media;
2. Conduct effective disaster response operations: respond to disasters in an effective and unified manner;"

3. Provide timely and appropriate disaster assistance: improve governmental, nongovernmental, and private-sector delivery of disaster assistance" (DHS, 2010).

Measures of effectiveness (MOEs) are critical for systems engineering processes and assessing how well the system performed against its goals. These measures "are an important management tool not only for the stakeholders who initiate the process input but for the systems engineers who must provide a solution for the stakeholders" (Sproles, 2000). Other related terms also include measures of performance (MOPs) and critical operational issues (COIs). The critical operational issues, or what some would call a showstopper, "refer to those emergent properties that the system must have in order to perform its function." "Performance measures of quantitative or qualitative levels against which achievement of a task or capability outcome can be assessed" (FEMA, 2007b).

Sproles surmises that "MOEs relate to how well any from a range of solutions may meet a need; MOPs are concerned with what a particular solution does regardless of its intended purpose" (Sproles, 2000). From a government perspective, Gawande and Wheeler (1999) address the lack of literature "about methods that quantify their effectiveness." Ozkan and others describe the "difficulty of formulating measures of effectiveness for complex socio-technical systems" (Ozkan, Cakir, & Bilgen 2008). One approach could start with breaking the system down into COIs, then developing MOEs for each COI. And finally, the MOEs would then be broken down further into measureable MOPs (Colombi & Cobb, 2009). Alternatively, Sproles outlines a streamlined process from the request to formulate the MOEs to its application, without the MOPs. This chapter follows this process of proposing measures of effectiveness for the Virtual Whole Community DSS (VWCDSS) Framework by analyzing the results from the Sandy case study and the overarching disaster response objectives.

Analyzing the areas from Figure 6 where social media added value to disaster response, below these are generalized to be the COIs for information systems in large-scale disasters. From here, 12 candidates for measures of effectiveness were developed:

1. Number of state, local, and federal groups connected to the VWCDSS
2. Number of private organizations connected to the VWCDSS
3. Number of survivors that received aid information on an official social media channel
4. Number of Emergency Support Functions supported by social media
5. Number of safety hazards relayed to on-scene personnel and survivors
6. Number of Mission Assignments coordinated over social media
7. Number of volunteers recruited
8. Number of posts, tweets, pictures, and videos aggregated by virtual JIC
9. Quantity of geo-based data uploaded to virtual JIC
10. Number of social media planning products created
11. Number of data sets cross-verified
12. "Likes" (or positive feedback comments) on official postings

These proposed MOEs are both system outputs and survivor outcomes. These measures assess the "connectivity" and effectiveness of the disaster response network. It will undoubtedly improve the Department of Homeland Security's mission to ensure disaster resilience.

FUTURE RESEARCH DIRECTIONS

More research can be explored deeper to fully leverage the capabilities of social media technologies for disaster response. Areas for future research include using a systems engineering design approach, redefining the role of the public information officer, applying social network analysis methods, social media organizational barriers, and social media for other phases of disaster management. As social media technologies continue to evolve, these recommended areas for future research should be considered.

The National Integration Center and the Science & Technology Directorate of DHS are best suited to oversee the development of new disaster response related technology (FEMA, 2008b). The VWCDSS framework should be explored further using the systems engineering design approach. There needs to be further analysis of the concept of operations, system requirements, and system architecture. Additionally, with respect to the online platform, principles of human-systems integration should be applied to prevent information overload, trust concerns with data, and interface design.

The Public Information Officer position for VWCDSS framework should be examined to determine how to best address gaps between the current role of the PIO and the level of support needed to manage a VOST type team and incorporate privacy and information security requirements. This expanded role may also present a culture shift as additional authorities may be needed. Supporting research would still be needed to determine organizational barriers to integrating social media practices and if it varied depending on the type of incident. This chapter emphasized natural disasters, but social media has also been demonstrated to be effective in man-made incidents and acts of terror, such as the 2013 Boston Marathon Bombing.

This chapter did not elaborate how social media could be effective beyond the basic ICS structure. Social network analysis has tremendous potential in studying incident command systems. Further research in this field will benefit the effectiveness of social media driven networks. Analysis techniques can also measure where there is power or influence in the network, which would be beneficial to detecting if there are sub-networks in the ICS structure that have influence in decision-making. Social media has also been demonstrated to aid in the preparedness phase of the disaster management lifecycle. For example, before a hurricane makes landfall, agencies and private organizations are already pushing information to the public on how they can best prepare or evacuate. Effective preparedness may have a direct impact on how an agency responds or how many lives are at risk.

CONCLUSION

The Department of Homeland Security continues to strive towards a more resilient nation which can withstand the most devastating disasters. Given the unpredictability of large-scale incidents, Hurricane Katrina serves as a permanent reminder of how people, in the face of despair and crisis, are resilient. The very essence of human nature is to survive. People create natural networks and come together from all backgrounds and talents to lend help when their communities have been overcome by events. The volunteer spirit of individuals, faith-based groups, schools, and corporations unite for a common purpose. Social networks and social media technologies enable and empower these groups and individuals to serve during crises. FEMA and DHS must continue to learn from the experiences of both official disaster responders and community volunteered. With a keen focus on serving disaster victims, the Department must continually adapt to new technologies and practices to fully harness the power of the network of problem-solvers that exists beyond the traditional boundaries of the Incident Command System.

Leveraging social media in large-scale natural disasters, especially hurricanes, has been incredibly effective in assisting incident responders. From trends of social media usage, it is evident that social media will continue to be seamlessly integrated in the lives of millions of Americans and others around the globe. Typical of any other form of media, however, social media is only as effective as the content of the information being inputted. Skeptics of social media advantages criticize the accuracy of the observer and the relevancy of their postings. Even though the government has only monitored and aggregated publically available social media postings, critics warn about the dangers of privacy, especially in era where there has been government mistrust following numerous National Security Administration debates over snooping allegations. This chapter presented numerous successes of how social media was effectively leveraged in disaster response as well as identifying areas where social media may not add value. From the fires in Texas to Hurricane Sandy, incident official across federal, state, county, and city levels exploited the information gleamed from social media channels. In all of the presented examples, incident leadership used the Public Information Officer to push vital disaster-related information to the community.

Systems engineering principles and approaches are exceedingly helpful in ensuring the monitoring, aggregating, synthesizing, and broadcasting of social media are applied in an efficient and systematic process. Recognizing that social media is not the preferred tool for every aspect of disaster response, the Virtual Whole Community Decision Support System framework was proposed to facilitate the best areas where social media would be most integrated in NIMS and ICS. Given the advancements of disaster response technologies, the Department of Homeland Security will need to closely examine how the National Integration Center can further address social media. Emergency response personnel will inevitably rely more on technology to help make better decisions with limited information. Systems engineering offers critical insight towards closing the gap of how to best integrate social media and disaster response to ensure a more resilient Nation, and provide support in the capacity to save lives and property.

REFERENCES

American Red Cross. (2012). *Superstorm Sandy: One month update*. Retrieved from http://www.redcross.org/news/press-release/Red-Cross-One-Month-Report-on-Superstorm-Sandy

Anonymous, . (2013). Ahead of the Storm: Engineering for Disaster. *Institute of Transportation Engineers. ITE Journal, 83*(12), 22–24.

Auer, M. R. (2011). The Policy Sciences of Social Media. *Policy Studies Journal: the Journal of the Policy Studies Organization, 39*(4), 709–736. doi:10.1111/j.1541-0072.2011.00428.x

Belardo, S., Karwan, K., & Wallace, W. (1984). An Investigation of System Design Considerations for Emergency Management Decision Support. *IEEE Transactions on Systems, Man, and Cybernetics, 14*(6), 795–804. doi:10.1109/TSMC.1984.6313308

Betts, S. C., & Stouder, M. D. (2004). The Network Perspective in Organization Studies: Network Organizations or Network Analysis. *Academy of Strategic Management Journal, 3*, 1–20.

Bhanot, S. (2012). Use of Social Media by Companies to Reach their Customers. *SIES Journal of Management*, *8*(1), 47–55.

Bharosa, N., Lee, J., & Janssen, M. (2010). Challenges and Obstacles in Sharing and Coordinating Information during Multi-agency Disaster Response: Propositions from Field Exercises. *Information Systems Frontiers*, *12*(1), 49–65. doi:10.1007/s10796-009-9174-z

Billington, M. G., & Billington, P. J. (2012). Social Media Tools for Leaders and Managers. *Journal of Leadership. Accountability and Ethics*, *9*(6), 11–19.

Buck, D. A., Trainor, J. E., & Aguirre, B. E. (2006). A Critical Evaluation of the Incident Command System and NIMS. *Journal of Homeland Security and Emergency Management*, *3*(3), 1252. doi:10.2202/1547-7355.1252

Centers for Disease Control and Prevention. (2013). Deaths associated with Hurricane Sandy - October-November 2012. (No. 20). Atlanta, GA: Author.

Chen, R., Sharman, R., Rao, H. R., & Upadhyaya, S. (2007). Design Principles for Critical Incident Response Systems. *Information Systems and eBusiness Management, 5*(3), 201-227. doi:10.1007/s10257-007-0046-0

Choi, S. O., & Kim, B. (2007). Power and Cognitive Accuracy in Local Emergency Management Networks. *Public Administration Review*, *67*(S1), 198–209. doi:10.1111/j.1540-6210.2007.00828.x

Colombi, J. M., & Cobb, R. G. (2009). Application of Systems Engineering to Rapid Prototyping for Close Air Support. *Defense AR Journal*, *16*(3), 284–303.

Crowe, A. (2010). The Elephant in the JIC: The Fundamental Flaw of Emergency Public Information within the NIMS Framework. *Journal of Homeland Security and Emergency Management*, *7*(1), 1662. doi:10.2202/1547-7355.1662

De Wever, S., Martens, R., & Vandenbempt, K. (2005). The Impact of Trust on Strategic Resource Acquisition through Interorganizational Networks: Towards a Conceptual Model. *Human Relations*, *58*(12), 1523–1543. doi:10.1177/0018726705061316

Department of Homeland Security. (2008). *National response framework*. Retrieved from http://www.fema.gov/pdf/emergency/nrf/nrf-core.pdf

Department of Homeland Security. (2010). *Quadrennial homeland security review report: A strategic framework for a secure homeland.* Retrieved from https://www.dhs.gov/xlibrary/assets/qhsr_report.pdf

Department of Homeland Security. (2013a). *DHS uses social media to enhance information sharing and mission operations, but additional oversight and guidance are needed* (No. OIG-13-115). Retrieved from http://www.oig.dhs.gov/assets/Mgmt/2013/OIG_13-115_Sep13.pdf

Department of Homeland Security. (2013b). *Disaster response and recovery*. Retrieved from https://www.dhs.gov/topic/disaster-response-and-recovery

Department of Homeland Security. (2013c). *Lessons learned: Social media and Hurricane Sandy*. Retrieved from http://www.oig.dhs.gov/assets/Mgmt/2013/OIG_13-115_Sep13.pdf

Department of Homeland Security. (2013d). *National response framework* (2nd ed.). Retrieved from http://www.fema.gov/media-library-data/20130726-1914-25045-1246/final_national_response_framework_20130501.pdf

Department of Homeland Security. (n.d.a). *If you see something, say something*. Retrieved from http://www.dhs.gov/if-you-see-something-say-something%E2%84%A2-campaign

Department of Homeland Security. (n.d.b). *Disasters overview*. Retrieved from https://www.dhs.gov/disasters-overview

Department of Homeland Security. (n.d.c). *Our mission*. Retrieved from https://www.dhs.gov/our-mission

Department of Homeland Security. (n.d.d). *Who joined DHS*. Retrieved from http://www.dhs.gov/who-joined-dhs

Department of Homeland Security Lessons Learned Information System. (2011). *Emergency public information: The Harris County, Texas, Office of Homeland Security and Emergency Management's implementation of a social media protocol*. Retrieved from https://www.llis.dhs.gov/

Department of Homeland Security Lessons Learned Information System. (2013). *Social media in emergency management: The virtual operations support team concept*. Retrieved from https://www.llis.dhs.gov/

Emergency Management Institute. (n.d.). *IS-42: Social media in emergency management course*. Retrieved from http://training.fema.gov/is/courseoverview.aspx?code=is-42

Federal Emergency Management Agency. (2003). *Terrorism incident annex*. Retrieved from http://www.fas.org/irp/offdocs/pdd39_frp.htm

Federal Emergency Management Agency. (2007a). *ICS all-hazard core competencies*. Retrieved from http://www.fema.gov/media-library-data/20130726-1617-20490-3897/ics_core_competencies_sep07.pdf

Federal Emergency Management Agency. (2007b). *Target capabilities list: A companion to the national preparedness guidelines*. Retrieved from http://www.fema.gov/pdf/government/training/tcl.pdf

Federal Emergency Management Agency. (2008a). *Emergency support function annexes: Introduction*. Retrieved from http://www.fema.gov/media-library-data/20130726-1825-25045-0604/emergency_support_function_annexes_introduction_2008_.pdf

Federal Emergency Management Agency. (2008b). *National incident management system*. Retrieved from http://www.fema.gov/pdf/emergency/nims/NIMS_core.pdf

Federal Emergency Management Agency. (2010). *FEMA independent study course IS208: State disaster management*. Retrieved from http://training.fema.gov/is/courseoverview.aspx?code=IS-208.a

Federal Emergency Management Agency. (2011). *A whole community approach to emergency management: Principles, themes, and pathways for action*. Retrieved from http://www.fema.gov/media-library-data/20130726-1813-25045-3330/whole_community_dec2011__2_.pdf

Federal Emergency Management Agency. (2013). *Hurricane sandy FEMA after-action report*. Retrieved from http://www.fema.gov/media-library-data/20130726-1923-25045-7442/sandy_fema_aar.pdf

Federal Emergency Management Agency. (2014). *Hurricane Sandy: Timeline*. Retrieved from https://www.fema.gov/hurricane-sandy-timeline

Federal Emergency Management Agency. (n.d.). *ICS Organization*. Retrieved from https://training.fema.gov/emiweb/is/icsresource/assets/icsorganization.pdf

Fugate, C. (2011a). *Evolution of emergency management and communication*. Hearing before the Subcommittee on Homeland Security on the U.S. Committee on Appropriations. House of Representatives. Retrieved from https://www.fema.gov/pdf/about/programs/legislative/testimony/2011/6_8_2011_evolution_of_emergency_management_and_communication.pdf

Fugate, C. (2011b). *Improving the nation's response to catastrophic disasters: How to minimize costs and streamline our emergency management programs*. Hearing before the Subcommittee on Economic Development, Public Buildings and Emergency Management on the Transportation and Infrastructure Committee. Retrieved from https://www.fema.gov/pdf/about/programs/legislative/testimony/2011/3_30_2011_improving_the_nations_response_to_catastrophic_disasters.pdf

Fugate, C. (2011c). *Understanding the power of social media as a communication tool in the aftermath of disasters*. Hearing before the Subcommittee on Disaster Recovery and Intergovernmental Affairs, Senate Committee on Homeland Security and Governmental Affairs. Retrieved from https://www.fema.gov/pdf/about/programs/legislative/testimony/2011/5_5_2011_use_of_ social_media_in_disasters.pdf

Gawande, K., & Wheeler, T. (1999). Measures of Effectiveness for Governmental Organizations. *Management Science*, *45*(1), 42–58. doi:10.1287/mnsc.45.1.42

Hémond, Y., & Benoît, R. (2012). Preparedness: The State of the Art and Future Prospects. *Disaster Prevention and Management*, *21*(4), 404–417. doi:10.1108/09653561211256125

Horsley, J. S., & Barker, R. T. (2002). Toward a Synthesis Model for Crisis Communication in the Public Sector. *Journal of Business and Technical Communication*, *16*(4), 406–440. doi:10.1177/105065102236525

Hughes, A. L., & Palen, L. (2012). The Evolving Role of the Public Information Officer: An Examination of Social Media in Emergency Management. *Journal of Homeland Security and Emergency Management*, *9*(1). doi:10.1515/1547-7355.1976

International Council on Systems Engineering. (2004). *What is systems engineering?* Retrieved from http://www.incose.org/practice/whatissystemseng.aspx

Jefferson, T. L. (2006). Evaluating the Role of Information Technology in Crisis and Emergency Management. *Vine*, *36*(3), 261–264. doi:10.1108/03055720610703542

Kapucu, N., Arslan, T., & Demiroz, F. (2010). Collaborative Emergency Management and National Emergency Management Network. *Disaster Prevention and Management*, *19*(4), 452–468. doi:10.1108/09653561011070376

Katims, L. (2013). *Virtual operations support teams harness social media: The teams of emergency managers and disaster volunteers lend virtual support to those on the site of a disaster*. Retrieved from http://www.emergencymgmt.com/disaster/Virtual-Operations-Support-Teams-Social-Media.html

Krueger, G. P. (2013). Human Systems Integration (HSI): Psychological Influences in Design Produce Exceptional Operator Performance. *Journal of the Washington Academy of Sciences*, *99*(1), 1.

Liebowitz, J. (2005). Linking Social Network Analysis with the Analytic Hierarchy Process for Knowledge Mapping in Organizations. *Journal of Knowledge Management*, *9*(1), 76–86. doi:10.1108/13673270510582974

Mendonca, D. J., & Wallace, W. (2007). A Cognitive Model of Improvisation in Emergency Management. *IEEE Transactions on Systems, Man, and Cybernetics. Part A, Systems and Humans*, *37*(4), 547–561. doi:10.1109/TSMCA.2007.897581

Moynihan, D. P. (2008). Learning Under Uncertainty: Networks in Crisis Management. *Public Administration Review*, *68*(2), 350–365. doi:10.1111/j.1540-6210.2007.00867.x

Nikolic, V., Savic, S., & Stankovic, M. (2007). Designing a Multimedia Platform for Emergency Management. *Management of Environmental Quality*, *18*(2), 198–210. doi:10.1108/14777830710725858

Ozkan, S., Cakir, M., & Bilgen, S. (2008). A Maturity Based Qualitative Information Systems Effectiveness Evaluation of a Public Organization in Turkey. *Journal of Cases on Information Technology*, *10*(3), 58–71. doi:10.4018/jcit.2008070106

Pellissier, R. (2011). The Implementation of Resilience Engineering to Enhance Organizational Innovation in a Complex Environment. *International Journal of Business and Management*, *6*(1), 145–164.

Perry, R. W. (2003). Incident Management Systems in Disaster Management. *Disaster Prevention and Management*, *12*(5), 405–412. doi:10.1108/09653560310507226

Saulais, P., & Ermine, J. (2012). Creativity and Knowledge Management. *Vine*, *42*(3/4), 416–438. doi:10.1108/03055721211267521

Selke, K. (2007). *Understanding social networks - A key requirement for systems engineers.* Retrieved from http://www.dtic.mil/ndia/2007systems/Thursday/AM/Track8/UnderstandingSocialNetworks_BriefingFINAL.pdf

Serino, R. (2011). *Accountability at FEMA: Is Quality Job #1?* Hearing before the Ad Hoc Subcommittee on Disaster Recovery and Intergovernmental Affairs. U. S. Senate Committee on Homeland Security and Government Affairs. Retrieved from https://www.fema.gov/pdf/about/programs/legislative/testimony/2011/10_20_2011_accountability_at_fema.pdf

Shaluf, I. M., Ahmadun, F., & Said, A. M. (2003). A Review of Disaster and Crisis. *Disaster Prevention and Management*, *12*(1), 24–32. doi:10.1108/09653560310463829

Sproles, N. (2000). Coming to Grips with Measures of Effectiveness. *Systems Engineering*, *3*(1), 50–58. doi:10.1002/(SICI)1520-6858(2000)3:1<50::AID-SYS4>3.0.CO;2-U

St. Denis, L. A., Palen, L., & Hughes, A. (2012). Trial by fire: The Deployment of Trusted Digital Volunteer in the 2011 Shadow Lake Fire. In *Proceedings of the 9th International ISCRAM Conference*. Vancouver, Canada: ISCRAM.

Su, Y. S., Wardell, C., & Thorkildsen, Z. (2012). *Social media in the emergency management field (No. IPP-2013-U-004984)*. National Emergency Management Association.

Tsai, J., & Chi, C. S. F. (2012). Cultural Influence on the Implementation of Incident Command System for Emergency Management of Natural Disasters. *Journal of Homeland Security and Emergency Management*, *9*(1). doi:10.1515/1547-7355.1970

Turoff, M., Chumer, M., Hiltz, S. R., & Klashner, R. (2004b). Assuring Homeland Security: Continuous Monitoring, Control & Assurance of Emergency Preparedness. *Journal of Information Technology Theory and Application*, *6*(3), 1–24.

Turoff, M., Chumer, M., Van de Walle, B., & Yao, X. (2004a). The Design of a Dynamic Emergency Response Management Information System (dermis). *Journal of Information Technology Theory and Application*, *5*(4), 1–35.

Van de Walle, B., & Turoff, M. (2008). Decision Support for Emergency Situations. *Information Systems and eBusiness Management*, *6*(3), 295-316. doi:10.1007/s10257-008-0087-z

Ward, R., Wamsley, G., Schroeder, A., & Robins, D. B. (2000). Network Organizational Development in the Public Sector: A Case Study of the Federal Emergency Management Administration (FEMA). *Journal of the American Society for Information Science*, *51*(11), 1018–1032. doi:10.1002/1097-4571(2000)9999:9999<::AID-ASI1004>3.0.CO;2-5

White House. (2003). *Homeland Security Presidential Directive-5: Management of domestic incidents*. Retrieved from http://www.dhs.gov/sites/default/files/publications/Homeland%20Security%20Presidential%20Directive%205.pdf

White House. (2003). *Homeland Security Presidential Directive-8: National preparedness*. Retrieved from http://www.fas.org/irp/offdocs/nspd/hspd-8.html

White House Office of Management and Budget. (2009). *Open government directive*. Retrieved from https://www.whitehouse.gov/omb/inforeg_infocoll

White House Office of Management and Budget. (2010a). *Guidance for agency use of third-party websites and applications*. Retrieved from https://www.whitehouse.gov/omb/inforeg_infocoll

White House Office of Management and Budget. (2010b). *Guidance for online use of web measurement and customization technologies*. Retrieved from https://www.whitehouse.gov/omb/inforeg_infocoll

White House Office of Management and Budget. (2010c). *Social media, web-based interactive technologies, and the paperwork reduction act*. Retrieved from https://www.whitehouse.gov/omb/inforeg_infocoll

Williams, M. D., & Smart, A. (2010). Patient Safety: A Casualty of Target Success? *International Journal of Public Sector Management*, *23*(5), 416–430. doi:10.1108/09513551011058457

Yang, L., Prasanna, R., & King, M. (2009). On-site Information Systems Design for Emergency First Responders. *Journal of Information Technology Theory and Application*, *10*(1), 5–27.

Zheng, L., Shen, C., Tang, L., Zeng, C., Li, T., Luis, S., & Chen, S. (2013). Data Mining Meets the Needs of Disaster Information Management. *IEEE Transactions on Human - Machine Systems*, *43*(5), 451-464. doi:10.1109/THMS.2013.2281762

Zimmermann, K. A. (2012). *Hurricane Katrina: Facts, damage & aftermath*. Retrieved from http://www.livescience.com/22522-hurricane-katrina-facts.html

KEY TERMS AND DEFINITIONS

Federal Emergency Management Agency (FEMA): The principal United States component under the Department of Homeland Security that is responsible for coordinating federal resources in support of large-scale disaster response.

Human Systems Integration: The process of integrating human related aspects throughout the Systems Engineering Life Cycle Process, including training, human factors, safety, and workforce capability.

Hurricane Sandy: The largest Atlantic hurricane in history, and the second costliest storm to affect the United States.

Incident Command System (ICS): A command and control component of the National Incident Management System which organizes emergency management officials and responders in a standard, scalable framework with a common hierarchy and reporting structure.

Information Systems: The study of networks of information and people that collects, organizes, and disseminates data in support of decision-making activities.

Public Information Officer (PIO): The officer in the Incident Command System framework responsible for providing relevant information to the general public and disaster response stakeholders in the community about the status of emergency management operations.

Resilience: As applied to organizations and communities, this concept is the inherent elasticity and strength to overcome or deflect crises.

Social Network Analysis: The study of the relationships between people, or actors, in a network. Social Network Analysis is grounded in network theory.

Systems Engineering: An interdisciplinary and macro field of engineering which deals with the design and maintenance of complex systems. Topics in project management, industrial psychology, industrial engineering, and other related disciplines are integrated into the appropriate aspects of the system analysis and design.

Section 3

The Law of Unintended Consequences:

Impacts of Social Media

Chapter 7
Cyberbullying Prevention:
Some Preventing Tips

Gilberto Marzano
Rezekne University of Applied Sciences (Rezekne Augstskola), Latvia

ABSTRACT

The Internet is a really wide, intriguing world, difficult to penetrate in depth, rich with dissimulations, full of useful and also evil things, that are continuously changing. Cyberbullying represents an actual risk, especially for the online generation which is often unable to distinguish between virtual and real-reality. The first step for cyberbullying prevention is the knowledge acquisition of what cyberbullying is and how it occurs within a specific context. This is not an easy task, since cyberbullying is a complex and quite new phenomenon, so much that researchers' opinion is often divided on its definition and there isn't agreement on the extent of its diffusion. This article presents and comments some cyberbullying preventing tips, inspired by the "top ten tips" that Hinduja and Patchin suggested to educators engaged in cyberbullying prevention.

INTRODUCTION

Many children, adolescents and youth have had exposure to educational programs aimed at preventing various kinds of violence perpetration and victimization. The most famous program was the Olweus Bullying Prevention Program, developed and evaluated over a period of almost 20 years (Olweus 1993, 1999). It is built on four key principles:

- warmth, positive interest, and involvement from adults;
- firm limits on unacceptable behavior;
- consistent application of non-punitive, nonphysical sanctions for unacceptable behavior or violations of rules;
- adults who act as authorities and positive role models.

DOI: 10.4018/978-1-4666-8556-7.ch007

Overall, several researchers agree that Olweus's program can reduce violence and aggression (Limber, 2013). However, in spite of the large violence prevention mobilization and indications of effectiveness, there are reasons to think that program dissemination has stalled (Finkelhor, Vanderminden, Turner, Shattuck, & Hamby, 2014).

But what about cyberbullying?

Some years ago, Hinduja and Patchin (2009) indicated the *top ten tips for educators* aimed to prevent cyberbullying. These tips can be grouped in three main classes:

- Contextualized knowledge acquisition (tip: *Formally assess*);
- Educational activities (tips: *Teach students that all forms of bullying are unacceptable; Specify clear rules regarding the use of the Internet and electronic devices; Use peer mentoring; Cultivate a positive school climate; Educate your community*);
- Technical support (tips: *Consult with your school attorney before incidents occur; Create a comprehensive formal contract; Implement blocking/filtering software; Designate a "Cyberbullying Expert"*).

The *educational activities* class includes five of the ten cyberbullying preventing tips, this way showing the importance that education occupies in promoting appropriate solution strategies.

In the following paragraphs the Hinduja and Patching's tips are analyzed and commented; though before dealing with cyberbullying prevention, we cannot leave aside the issue of what is cyberbullying.

WHAT CYBERBULLYING IS

Currently there is debate about what constitutes cyberbullying and how this phenomenon is similar or different to traditional forms of bullying (Thomas, Connor, & Scott, 2014; Menesini et al., 2012).

Beran and Li (2005) used for cyberbullying the expression "old wine in new bottles".

The hypothesis of contiguity between bullying and cyberbullying seems to be confirmed by some researches' data (Ybarra & Mitchell, 2004; Beran & Li, 2008; Kowalski, Limber, & Agatston, 2008; Del Rey, Elipe, & Ortega-Ruiz, 2012). These researches show that there is an overlap of 44% between victims of "on line aggression" in cyberspace and real life victims. In particular, Beran and Li, as well as Kowalski and Limber, found an overlap of about 60% between being a victim in real life and being a cyber-victim – although, if 60% of cyber-victims are also bullied in real life, that leaves an amount of 40% who are not victimized in real life. Data from a German study (Riebel, Jaeger, & Fischer, 2009) confirms that more than 80% of cyberbullies bully their fellow students in real life as well; although in most cases it seems that cyberbullying is a further strategy in the repertoire of a typical bully. There are also some recent studies that indicate a significant overlap between traditional bullying and cyberbullying (Perren, Dooley, Shaw, & Cross, 2010; Sourander et al., 2010).

Literature offers a wide variety of cyberbullying definitions (Smith, Cowie, Olafsson, & Liefooghe, 2002; Hinduja & Patchin, 2006, 2007, 2008; Ybarra, Mitchell, Wolak, & Finkelhor, 2006; Ybarra & Mitchell, 2007; Dooley, Pyżalski, & Cross, 2009; Lenhart, 2010; Ortega-Ruiz & Nunez, 2012; Heirman & Walrave, 2012; Wachs, Wolf, & Pan, 2012; Palladino, Nocentini, & Menesini, 2012; Paul, Smith, & Blumberg, 2012; Vandebosch & Van Cleemput, 2008; Vandebosch, Beirens, D'Haese, Wegge, & Pabian, 2012; Franco, Bellerín, Borrego, Díaz, & Molleda, 2012). However, according to Kowalski and Limber's opinion (2007), many insiders consider cyberbullying a particular form of bullying, and described it as bullying that occurs through media and communication devices such as mobile phones, email, and the Internet (e.g. social networking sites, web pages, chat rooms and blogs). Cyberbullies threat, scare, libel, humiliate, and offend their victims from a distance. They use pictures, movies, emails, messages and comments to harass and pick on their targets on the Internet.

Some of the most cited definitions of cyberbullying are:[1]

An aggressive, intentional act carried out by a group or individual, using electronic forms of contact, repeatedly and over time against a victim who cannot easily defend him or herself (Smith et al., 2008, 376).

[cyberbullying is when someone] repeatedly makes fun of another person online or repeatedly picks on another person through e-mail or text messages or when someone posts something online about another person that they don't like (Hinduja & Patchin, 2012, 540)

Cyberbullying involves the use of information and communication technologies to support deliberate, repeated, and hostile behavior by an individual or group, that is intended to harm others (Belsey, 2013)

Although there is a high degree of empirical and conceptual overlap between traditional bullying and cyberbullying, specific, distinct features have been identified in cyberbullying (Dooley, Pyżalski, & Cross, 2009; Smith et al., 2008). Cyberbullies can reach particularly large audiences in a peer group as opposed to the small groups who are the usual audience in traditional bullying. Compared to the most traditional forms of bullying, the person carrying out cyberbullying may be less aware or is completely unaware of the consequences of his or her actions.

In our opinion, two issues, which are specific of cyberbullying, continue to create many theoretical and practical difficulties. The first is related to the lack of complete control which is due to the technology being used. For example, a harmful digital content (sms, text, picture or video) once is sent out or uploaded on the Internet, can be downloaded, forwarded and diffused by anyone else, not only by the initial perpetrator, and the flux cannot be stopped. Furthermore, the victim can experience the same aggression many times, accessing and running the same toxic digital item repeatedly. Internet memory is out of the perpetrators and victims' control, and cyberbullies take advantage from the specific features of online communication (anonymity, asynchronicity, and accessibility) to torment their targets (Valkenburg & Peter, 2011).

The second issue is related to the notion of *power imbalance*, which can involve different spheres of domination and weakness (Law, Shapka, Hymel, Olson, & Waterhouse, 2012). In fact one can presume that cyberbullies have a very deep knowledge of technologies and profit from their anonymity. But things could go in a different way: cyberbullies might be less technologically skilled than their victims, using technologies only as an extension of their face to face aggression or *to document* it. Conversely, it has been observed that the hidden nature of electronic communication can offer to victims, or self-presumed victims, of traditional bullying the opportunity to retaliate on line or to attack their real-world perpetrators (Kowalski, Limber, & Agatston, 2008). An online survey conducted on 473 students (König, Gollwitzer, & Steffgen, 2010) investigated the role of revenge and retaliation as a motive to engage in acts of cyberbullying. Data showed that 149 respondents were identified as traditional victims/cyberbullies, and that, although individual differences play a role in the choice of their victims, *traditionally* bullied students indeed tend to choose their former perpetrators as cyber victims.

In our opinion, cyberbullying is an extremely complex issue. In general, it can assume two different faces due to different use of communication technology. Online bullying can be only a simple, but really serious, extension of face to face bullying: "new wine in old bottles". Victims know their perpetrators and the dynamics between perpetrators and victims are the same ones as in traditional bullying. But there is another aspect of cyberbullying that is new and more difficult to probe, due to the nature of virtual relationships and the significance of perception of these kinds of relationships. This type of bullying is sneaky and dangerous since it involves the sphere of intimacy and especially the image that young people have of themselves. Furthermore, young people are digital natives and are using digital technology as the dominant medium for communication among them. But between these two different patterns of cyberbullying there are many other forms which depend on the context and involve the individual personality of the various implicated actors (perpetrators, victims, bystanders, educators, parents), and also cultural factors, since culture and historical heritage could influence cyberbullying or at least its perception:

[An] Important consideration to keep in mind when comparing studies are cultural similarities and differences in both the prevalence of research on the topic of cyberbullying and the prevalence of cyberbullying itself within a particular culture (Kowalski, Limber, & Agatston, 2012, p. 89).

In developing an effective educational *no-cyberbullying* program all these aspects of the problem cannot be ignored.

In the following paragraphs the Hinduja and Patchin's tips are commented, and some educational suggestions are introduced. Some comments come from the experience of a two year bilateral EU project, "Threats of cyberspace: new qualification of a social worker", conducted by Janusz Korczak Pedagogical University in Warsaw in collaboration with Rezekne University of Applied Sciences (2012-2014).

CONTEXTUALIZED KNOWLEDGE ACQUISITION

The first tip indicated by Hinduja and Patchin is the knowledge acquisition about the extent and the scope of the problem within the school district. They suggest collecting survey and/or interview data from students in order to acquire a baseline measure of what is happening in the school. Acquiring contextualized knowledge is the necessary premise for implementing specific countermeasures against cyberbullying. The surveys should be primarily aimed to collect information useful to educate students and staff about online safety and the use of the Internet in a creative and powerful way.

However, in our opinion, understanding the school's context should not be limited to gather data on off-line and on-line bullying among students, but should also be extended to acquire information about the staff awareness of violent conducts at their school and their competences in tackling them. Although teachers are aware and well prepared on *critical media literacy*, often they are considered incompetent on computing and new media use by their students. More than 90% of today's students are using computer and smartphones to access and disseminate media.

Teachers and educators' awareness as to what constitutes bullying/cyberbullying can vary (Kochenderfer-Ladd & Pelletier, 2008) and requires a wide basic knowledge of different aggressive features; many of them are subtle, hidden and undetectable but having devastating consequences. In fact, the over violence is a key factor easy to recognize, but there are also forms of relational aggression which are covert, such as exclusion and sneaky mocking (Ryan et al., 2011; Sahin, 2010). A lack of knowledge of bullying-type behaviors can cause that teachers to underestimate relational aggression, and are unlikely to intervene if such problems occur. For this reason, assessing the school staff capability in recognizing the spectrum of bullying behaviors is a primary step to deal with bullying issues (Craig, Henderson, & Murphy, 2000).

Strategies and techniques used for data gathering about students' behavior and teachers' awareness and skills as to off-line and on-line bullying are also important.

The most suitable and used means to gather data are questionnaires. In the case of bullying and cyberbullying it is necessary to prepare the questionnaire carefully and provide some preliminary actions. It is crucial that school staff is involved in preparing and submitting the questionnaire with detailed instruction and appropriate suggestions. Students should be informed that it is not mandatory filling out the questionnaire, but that school would be grateful if they did. Furthermore, they should be assured that anything that they write will be treated as most confidential: questionnaire's anonymity will be preserved, and teachers, head teacher and classmates will not be shown their answers. It is of the utmost importance underline that the answers ought to be truthful, explaining that the reason of the questionnaire is to pursue the students' interest, protecting them and making school safer. Students should be involved in the survey and motivated to share their advice, following a participating process and a responsible engagement.

There exist many good templates that can be used for gathering data about the presence of bullying and cyberbullying at school and to measure the awareness and skill of school staff in facing these phenomena.

Generally, questionnaires focus on students' life in and out of school in the past 2 or 3 months; students are invited to answer the questions thinking of how their live has been during the past couple of months and not only how it is just now. Often designed for scientific aims, many questionnaires premise the definition or explanation of the word *bullying*:

We say a student is being bullied when another student, or several other students

- say mean and hurtful things or make fun of him or her and call him or her mean and hurtful names;
- completely ignore or exclude him or her from their group of friends or leave him or her out of things on purpose;
- hit, kick, push, shove around, or lock him or her inside a room;
- tell lies or spread false rumors about him or her or send mean notes and try to make other students dislike him or her;
- and other hurtful things like that. (Smith, Mahdavi, Carvalho, & Tippett, 2006)

This bullying's definition is generally acknowledged by researchers and it is in accordance with the Olweus's definition (Olweus, 1993).

Therefore, questionnaires introduce the *cyberbullying* definition. But cyberbullying definitions can differ, since there are divergent opinions among researchers. Following, there is an example of a cyberbullying definition from a well-known investigation on the subject which aimed at exploring its forms, awareness and impact, and the relationship between age and gender in cyberbullying:

[…] Today we would like to look at a special kind of bullying: Cyberbullying.

This include bully

- through text messaging
- through picture/photos or video clip
- through phone calls (nasty, silent etc.)
- through email

- in Chat rooms,
- through Instant Messaging
- through Websites (Smith et al., 2006)

Even in the questionnaire's premise, students are invited to keep in mind that bullying is when the above mean things happen repeatedly, and it is difficult for the student being bullied to defend him or herself. In fact it is crucial that students understand that one is bullied when him/her is teased repeatedly in a mean and hurtful way, but one isn't bullied when the teasing is done in a friendly and playful way. Also, it is not bullying when two students of about equal strength or power argue or fight.

We note that a circumscribed definition of bullying and cyberbullying is a necessary condition in a scientific survey. Nevertheless, it is not necessary to embark on a scientific discussion regarding cyber-bullying definition when one would like to acquire information in the school district before planning a no-cyberbullying program. One needs more general information about students' habits and their family context: the use of the Internet and mobiles, the hours spent on line, the presence of violent conducts in their family context, their involvement in harmful events, the perceived hostility by class-mates, their confidence in new technologies and their presumed ability to use them, the awareness about cyber risks, the importance attributed to have be present with an on-line profile, the importance of virtual relation-ships, and so on.

Combating cyberbullying is impossible if one doesn't know the context where cyberbullying takes place. The Internet is a really wide world, difficult to probe, rich with dissimulations, full of useful and evil things, continuously changing. One can observe that when young people are together in face to face meeting and talks, they continue using their mobiles and smartphones, simultaneously communicating with present mates, with others that are in distant places, and with virtual friends who they know only through the Internet. An endeavor must be made understanding the world of young people and deal with the presence of virtual reality[2] in their life, recognizing their concern and significance about their on-line reputation. Moreover, victimization can begin at school and then extends into the home and community through the use of technology, but it is also possible that victimization via computers and mobiles can leads to face-to-face victimization. The most troubled ones are those who are both bullies and victims. This implies the need for new strategies for prevention and intervention. There are researchers in psycho-logical and psychiatric fields who argue that both cyberbullying and cybervictimization are associated with psychiatric and psychosomatic problems, and the most troubled individuals are those who are both cyberbullies and cybervictims (Sourander et al., 2010).

School staff, willy-nilly, ought to take awareness of this topical situation.

An alarming murderous episode which happened in Italy on 26th of December 2013 demonstrates that topicality of close links between real and virtual life is not only among young people. Creating a fake profile on Facebook, a 42 years old woman unleashed the mad jealousy of her partner who brutally killed her.[3]

EDUCATE THAT ALL FORMS OF BULLYING ARE UNACCEPTABLE

The most important educational tip by Hinduja and Patchin is that one ought to "teach students that all forms of bullying are unacceptable, and that cyberbullying behaviors are potentially subject to discipline". They suggest to educators to have a conversation with students about what "substantial disruption" means. Students need to know that even a behavior that occurs miles away from the school could be subject to school sanction if it disrupts the school environment. This tip is tied to another tip which we have included into the *technical support* class. The authors suggest to "create a comprehensive formal contract" specific to cyberbullying in the school's policy manual, or introduce clauses within the formal "honor code" which identify cyberbullying as an example of inappropriate behavior. Following this tip many schools in USA have adopted the solution of *zero tolerance* against cyberbullying.

We concur that teaching students about the cyberbullying illegitimacy is a primary action in combating it. We agree that a school must openly declare that cyberbullying is not an allowed and admissible behavior and rules about this phenomenon must be taken. We suggest to define these rules, which should be periodically reviewed, involving schools' staff, parents and students' representatives. No-cyberbullying rules definition and updating is an opportunity to illustrate the consequences related to cyberbullying acts, also under a legal point of view.

However, teachers and educators must keep in mind that current regulations have a low impact in combatting and punishing cyberbullying. Social networking such as MySpace and Facebook allow any users older than fourteen to become members of their sites once they register. Registration requires users to submit personal information, choose a password, and agree to the Terms of Service and Privacy Policy (TOS) by checking a box or doing other operations for accepting the service. But about TOS the case of Lori Drew is an instructive and illuminating one.[4] This case has been widely reported on the international media, becoming a cyberbullying icon. Lori Drew helped her daughter creating a false profile on MySpace.com in order to contact a class-mate and a neighbor of hers, thirteen-year-old Megan Meier. The profile falsely claimed to belong to an attractive sixteen-year-old boy named Josh Evans. Lori Drew declared the fake profile purpose was to get confidential information on what Meier was saying about her daughter. A month later Meier received an instant message from "Evans" saying that he no longer liked her and that "the world would be a better place without her in it." That same day, Meier committed suicide. Lori Drew was accused of breaking federal laws, since she orchestrated a MySpace hoax that led to her neighbor child's dead. The government's theory in the Drew case was that the creation of the "Josh Evans" profile had violated the TOS of MySpace.com, and that the TOS violation rendered the access to MySpace's computers either without authorization or exceeded authorization. The first guilty verdict was overturned by the federal judge handling the case. The judge accepted what defense argued. As a matter of statutory construction, the TOS did not govern authorization: in fact MySpace had given authorization to access its computers by creating the Josh Evans account and allowing the group to send and receive messages using it (Kerr, 2009). At trial, the vice president of customer care at MySpace testified that the sheer volume of 400 million MySpace accounts made it nearly impossible to determine which accounts were in violation of the TOS.

As Meredith observes:

Practically speaking, the repercussions for violating the MySpace TOS involve MySpace contacting law enforcement directly only in rare circumstances. More likely, MySpace would simply warn the violative users that their actions might warrant involvement of law enforcement or the removal of the offensive profile from the MySpace site (Meredith, 2010, p. 321).

The other Hinduja and Patchin's tips for a no-cyberbullying educational program are referred to a crucial aspect: the participation of all involved subjects to the development of same no-cyberbullying program. They suggest using peer mentoring to promote positive online interactions, e. g. organizing lessons by older students who informally teach and share learning experiences with younger students. Another tip is to cultivate a positive school climate. They argue that it is crucial to establish and maintain a school climate of respect and integrity where violations result in informal or formal sanction. They also recommend to utilize specially-created cyberbullying curricula, or general information sessions such as assemblies and in-class discussions to raise awareness among youth. Finally, they underline the importance of inviting specialists to come to talk to staff and students, sending information out to parents and organizing educational events involving parents, relatives and any other relevant adult. Specialist and educators should emphasize the advantages offered by new technologies. They can be used to support not only learning activities, but also to enhance positive social relationships. Moreover, social networking and virtual relationships are useful for people who experience difficulties in engaging in social relationships, and online anonymity may lower the barriers in meeting new friends. Social networking sites help to reach large audiences to tell what's going on in their lives and how they feel (Heirman & Walrave, 2008). Social networking can help finding others who share mainstream but specific interests or those who share culturally devalued identity aspects that could be quite difficult in one's everyday social world (McKenna & Green, 2002).

Also, from what Meredith (2010) writes, we add to the above suggestions a consideration about overcriminalization of cyberbullying. We agree with her observations that one ought to avoid overcriminalization and carefully criminalize only criminal acts. In this regard, Meredith reports about the failure to distinguish cyberbullying done by minors from that caused by adult perpetrators, and underlines the difficulty to regulate unpleasant speech that doesn't reach the level of a true threat. She cites (p. 328) some cases where law considered cyberbullying to be a *juvenile issue* and thus required the schools to create policy to prevent it. This is some sort of shifting the blame which charges the schools with extra responsibility.

In this perspective, efforts should be concentrated in increasing the Internet safety education. Educators ought to be empowered with the proper tools to inform students and parents about how to use technology in a wise way. It is crucial to provide information that increases awareness about the cyberspace's risks. In the first place, children should be taught to utilize the positive aspects of the Internet without becoming victims of the threats hidden inside it.

However, attempts of combating cyberbullying through legislation should be pursued. The first US legislation related to the prevention of cyberbullying was introduced in April 2009, but, at the moment, only few states in the United States have enacted antibully legislation, while some states have either antibullying policies, laws pending, or comments about the need for or status of their policies or laws.

Cyberbullying legislation is not an easy issue, it is sometime difficult to extricate and distinguish what is really on-line harassment, defamation and hateful speech at a legal level. Although they represent a significant problem, there are few legal remedies for victims (Marwick & Miller, 2014). Accordingly, one should tackle this issue keeping in mind a Linda Sanchez's consideration, cited by Meredith (2010, p. 330):[5]

I want the law to be able to distinguish between an annoying chain email, a righteously angry political blog post, or a miffed text to an ex-boyfriendall of which are and should remain legal; and serious, repeated, and hostile communications made with the intent to harm.

Sanchez alludes to a legislative effort, but her aim should be extended to the possibility of preventing cyberbullying through new and sophisticated software tools. Software producers should provide solutions for safety use of their products. But also the role of media should not be ignored.

THE MEDIA ROLE: IS CYBERBULLYING AN EXAGGERATED THREAT?

In an article which reviews and discusses the current state of art of international research on cyberbullying, Cassidy, Faucher and Jackson (2013) observe that, increasingly, cyberbullying is mentioned alongside traditional bullying, as an exploding phenomenon with which researchers and practitioners must contend. Researchers who share the opinion of cyberbullying increasing are very numerous (Ybarra & Mitchell, 2004; Campbell, 2005; Cowie, Hutson, Jennifer, & Myers, 2008; Smith & Slonje, 2010; Bulut & Gündüz, 2012; Demaray, Malecki, Jenkins, & Westermann, 2012; Hatzchristou, Polychroni, Issari, & Yfanti, 2012; Kowalski et al., 2012a; Kowalski, Morgan, & Limber, 2012; Soto, Campos, & Morales, 2012; von Mare´es & Petermann, 2012). International quantitative researches show a significant magnitude of cyberbullying among pupils, but we cannot help observing that specific results vary considerably across cultures and across studies (David-Ferdon & Hertz, 2007; Smith et al., 2008). This can be partially explained by the disparate definitions and methods utilized so far by current studies. Therefore it seems problematic to compare prevalence rates. In Li's Canadian study, it is found that 25% of pupils have never been a victim of cyberbullying and 17% have cyberbullied others (Li, 2006). But data is not so definite. In a less recent survey carried out in the USA, Ybarra and Mitchell (2004) challenged the dichotomized perception of cyberbullies/online -victims, and compared the prevalence rates of four groups: 4% of which were a target of online aggression, 12% were online aggressors, 3% of the youngsters were both aggressors and targets of online bullying, whilst 81% were not involved in online bullying. Some researchers (e.g. Cassidy, Brown, & Jackson, 2011; Kowalski et al., 2012a; Rivers & Noret, 2010) have found an increased rate over the past five years.

However, there are those who don't agree with the idea of a cyberbullying deluge. Some researchers (Olweus, 2012a, 2012b; Li et al., 2012a; Hinduja & Patchin,2012) claim that the rate of cyberbullying has not increased since it first emerged as a problem in the middle of the last decade. But, a fact is certain, in this last years, there was a growing attention of media about cyberbullying. Media have stimulated the awareness of public authorities and also the attention of ordinary citizens (Dooley et al., 2009; Patchin & Hinduja, 2011; Li, Smith, & Cross., 2012).

Recently, newspapers and television programs are beginning to deal with cyberbullying, especially when an extreme case occurs, e. g. the suicide of a targeted adolescent. In general, media report about victims' parents who angrily accuse classmates, school authorities and the government, since in their opinion, they didn't anything for combating and arresting cyberbullies. On the other hand, classmates and school authorities always claim that they didn't notice anything, while police complain that there are huge amount of grossly offensive, indecent and obscene comments made every day on social media. Police object that they cannot analyse all malicious/aggressive contents and claim that many of them don't meet the threshold that would justify their involvement. However, all the above mentioned subjects confirm that social networking providers should do more, taking countermeasures, controlling and filtering sites' contents. There is no doubt that newspapers and media interpret the people disquiet about cyberbullying and present it as an alarming and increasing phenomenon. In this regard, one must keep in mind that media often exaggerate, as in the following example.

"Quando i social network uccidono: adolescenti vittime di cyberbullismo [When social networks kill: adolescent victims of cyberbullying]" [6] this is an on line article, published on 3th June 2013 by *Repubblica*, a largely read Italian newspaper. The same article was re-published on 12th August of the same year, when a fourteen years old boy who was presumed to be gay committed suicide. But, after some days of flamed articles and recriminations against homophobia, police excluded the cyberbullying hypothesis as the cause of his suicide. It shouldn't surprise anyone if immediately after the incident, gay, civil rights and left associations had invoked a new, stricter law to contrast homophobia and cyberbullying. But we have showed that improvement of legal options is not an easy matter, because there isn't a completely agreed definition of cyberbullying.

However, a question remains opened about media and cyberbullying. The increasing of cyberspace risks corresponds to reality or is an overestimated projection of people's fear?

In this end, it is useful to report on Olweus's considerations about cyberbullying.

OLWEUS'S CONSIDERATIONS

Olweus, the pioneering researcher in the field of bullying prevention and the creator of the *Olweus Bullying Prevention Program*, observes that cyberbullying has been studied ''in isolation'', that is, outside a general context of traditional bullying, and it often occurs without a general student-friendly definition of what is meant by bullying. For Olweus, a student is being bullied when another student, or several other students:

- say mean and hurtful things or make fun of him or her or call him or her mean and hurtful names;
- completely ignore or exclude him or her from their group of friends or leave him or her out of things on purpose;
- hit, kick, push, shove around, or lock him or her inside a room;
- tell lies or spread false rumours about him or her or send mean notes and try to make other students dislike him/ her;
- and other hurtful things like that.

These characterizing traits of bullying are the same reported in the above mentioned example questionnaire on cyberbullying by Smith et al. (2006).

Olweus underlines that the general picture created in the media—and also by researchers and authors of books on cyberbullying— is that cyberbullying is very frequent, that it has increased dramatically over time and that this new form of bullying has created many new victims and bullies. More precisely, he argues that claims about cyberbullying made in the media and elsewhere are often greatly exaggerated and, by and large, they have very little scientific support. Furthermore, he notes that such claims may have some unfortunate consequences, confirming that his arguments are based on empirical analyses of several large-scale studies. He concludes that a presumably quite effective measure an individual school or community can take in counteracting and preventing cyberbullying is to invest time and technical competence in disclosing thoroughly a few identified cases of cyberbullying—and then communicate clearly and openly but anonymously the results to the students.

In our opinion, Olweus's considerations only confirm the novelty, the multifarious nature of cyberbullying and the difficulties to embrace all its facets. However, there is an uncomfortable evidence: nobody who is sane of mind would put a gun in the hand of a child: but new technology can be more powerful and lethal than a hand-held weapon.

CYBERBULLYING AND NEW MEDIA KNOWLEDGE

Fully 95% of teens are online, a percentage that has been consistent since 2006 (Figure 1).[7] Yet, the nature of teens' internet use has transformed dramatically during that time — from stationary connections tied to desktops in the home to always-on connections that move with them throughout the day. In many ways, teens represent the leading edge of mobile connectivity, and the patterns of their technology use often signal future changes in the adult population. Teens are just as likely to have a cell phone as they are to have a desktop or laptop computer.

Teachers, educators and parents are generally less skilled than youth in new technologies. This is a serious problem which leads to inadequate supervision at school and at home and limits cyberbullying prevention (Popovic´-C´ itic´, Djuric´, & Cvetkovic´, 2011). Moreover, although younger educators

Figure 1. Internet use over time by teen and adults

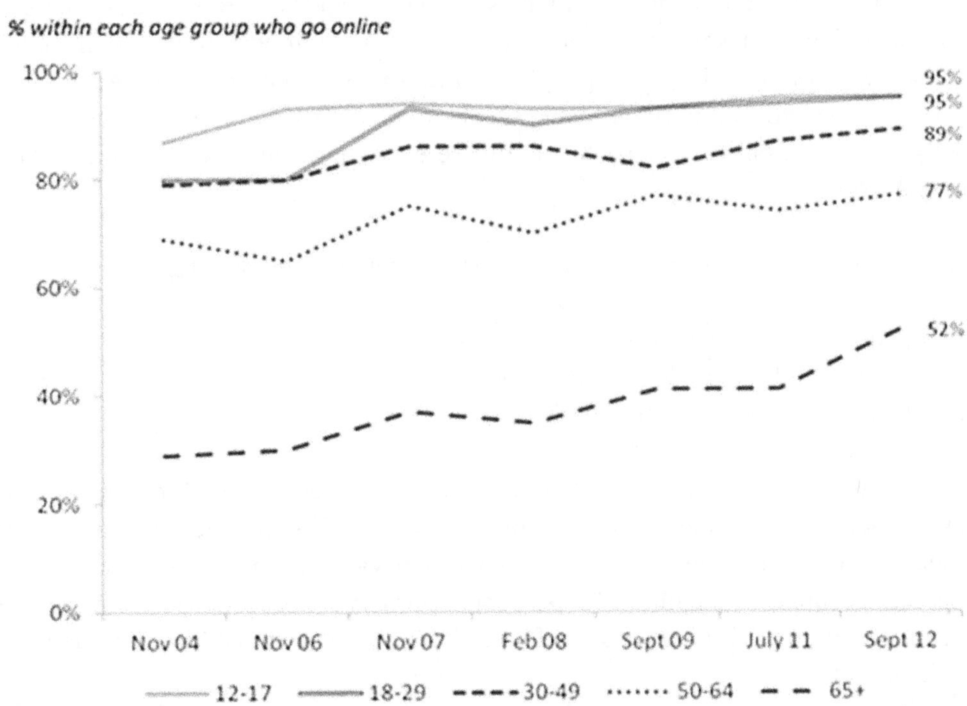

Source: The Pew Research Center's Internet & American Life Project surveys. All teen data comes from separate surveys of teens and their parents. Methodological info for each survey is available at: http://pewinternet.org/Data-Tools/Download-Data

show more familiarity with technology and they are concerned about cyberbullying, they are not able to translate their knowledge into policy or programs within the school.

In general, teachers are more likely to focus on preventing face-to-face bullying than cyberbullying. However, the lack of knowledge of technology is not the only obstacle in preventing cyberbullying. The main problem is that cyberbullying requires a multi-disciplinary approach in both policy development and teacher training. Cyber-themed school assemblies and curriculum lessons can be useful to spread knowledge among school staff and recover the technological gap (Paul, Smith, & Blumberg, 2010). But school staff professional background should also be enriched by other notions related to the psychological, pedagogical and social themes being present in cyberbullying, such as reviewing traditional concepts about self-esteem, peer relationships, empathy, communication, power imbalance, etc.

In regard of technology, it needs a thorough knowledge of the various media used for online bullying. Some authors classify cyberbullying by the types of medium, distinguishing cyberbullying via SMS, via e-mail, via instant messaging and so on (Patchin & Hinduja, 2007; Smith et al., 2008; Limber, 2012).

Ortega, Mora-Merchán and Jäger (2007) recommend the categorization by type of action, such as the taxonomy by Willard (2007), who defined eight subcategories of cyberbullying. Willard herself acknowledges that some of the phenomena she lists should rather be called "online social cruelty".

The classical forms of cyberbullying are:

- *Harassment* can be defined as repetitiously sending insulting or threatening messages to another person by e-mail, SMS, instant messaging or in chatrooms.
- *Denigration* is the spreading of rumours via electronic communication devices. Unlike gossip in real life, by means of the internet, information can be sent to thousands of people within seconds.
- *Outing & trickery* is similar: a message revealing personal information, which the victim sent to someone in confidence, is forwarded to other people in order to compromise the victim.
- *Exclusion* is equivalent to exclusion in real life and means withholding the opportunity of taking part in social activities. In an online context this could be excluding someone from multiplayer games, chats, or platforms.

Technology affects the relationship between victim and perpetrator creating new types of bullying:

- Bully knows victim and victim knows bully;
- Bully doesn't know victim and victim doesn't know bully;
- Bully knows victim and victim doesn't know bully.

In cyberspace, bullies can choose unknown targets to harass sifting social networking profiles, but there are frequent cases where an anonymous post in a blog provokes aggressiveness against his/her author. On the Internet, episodes which mutate into virtual group violence are not unusual: people participating to a blog begin to insult the author of a post, but later they look for him/her on the Web continuing to offend and threat the victim. Level of violence also in some political blogs is often very high (see the blog of Grillo in Italy).[8]

Harmful use of new media is not only specific of young people and circumscribed to the private or political sphere; it can also affect public institutions. This is the case of an Italian public university which has used Facebook facilities with the aim of giving students a place where posting their questions, exchanging comments/complaints and vent their protest, or to make a compliment. Students are free to enter their posts without any filter, and recently, in 2014, a big problem arose, when a student posted unfair accusations, offending an employee. Incredible, but for more than 20 days, the responsible of the site continued to exchange messages with the student, expressing solidarity, before alerting the unaware colleague about the situation. The responsible also asked the student to send the emails received by the employee. When the student sent them, apologizing for the trouble engendered, the responsible wrote: "You don't have to apologize: we should do it ".

The messages were visible to a community of about 6.000 members, including many persons out of the University context.

One may think the incident should inspire Sutton to enrich his famous book, *The no assole rule* (Sutton, 2007), but it is emblematic of the risks of uncontrolled or personal use of social media in a public environment. At the moment, the Internet is an uncontrollable and unregulated world, and technology offers to backstabbing and treachery a much wider audience than it did in the past.

TECHNICAL SUPPORT

The Hinduja and Patchin's ten tips include some suggestions which we have defined as *technical support*. For instance, they suggest implementing blocking/filtering software on computer network to prevent access to certain Web sites and software. These preventive measures are necessary, but they are not decisive. In fact, there are many commercial, open source and freeware software which provide functions for parental control. Many of these allow to control children´s Internet activities, restrict certain websites´ access and limit the time of surfing. Nevertheless, parental control can be bypassed; one can use a different computer where no filters are configured, while the proliferation of mobile technologies offers new opportunities of accessing the Internet.

Hinduja and Patchin also underline the importance of consulting school attorney before cyberbullying incidents occur to find out what actions you can or must take in various situations. It is certain that knowledge of law can help to tackle cyberbullying incidents, it is especially useful for avoiding conflict and long and expensive lawsuits. At the moment, we are sceptic about the effectiveness of law in preventing cyberbullying. The debate on this point demonstrates the large divergence between insiders and experts. A delicate underlying issue concerns the Internet censorship. Are cyberbullying law a form of censorship? Edward Snowden's revelations on US global surveillance operations are alimenting the debate on this issue.

However, some practical ideas about this issue are contained in the book of C. Zuchora-Walske (2010). The author sustains that the Internet censoring debate should be tackled considering that is made up of several smaller debates and involves many sticky questions:

- What is the best way to keep children safe on line?
- Should we use laws to protect children from inappropriate online contents?
- Is a combination of education and parental involvement a better strategy?
- It is possible - and even desirable – to demand civility (polite behavior) on the Internet?
- Does online civility help preserve a peaceful society?
- Should Internet users adhere to traditional

Finally, about technical support we note that effective technological help for combating cyberbullying could come from *machine learning* applications (Moore, Nakano, Suda, T., & Enomoto, 2013). Automatic text analysis of social networking contents is a new field of research and the results obtained encourage the insiders to continue in this direction (Dinakar, Reichart, & Lieberman, 2011; Garaigordobil & Berrueco, 2011; Reynolds, Kontostathis, & Edwards, 2011; Kontostathis, Reynolds, Garron & Edwards, 2013). On this regards, 179 experts on cyberbullying have been contacted and asked on three open-ended questions relating to the desirability of automatic monitoring, and 50 (28%) responded (Van Royen, Poels, Daelemans, & Vandebosch, 2015). Most of these experts favoured automatic monitoring, but specified clear conditions under which such systems should be implemented, including effective follow-up strategies, protecting the adolescents' privacy and safeguarding their self-reliance.

PREVENTING STRATEGIES

Many advices about preventing cyberbullying concern training children and their parents in e-safety and implementing technological tools to counteract the cyberbullies' behaviour, such as blocking bullying behaviour online or creating panic buttons for cyber victims to use when feeling under threat.

Prevention through education is a strategy largely agreed on researchers and practitioners. All insiders concord with the opinion that criminalization is not the right answer to combating cyberbullying, since its negative consequences outweigh the possible positive effects. Internet safety education has proven effective, and empowering educators with theoretical and practical tools is an extremely valid key.

Peer education programs and peer support models are also deemed useful. They are based on the assumption that peers learn from one another, having significant influence on each other. Conducts and behaviours are most likely to change when liked and trusted group members take the lead in changing the situation (Naylor & Cowie 1999; Turner & Shepherd, 1999; Maticka-Tyndale & Barnett, 2010).

There are several curriculum based programs available to educators that strive to help in preventing cyberbullying (Snakenborg, 2011). Some examples of cyberbullying combating programs are also available on the Internet, while prevention programs focus at school level appear to be more efficient (Couvillon & Ilieva, 2011).

It seems that traditional anti-bullying programs aimed at systematically tackling cyberbullying, including the participation of students and their parents, might yield fruitful results in terms of preventing cyberbullying or handling it (Perren et al., 2012; Mc Guckin et al. 2014). However, special programs which take into account the specific characteristics of cyberbullying may be needed. In this respect, Riebel, Jaeger and Fischer (2009) identified four major copying strategies:

- social coping: seeking help from family, friends, teachers, peer supporters;
- aggressive coping: retaliation, physical attacks, verbal threats;
- helpless coping: hopelessness, passive reactions, such as avoidance; displays of emotion;
- cognitive coping: responding assertively, using reason; analysing the bullying episode and the bully's behaviour.

However, designing preventive strategies is not simple at all. First, an effective preventing program should ground on the knowledge of new media and the understanding of their use by the different typologies of users, especially by adolescents and young people. This implies trainers ought to be skilled on new media, and should show their mastery to learners, otherwise their image of expert and authority could be discredited.

Another important aspect concerns the virtual identity issue and its reciprocal effects on physical identity. It has been observed that at the heart of the explosion in online communication is the desire to construct a valued representation of oneself which affirms and is affirmed by one's peers (Livingstone & Brake, 2010). Accordingly, exploiting new media people can constitute and experiment their identity in the online domain. If identities are constituted through interaction with others, what is the consequence on identity constitution if interaction is a mix of physical and virtual interactions? And what about virtual interactions are not only with real persons but also with electronic programs and with virtual fake profiles built by physical persons?

On this regards, what Leanhart[9] wrote in 2008 on Facebook about offline identities sounds still topical:[10]

In the offline world, we don't present ourselves in the same way to all people in our lives - we show different sides of ourselves to our mothers, our friends, our employers. And even in the age of fine-grained privacy tools, those tools do not eliminate the complexity of figuring out how to best present oneself in a multi-use public space, particularly for those who have personal, professional and family contacts on these sites.

Finally, cultural aspects should not be undervalued. They can play a role that could not be secondary if one adopts preventing programs created and experimented in other countries. The cultural sensitivity of cyberbullying is an issue not largely explored by researchers. Some researches try to test the relation between culture and cyberbullying comparing questionnaire results submitted in different countries, e. g. in United States and Japan (Barlett et al., 2014).

Indeed, the new cultural attitudes about new media open to an intriguing question: is really cyber bullying an alarming issue across various age groups regardless of cultural background (Mura, Topcu, Erdur-Baker, & Diamantini, 2011)? Does cyberbullying depend only on the spread of new technologies or some factors, such as culture and historical heritage, can influence it or at least its perception?

In this regard, a research conducted on teenagers in Latvia and Russia (Boronenko, Menshikov & Marzano, 2013) seems to show that cyberbullying is culturally influenced, and it matches the following observation:

Important consideration to keep in mind when comparing studies are cultural similarities and differences in both the prevalence of research on the topic of cyberbullying and the prevalence of cyberbullying itself within a particular culture (Kowalski et al., 2012a, p. 89).

Nevertheless, investigating about cultural influences affecting cyberbullying presents many methodological problems. Analysis can be expensive and take long time. Questionnaires should be integrated with other forms of data gathering, such as structured interviews, direct observation and web analysis. The above mentioned comparative research on cyberbullying in US and in Japan and the research on cyberbullying in Russia and Latvia have used different methodological approaches, the first a classic quantitative approach, while the second a mix one, socio anthropological qualitative-quantitative. It is interesting that, even though the differences, the two researches share the similar hypothesis that cy-

Figure 2. Online Bullying Among Youth 8-17 Years Old – Worldwide[13]

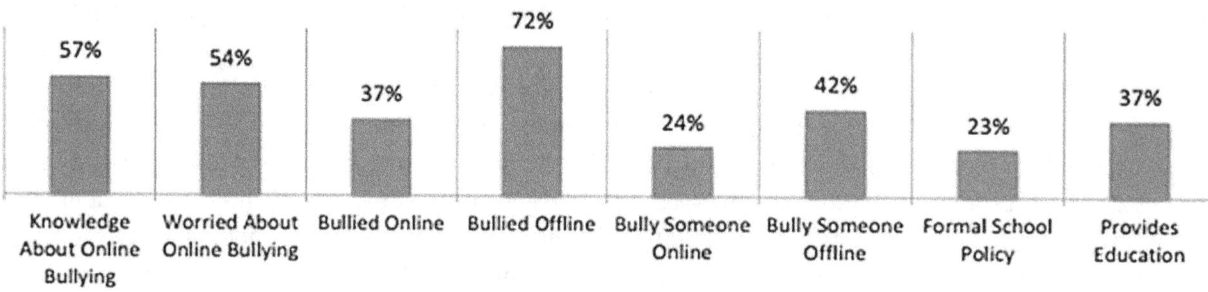

berbullying is related on how individuals perceive, comprehend, and interpret the world around them (Individualism and Uncertainty Avoidance[11] in the Latvia-Russia research, independent self-construal in US-Japan one).

CONCLUSION

At the moment, the worldwide available data about cyberbullying is that of a study commissioned by Microsoft (2012)[12] on pervasiveness of online bullying among youth 8-17 years old in 25 countries (Figure 2). To overcome the problem of cyberbullying definition, and that cyberbullying can vary between different cultures, and even among different individuals, the Microsoft study explored the issue by asking children about negative experiences they've had online - from their point of view (i.e., being called mean names, being teased, etc.). 37% (25 country average) of children age 8-17 who responded to the survey said they have been subjected to a range of online activities that some may consider to be online bullying or to have adverse effects:

- 19% - Mean or unfriendly treatment
- 20% - Made fun of or teased
- 18% - Called mean name

The Ditch the Label *Annual Cyberbullying Survey* (2013)[14] conducted in UK on a sample of 10,008 young people shows that 69% of young people in the UK experience bullying before the age of 18 and that over 2.5 million youths are bullied each year and 550,547 are bullied every single day.

A survey on 23,420 children and young people across Europe (Livingstone, Haddon, Görzig, & Ólafsson, 2011) found that 5% of respondents were being cyberbullied more than once a week, 4% once or twice a month and 10% less often; the vast majority were never cyberbullied. Different results came from a recent meta-analysis which shows a high prevalence of the problem in all countries (Garaigordobil, 2011): approximately, 40% and 55% of students are involved in some way (as victims, perpetrators, or observers). These discordances are in part due to the novelty of cyberbullying research and in part to the different approaches used by researchers:

Los diferentes estudios realizados varían mucho en cuanto a la edad (10-18 años), a la técnica o instrumento de evaluación empleado, al tipo de conductas de CB estudiadas o al intervalo de tiempo con-

siderado [Different studies vary widely in age (10-18 years old), technical or assessment of tools used, type of cyberbullies behaviour studied, and time interval considered] (Garaigordobil, 2011, p. 243).

However, cyberbullying phenomenon is an evil reality and it is a serious form of misbehavior among youth, and its spreading among early adolescents is alarming (Kowalski, Giumetti, Schroeder, & Lattanner, 2014).

Moreover, about cyberbullying, there are some evidences that should be carefully considered. Within the last decade an impressive expansion of social media technologies has occurred. Over 65% of people 11-16-year old have a profile on a social networking site and information exposure is more and more diffused among youth. Information exposure is a seemingly bizarre phenomenon whereby individuals freely and deliberately disseminate confidential or personal damaging information (including incriminating facts) to the widest possible audience, apparently without being concerned about any consequences. It is well known that adolescents have the tendency to feel constantly watched or "on stage" ("imaginary audience"), but information exposure is not just a phenomenon affecting adolescence. About Facebook, Leanhart observed (2008):[15]

And now, Facebook is adding yet another layer of complexity on top of this, yielding a new round of personal behavioral scrutiny: is it OK for my coworker or professional colleague to know that I was watching a video yesterday? Or that I shopped at the Discovery Kids website? Do I want them to know that about me? And what about my child who uses these services?

Finally, there is a significant rise in gossip sites, which, in essence, are websites where individuals strive against each other to produce the most tantalizing piece of gossip. The sites "rate" gossip based on the number of people who click on (and presumably read) the contents.

In this context, really complex and fluid at the same time affected by technology fast evolution and consumerism, what are the best strategies for cyberbullying prevention?

Analysing the ten tips proposed by Hinduja and Patchin new questions regarding the best strategies for combating cyberbullying have been presented. Indeed, prevention activities should follow the progress of research, paying attention to both the current evolution of technology, and the emerging aspects and harmful effects of cyberbullying. Empirical research should be increased and common initiatives on implementing preventative measures, testing their applicability and benchmarking their results should be multiplied.

But, above all it is crucial advocating the role of school in this purpose, otherwise one may run the risk of closing the barn door when the cow has bolted.

REFERENCES

Barlett, C. P., Gentile, D. A., Anderson, C. A., Suzuki, K., Sakamoto, A., Yamaoka, A., & Katsura, R. (2014). Cross-Cultural Differences in Cyberbullying Behavior A Short-Term Longitudinal Study. *Journal of Cross-Cultural Psychology*, *45*(2), 300–313. doi:10.1177/0022022113504622

Belsey, B. (2005). *Cyberbullying: An emerging threat to the 'always on' generation.* Retrieved March 2 2015 from http://www.inukshuk.com/pdf/Cyberbullying_Presentation_Description.pdf

Beran, T., & Li, Q. (2005). Cyber-harassment: A study of a new method for an old behavior. *Journal of Educational Computing Research*, *32*(3), 265–277. doi:10.2190/8YQM-B04H-PG4D-BLLH

Beran, T., & Li, Q. (2008). The relationship between cyberbullying and school bullying. *The Journal of Student Wellbeing*, *1*(2), 16–33.

Boronenko, V., Menshikov, V., & Marzano, G. (2013). Topicality of cyberbullying among teenagers in Russia and Latvia. *Social Sciences Bulletin*, *2*(17), 84–104.

Bulut, S., & Gündüz, S. (2012). Exploring violence in the context of Turkish culture and schools. In S. R. Jimerson, A. B. Nickerson, M. J. Mayer, & M. J. Furlong (Eds.), *Handbook of school violence and school safety: International research and practice* (2nd ed.; pp. 165–174). New York, NY: Routledge.

Campbell, M. A. (2005). Cyber bullying: An old problem in a new guise? *Australian Journal of Guidance & Counselling*, *15*(1), 68–76. doi:10.1375/ajgc.15.1.68

Cassidy, W., Brown, K., & Jackson, M. (2011). Moving from cyber-bullying to cyber-kindness: What do students, educators and parents say? In E. Dunkels, G.-M. Franberg, & C. Hallgren (Eds.), *Youth culture and net culture: Online social practices* (pp. 256–277). Hershey, PA: Information Science Reference. doi:10.4018/978-1-60960-209-3.ch015

Cassidy, W., Faucher, C., & Jackson, M. (2013). Cyberbullying among youth: A comprehensive review of current international research and its implications and application to policy and practice. *School Psychology International*, *34*(6), 575–612. doi:10.1177/0143034313479697

Couvillon, M. A., & Ilieva, V. (2011). Recommended practices: A review of schoolwide preventative programs and strategies on cyberbullying. *Preventing School Failure: Alternative Education for Children and Youth*, *55*(2), 96–101. doi:10.1080/1045988X.2011.539461

Cowie, H., Hutson, N., Jennifer, D., & Myers, C. A. (2008). Taking Stock of Violence in UK Schools Risk, Regulation, and Responsibility. *Education and Urban Society*, *40*(4), 494–505. doi:10.1177/0013124508316039

Craig, W. M., Henderson, K., & Murphy, J. G. (2000). Prospective teachers' attitudes toward bullying and victimization. *School Psychology International*, *21*(1), 5–21. doi:10.1177/0143034300211001

David-Ferdon, C., & Hertz, M. F. (2007). Electronic media, violence, and adolescents: An emerging public health problem. *The Journal of Adolescent Health*, *41*(6), S1–S5. doi:10.1016/j.jadohealth.2007.08.020 PMID:18047940

Del Rey, R., Elipe, P., & Ortega-Ruiz, R. (2012). Bullying and cyberbullying: Overlapping and predictive value of the co-occurrence. *Psicothema*, *24*(4), 608–613. PMID:23079359

Demaray, M. K., Malecki, C. K., Jenkins, L. D., & Westermann, L. D. (2012). Social support in the lives of students involved in aggressive and bullying behaviours. In Handbook of school violence and school safety: International research and practice (pp. 57-67). Academic Press.

Dinakar, K., Reichart, R., & Lieberman, H. (2011, July). Modeling the Detection of Textual Cyberbullying. In *The Social Mobile Web*. Retrieved from http://web.media.mit.edu/~kdinakar/3841-16937-1-PB.pdf

Dooley, J. J., Pyżalski, J., & Cross, D. (2009). Cyberbullying versus face-to-face bullying. *Zeitschrift für Psychologie. The Journal of Psychology*, *217*(4), 182–188.

Finkelhor, D., Vanderminden, J., Turner, H., Shattuck, A., & Hamby, S. (2014). Youth exposure to violence prevention programs in a national sample. *Child Abuse & Neglect*, *38*(4), 677–686. doi:10.1016/j.chiabu.2014.01.010 PMID:24630440

Franco, L. R., Bellerín, M., Borrego, J., Díaz, F., & Molleda, C. (2012). Tolerance Towards Dating Violence in Adolescents. *Psicothema*, *24*(2), 236–242. PMID:22420351

Garaigordobil, M. (2011). Prevalencia y consecuencias del cyberbullying: Una revisión. *International Journal of Psychology & Psychological Therapy*, *11*(2), 233–254.

Garaigordobil, M., & Berrueco, L. (2011). Effects of a play program on creative thinking of preschool children. *The Spanish Journal of Psychology*, *14*(2), 608–618. doi:10.5209/rev_SJOP.2011.v14.n2.9 PMID:22059307

Hatzchristou, C., Polychroni, F., Issari, P., & Yfanti, T. (2012). A synthetic approach for the study of aggression and violence in Greek schools. In S. R. Jimerson, A. B. Nickerson, M. J. Mayer, & M. J. Furlong (Eds.), *Handbook of school violence and school safety: International research and practice* (2nd ed., pp. 141–152). New York, NY: Routledge.

Heirman, W., & Walrave, M. (2008). Assessing Concerns and Issues about the Mediation of Technology in Cyberbullying. *Cyberpsychology: Journal of Psychosocial Research on Cyberspace, 2*(2), article 1. Retrieved February 12 2015 from http://cyberpsychology.eu/view.php?cisloclanku=2008111401&article=1

Heirman, W., & Walrave, M. (2012). Predicting adolescent perpetration in cyberbullying: An application of the theory of planned behavior. *Psicothema*, *24*(4), 614–620. PMID:23079360

Hinduja, S., & Patchin, J. (2006). Bullies move beyond the schoolyard: A preliminary look at cyberbullying. *Youth Violence and Juvenile Justice*, *4*(2), 148–169. doi:10.1177/1541204006286288

Hinduja, S., & Patchin, J. (2007). Offline consequences of online victimization: School violence and delinquency. *Journal of School Violence*, *6*(3), 89–112. doi:10.1300/J202v06n03_06

Hinduja, S., & Patchin, J. (2008). Cyberbullying: An exploratory analysis of factors related to offending and victimization. *Deviant Behavior*, *29*(2), 129–156. doi:10.1080/01639620701457816

Hinduja, S., & Patchin, J. W. (2009). *Preventing cyberbullying: Top ten tips for educators*. Retrieved December 28. 2013 from http://www.nyasp.biz/conf_2013_files/Kelly_Top_Ten_Tips_Educators_Cyberbullying_Prevention.pdf

Hinduja, S., & Patchin, J. W. (2012). Cyberbullying: Neither an epidemic nor a rarity. *European Journal of Developmental Psychology*, *9*, 539–543, doi:.(2012).706448.10.1080/17405629

Hofstede, G., & Hofstede, G. J. (2005). *Cultures and Organizations: Software of the Mind: Revised and expanded* (2nd ed.). New York: McGraw-Hill.

Kerr, O. S. (2009). Vagueness Challenges to the Computer Fraud and Abuse Act. *Minnesota Law Review, 94*, 1561–1587.

Kochenderfer-Ladd, B., & Pelletier, M. E. (2008). Teachers' views and beliefs about bullying: Influences on classroom management strategies and students' coping with peer victimization. *Journal of School Psychology, 46*(4), 431–453. doi:10.1016/j.jsp.2007.07.005 PMID:19083367

König, A., Gollwitzer, M., & Steffgen, G. (2010). Cyberbullying as an act of revenge? *Australian Journal of Guidance & Counselling, 20*(02), 210–224. doi:10.1375/ajgc.20.2.210

Kontostathis, A., Reynolds, K., Garron, A., & Edwards, L. (2013, May). Detecting cyberbullying: query terms and techniques. In *Proceedings of the 5th Annual ACM Web Science Conference* (pp. 195-204). ACM. doi:10.1145/2464464.2464499

Kowalski, R. M., Giumetti, G. W., Schroeder, A. N., & Lattanner, M. R. (2014, February 10). Bullying in the Digital Age: A Critical Review and Meta-Analysis of Cyberbullying Research Among Youth. *Psychological Bulletin, 140*(4), 1073–1137. doi:10.1037/a0035618 PMID:24512111

Kowalski, R. M., & Limber, S. P. (2007). Electronic bullying among middle school students. *The Journal of Adolescent Health, 41*(6), S22–S30. doi:10.1016/j.jadohealth.2007.08.017 PMID:18047942

Kowalski, R. M., Limber, S. P., & Agatston, P. W. (2008). *Cyber bullying*. Malden, MA: Blackwell. doi:10.1002/9780470694176

Kowalski, R. M., Limber, S. P., & Agatston, P. W. (2012a). *Cyberbullying: bullying in the digital age* (2nd ed.). Malden, MA: Wiley-Blackwell.

Kowalski, R. M., Morgan, C. A., & Limber, S. P. (2012b). Traditional bullying as a potential warning sign of cyberbullying. *School Psychology International, 33*(5), 505–519. doi:10.1177/0143034312445244

Law, D. M., Shapka, J. D., Hymel, S., Olson, B. F., & Waterhouse, T. (2012). The changing face of bullying: An empirical comparison between traditional and internet bullying and victimization. *Computers in Human Behavior, 28*(1), 226–232. doi:10.1016/j.chb.2011.09.004

Lenhart, A. (2010). Cyberbullying 2010: what the research tells us. *Youth Online Safety Working Group*. Retrieved March 1 2015, from http://www.pewinternet.org/2010/05/06/cyberbullying-2010-what-the-research-tells-us/

Li, Q., Smith, P. K., & Cross, D. (2012). Research into cyberbullying: Context. In Q. Li, D. Cross, & P. K. Smith (Eds.), *Cyberbullying in the global playground: Research from international perspectives* (pp. 3–12). West Sussex, UK: Wiley-Blackwell. doi:10.1002/9781119954484.ch1

Limber, S. P. (2013). Best Practices in the Prevention of Bullying: An Example of the Olweus Bullying Prevention Program. In M. G. Masiello & D. Schroeder, (Eds.), A Public Health Approach to Bullying Prevention (pp. 127-148). Washington, DC: American Public Health Association.

Livingstone, S., & Brake, D. R. (2010). On the rapid rise of social networking sites: New findings and policy implications. *Children & Society, 24*(1), 75–83. doi:10.1111/j.1099-0860.2009.00243.x

Livingstone, S., Haddon, L., Görzig, A., & Ólafsson, K. (2010). Risks and safety on the internet. In *The perspective of European children: Final findings from the EU Kids online survey of* (pp. 9-16). Academic Press.

Marwick, A. E., & Miller, R. W. (2014). Online Harassment, Defamation, and Hateful Speech: A Primer of the Legal Landscape. *Fordham Center on Law and Information Policy Report*, (2). Retrieved February 2 2015, from http://ir.lawnet.fordham.edu/cgi/viewcontent.cgi?article=1002&context=clip

Maticka-Tyndale, E., & Barnett, J. P. (2010). Peer-led interventions to reduce HIV risk of youth: A review. *Evaluation and Program Planning*, *33*(2), 98–112. doi:10.1016/j.evalprogplan.2009.07.001 PMID:19647874

Mc Guckin, C., Perren, S., Corcoran, L., Cowie, H., Dehue, F., Ševčíková, A., . . . Völlink, T. (2014). Coping with cyberbullying: How can we prevent cyberbullying and how victims can cope with it. In P. K. Smith & G. Steffgen (Eds.), Cyberbullying through the new media: Findings from an international network. New York, NY: Psychology Press.

McKenna, K. Y., & Green, A. S. (2002). Virtual group dynamics. *Group Dynamics*, *6*(1), 116–127. doi:10.1037/1089-2699.6.1.116

Menesini, E., Nocentini, A., Palladino, B. E., Frisén, A., Berne, S., Ortega-Ruiz, R., & Smith, P. K. et al. (2012). Cyberbullying definition among adolescents: A comparison across six European countries. *Cyberpsychology, Behavior, and Social Networking*, *15*(9), 455–463. doi:10.1089/cyber.2012.0040 PMID:22817693

Meredith, J. P. (2010). Combating cyberbullying: Emphasizing education over criminalization. *Federal Communications Law Journal*, *63*(1), 311–340.

Moore, M. J., Nakano, T., Suda, T., & Enomoto, A. (2013). Social interaction and automated detection: Tools in Cyberbullying. In A. Merlo, M. Coccoli, & L. Caviglione (Eds.), Social Network Engineering for Secure Web Data and Services (pp. 67-87). IGI Global.

Mura, G., Topcu, C., Erdur-Baker, O., & Diamantini, D. (2011). An international study of cyberbullying perception and diffusions among adolescents. *Social and Behavioral Sciences*, *15*(2), 3805–3809.

Naylor, P., & Cowie, H. (1999). The effectiveness of peer support systems in challenging school bullying: The perspectives and experiences of teachers and pupils. *Journal of Adolescence*, *22*(4), 467–479. doi:10.1006/jado.1999.0241 PMID:10469511

Olweus, D. (1993). *Bullying at school: What we know and what we can do*. Malden, MA: Wiley-Blackwell.

Olweus, D. (1999). Norway. In P. K. Smith, Y. Morita, J. Junger-Tas, D. Olweus, R. Catalano, & P. Slee (Eds.), *The nature of school bullying: A cross-national perspective* (pp. 28–48). London: Routledge.

Olweus, D. (2012a). Cyberbullying: An overrated phenomenon? *European Journal of Developmental Psychology*, *9*(5), 520–538. doi:10.1080/17405629.2012.682358

Olweus, D. (2012b). Comments on cyberbullying article: A rejoinder. *European Journal of Developmental Psychology*, *9*(5), 559–568. doi:10.1080/17405629.2012.705086

Ortega, R., Mora-Merchán, J. A., & Jäger, T. (2007). *Acting against school bullying and violence. The role of media, local authorities and the internet*. Landau: Verlag Empirische Pädagogik. Retrieved January 40 2015, from http://iamnotscared.pixel-online.org/data/database/publications/618_Acting_against_school_bullying_and_violence.pdf

Ortega-Ruiz, R., & Nunez, J. C. (2012). Bullying and cyberbullying: Research and intervention at school and social contexts. *Psicothema*, *24*(4), 603–607. PMID:23079358

Palladino, B. E., Nocentini, A., & Menesini, E. (2012). Online and Offline peer led models against bullying and cyberbullying. *Psicothema*, *24*(4), 634–639. PMID:23079363

Patchin, J. W., & Hinduja, S. (2007). Cyberbullying: An exploratory analysis of factors related to offending and victimization. *Deviant Behavior*, *29*(2), 1–29.

Paul, S., Smith, P. K., & Blumberg, H. H. (2010). Addressing cyberbullying in school using the quality circle approach. *Australian Journal of Guidance & Counselling*, *20*(02), 157–168. doi:10.1375/ajgc.20.2.157

Paul, S., Smith, P. K., & Blumberg, H. H. (2012). Investigating legal aspects of cyberbullying. *Psicothema*, *24*(4), 640–645. PMID:23079364

Perren, S., Corcoran, L., Cowie, H., Dehue, F., Garcia, D. J., Mc Guckin, C., & Völlink, T. et al. (2012). Tackling cyberbullying: Review of empirical evidence regarding successful responses by students, parents, and schools. *International Journal of Conflict and Violence*, *6*(2), 283–292.

Perren, S., Dooley, J., Shaw, T., & Cross, D. (2010). Bullying in school and cyberspace: Associations with depressive symptoms in Swiss and Australian adolescents. *Child and Adolescent Psychiatry and Mental Health*, *4*(28), 1–10. PMID:21092266

Popovic´-C´ itic´, B., Djuric´, S., & Cvetkovic´, V. (2011). The prevalence of cyberbullying among adolescents: A case study of middle schools in Serbia. *School Psychology International*, *32*, 412–424.

Reynolds, K., Kontostathis, A., & Edwards, L. (2011, December). Using machine learning to detect cyberbullying. In *Machine Learning and Applications and Workshops (ICMLA), 2011 10th International Conference on* (Vol. 2, pp. 241-244). IEEE. doi:10.1109/ICMLA.2011.152

Riebel, J., Jaeger, R. S., & Fischer, U. C. (2009). Cyberbullying in Germany–an exploration of prevalence, overlapping with real life bullying and coping strategies. *Psychology Science Quarterly*, *51*(3), 298–314.

Rivers, I., & Noret, N. (2010). 'I h8 u': Findings from a five-year study of text and email bullying. *British Educational Research Journal*, *36*(4), 643–671. doi:10.1080/01411920903071918

Ryan, T., Kariuki, M., & Yilmaz, H. (2011). A Comparative Analysis of Cyberbullying Perceptions of Preservice Educators: Canada and Turkey. *Turkish Online Journal of Educational Technology-TOJET*, *10*(3), 1–12.

Sahin, M. (2010). Teachers' Perceptions of Bullying in High Schools: A Turkish Study. *Social Behavior and Personality: an International Journal*, *38*(1), 127–142. doi:10.2224/sbp.2010.38.1.127

Smith, P. K., Cowie, H., Olafsson, R., & Liefooghe, A. (2002). Definitions of bullying: A comparison of terms used, and age and gender differences, in a fourteen-country international comparison. *Child Development*, *73*(4), 1119–1133. doi:10.1111/1467-8624.00461 PMID:12146737

Smith, P. K., Mahdavi, J., Carvalho, M., Fisher, S., Russell, S., & Tippett, N. (2008). Cyberbullying: Its nature and impact in secondary school pupils. *Journal of Child Psychology and Psychiatry, and Allied Disciplines*, *49*(4), 376–385. doi:10.1111/j.1469-7610.2007.01846.x PMID:18363945

Smith, P. K., Mahdavi, J., Carvalho, M., & Tippett, N. (2006). *An investigation into cyberbullying, its forms, awareness and impact, and the relationship between age and gender in cyberbullying.* Research Brief No. RBX03-06. London: DfES.

Smith, P. K., & Slonje, R. (2010). Cyberbullying: The nature and extent of a new kind of bullying, in and out of school. In S. R. Jimerson, S. M. Swearer, & D. L. Espelage (Eds.), *Handbook of bullying in schools: An international perspective* (pp. 249–262). New York, NY: Routledge.

Snakenborg, J. A., Van Acker, R., & Gable, R. A. (2011). Cyberbullying: Prevention and intervention to protect our children and youth. *Preventing School Failure*, *55*(2), 88–95. doi:10.1080/1045988X.2011.539454

Soto, C. M., Campos, J. C., & Morales, L. B. (2012). Bullying in Peru: A code of silence. In S. Jimerson, A. Nickerson, M. J. Mayer, & M. J. Furlong (Eds.), Handbook of school violence and school safety: International research and practice. New York: Routledge.

Sourander, A., Brunstein Klomek, A., Ikonen, M., Lindroos, J., Luntamo, T., Koskelainen, M., & Helenius, H. et al. (2010). Psychosocial risk factors associated with cyberbullying among adolescents: A population-based study. *Archives of General Psychiatry*, *67*(7), 720. doi:10.1001/archgenpsychiatry.2010.79 PMID:20603453

Sutton, R. I. (2007). *The no asshole rule: Building a civilized workplace and surviving one that isn't.* New York: Business Plus.

Thomas, H. J., Connor, J. P., & Scott, J. G. (2014). Integrating Traditional Bullying and Cyberbullying: Challenges of Definition and Measurement in Adolescents–a Review. *Educational Psychology Review*, *27*(1), 135–152. doi:10.1007/s10648-014-9261-7

Turner, G., & Shepherd, J. (1999). A method in search of a theory: Peer education and health promotion. *Health Education Research*, *14*(2), 235–247. doi:10.1093/her/14.2.235 PMID:10387503

Valkenburg, P. M., & Peter, J. (2011). Online communication among adolescents: An integrated model of its attraction, opportunities, and risks. *The Journal of Adolescent Health*, *48*(2), 121–127. doi:10.1016/j.jadohealth.2010.08.020 PMID:21257109

Van Royen, K., Poels, K., Daelemans, W., & Vandebosch, H. (2015). Automatic monitoring of cyberbullying on social networking sites: From technological feasibility to desirability. *Telematics and Informatics*, *32*(1), 89–97. doi:10.1016/j.tele.2014.04.002

Vandebosch, H., Beirens, L., D'Haese, W., Wegge, D., & Pabian, S. (2012). Police actions with regard to cyberbullying: The belgian case. *Psicothema*, *24*(4), 646–652. PMID:23079365

Vandebosch, H., & Van Cleemput, K. (2008). Defining cyberbullying: A qualitative research into the perceptions of youngsters. *Cyberpsychology & Behavior, 11*(4), 499–503. doi:10.1089/cpb.2007.0042 PMID:18721100

von Mare´ es, N., & Petermann, F. (2012). Cyberbullying: An increasing challenge for schools. *School Psychology International, 33*(5), 467–476.

Wachs, S., Wolf, K. D., & Pan, Ch.-Ch. (2012). Cybergrooming: Risk factors, coping strategies and associations with cyberbullying. *Psicothema, 24*(4), 628–633. PMID:23079362

Willard, N. E. (2007). *Cyberbullying and cyberthreats: Responding to the challenge of online social aggression, threats, and distress*. Champaign, IL: Research Press.

Ybarra, M., & Mitchell, K. (2007). Prevalence and frequency of Internet harassment instigation: Implications for adolescent health. *Adolesc Health, 41*(2), 89–95. doi:10.1016/j.jadohealth.2007.03.005 PMID:17659224

Ybarra, M., Mitchell, K., Wolak, J., & Finkelhor, D. (2006). Examining characteristics and associated distress related to Internet harassment: Findings from the Second Youth Internet Safety Survey. *Pediatrics, 118*(4), 69–77. doi:10.1542/peds.2006-0815 PMID:17015505

Ybarra, M. L., & Mitchell, K. J. (2004). Online aggressor/targets, aggressors, and targets: A comparison of associated youth characteristics. *Journal of Child Psychology and Psychiatry, and Allied Disciplines, 45*(7), 1308–1316. doi:10.1111/j.1469-7610.2004.00328.x PMID:15335350

Zuchora-Walske, C. (2010). *Internet Censorship: Protecting Citizens Or Trampling Freedom?* Minneapolis, MN: Twenty-First Century Books.

KEY TERMS AND DEFINITIONS

Bullying: The repeated use of force, threat, or coercion to abuse, intimidate, or aggressively dominate others.

Critical Media Literacy: An educational response aimed to expand the notion of literacy to include different forms of mass communication, popular culture, and new technologies.

Cyberbullying: Repeated aggressive, intentional acts carried out by a group or individual using electronic media.

Empathy: The capacity to feel and share another person's emotions.

Information Exposure: The intentional or unintentional disclosure of information to someone who is not explicitly authorized to have access to that information.

On Line Denigration: The spreading of rumours via electronic communication devices.

Olweus Bullying Prevention Program: The most researched and best-known bullying prevention program available today, designed to improve peer relations and make schools safer.

Machine Learning: A subfield of artificial intelligence that focuses on the development of computer programs that can teach themselves without being explicitly programmed.

Power Imbalance: A situation of inequality between the parties which leaves one relatively unable to defend its own interests.

Social Networking Site: An online place where a user can create a profile and build a personal network that connects him or her to other users.

ENDNOTES

1 See also Robert S. Tokunaga, 2010 (p. 278) for conceptual definitions of cyberbullying used in research.

2 In Italy, two girls 15 years old, who, in 2013, murdered a man 67 years aged, so told about it to an unknown boy they casually encountered on the train: "It was like we were within GTA, the video game. We felt like the hero of the game". Retrieved January 1 2014, from http://www.repubblica.it/cronaca/2013/04/10/news/delitto_udine_sequestrate_pagine_fb_delle_ragazze-56337211/

3 Retrieved May 10 2014, from http://bari.repubblica.it/cronaca/2013/12/28/news/cadavere_di_donna_nelle_campagne_l_amante_confessa_sono_stato_io-74647487/?ref=HREC1-8

4 United States v. Drew, 2008 WL 2078622 (C.D. Cal. May 15, 2008).

5 In 2009, Linda T. Sanchez, a Member of Congress from California, introduced the Megan Meier Cyberbullying Prevention Act.51. The proposed statute sought to criminalize "any communication, with the intent to coerce, intimidate, harass, or cause substantial emotional distress to a person, using electronic means to support severe, repeated, and hostile behavior." The Megan Meier act was not approved by congress in 2009 and there have been no new proposed acts in congress since.

6 Retrieved Cecember 20 2015, from http://inchieste.repubblica.it/it/repubblica/rep-it/2013/06/03/news/quando_i_social_network_uccidono_gli_adolescenti_vittima_di_cyberbullismo-60257903/

7 Retrived February 2 2015, from http://www.pewinternet.org/fact-sheets/social-networking-fact-sheet/

8 Beppe Grillo is an Italian comedian and politician. He launched Italy's Five Star Movement, which just won 109 seats in the country's lower chamber of deputies, as well as 54 seats in the senate. His blog is characterized by verbal violence and boorishness.

9 Amanda Lenhart is a senior research specialist at the *Pew Internet & American Life Project* who resides in Washington, D.C; she has published numerous articles and research reports, many of which focus on teenagers and their interactions with the internet and other new media technologies.

10 http://www.pewinternet.org/2008/12/02/facebook-connect-and-a-failure-to-understand-online-identity-management/

11 The uncertainty avoidance dimension expresses the degree to which the members of a society feel uncomfortable with uncertainty and ambiguity (Hofstede & Hofstede, 2005).

12 Retrieved April 10 2014, from http://www.microsoft.com/en-us/download/details.aspx?id=30148

13 Retrieved October 4 2014, from file:///C:/Documents%20and%20Settings/gilberto%20marzano/Documenti/Downloads/WW%20Online%20Bullying%20Survey%20-%20Executive%20Summary%20-%20WW_Final.pdf

14 Retrieved April 10 2014, from http://www.ditchthelabel.org/.

15 Retrieved May 10 2014, from http://www.pewinternet.org/2008/12/02/facebook-connect-and-a-failure-to-understand-online-identity-management/

Chapter 8
Business and Social Media: Collaboration for the Sixth Discipline

Kate Andrews
University of Phoenix, USA

Bethany Mickahail
University of Phoenix, USA

ABSTRACT

During the 21st century, businesses benefit from two key components of effective innovation: social media and collaboration. This chapter provides an in-depth analysis to study the interrelatedness of these components. Featured in this chapter are classroom exercises to deepen the learning of students. By providing multiple recent social media examples, the reality of the integral ways in which social media permeates our lives is delivered. Introduced in the chapter is a sixth discipline, an extension of Senge's five disciplines of collaboration. Through the presentation of a new leadership model based upon the six disciplines, the impact of social media is examined. The conclusion of the chapter contains definitions of the concepts introduced. The use of social media has been and still remains a strategically keen tool in business effectiveness.[*]

INTRODUCTION

Social Media and Business

During the 21st century, businesses have benefited from the use of two key interventions: social media and collaboration. What is included in the context of social media? We can only say what is included today because tomorrow will bring new innovations. Social media uses any form of electronic communication connectivity used to enhance collaboration and sharing of information through the internet. Technical support for social media are electronic tools such as, smartphones, tablets, laptops, and desktops. Technical tools that are digitally connected to social media venues are those such as Facebook, Twitter, LinkedIn, Instagram, Google +, Pinterest, Clouds, and Reddit feeds (Mickahail & Andrews, 2015). These are the well-known social media of today but for tomorrow, the trends reported are wearables,

DOI: 10.4018/978-1-4666-8556-7.ch008

drivables, flyables, and scannables. Present day examples are Google glasses and watches (the purchase and use of wearables has doubled every month since 2012), self-propelled cars (prototypes are running in California), personal drones, and bar code futures (Meeker & Wu, 2013).

There is no question that the use of social media is a strategic direction for business. As the latest innovations enter the market during each business cycle, the reasons for business utilization of social media proliferate. "Consider recent statistics, the number of mobile connected technological devices count more than the number of the entire population of the earth (Cisco, 2013). The average users of cell phones check their text, phone, and Internet contacts on their phones 150 times per day (Meeker & Wu, 2013). Fifty billion devices will be connected to the Internet by 2020 (Vestberg as cited in GIGAcom, 2013). The average North American uses a mobile device 141 minutes a day and 90% of all Internet traffic in 2017 will be video (Advertising Age Mobile Fact Pack, 2013). Globally, the US ranks 10 in the number of new Internet users from 2008-2012 while China, India, Indonesia, Iran, Russia, Nigeria, are all above the US (Meeker & Wu, 2013). From these predictions and facts, it is clear that social media as a corporate tool for profit, is growing exponentially every year" (Mickahail & Andrews, 2015).

Social Media for Business Innovation-Senge's Disciplines and Collaboration

Senge first brought the terminology of disciplines as they relate to ***learning organizations*** to the forefront as the preeminent guru in his 1990 work, *The Fifth Discipline.* Senge's five disciplines are needed for developing modern collaborations for innovation and fostering essential change in business delivery models through social media.

The five disciplines cultivate what Senge (1990) called learning organizations. The disciplines are personal mastery, mental models, shared vision building, team learning that together form systems thinking, which is needed for necessary change. "All five of Senge's core learning capabilities are actionable collaboration in culture, dialogue, and context revealing the potential of globalization in how we interact with social media (Mickahail & Andrews, 2015).

Because Senge's (2013) core learning capabilities are tied together with collaboration, a case can be made that collaboration is the ***sixth discipline*** for 21st century business innovation. Caldecott (2013) wrote that the success of Inventor Thomas Edison's "collaborative culture" nurtured 20th century innovation. Senge's ideas support Edison's in promoting a growing culture of team communication and collaboration and subsequent business innovation that meets consumer needs of the 21st century and beyond.

SECTION ONE: RECENT INNOVATION FOR COLLABORATION, CONNECTIVITY, AND PRODUCTIVITY

Social Media from Telephones to Media Convergence

The US has been a leader in Internet use internationally since 2008 with social media securely anchored as a profitable American business industry. From 2012 to the present, Internet use in the US was overshadowed by the rapid use of the Internet by citizens in China, India, Indonesia, Iran, Russia, and Nigeria on smartphones so that now the US ranks 10th in the world for Internet use (Meeker & Wu, 2013). Media convergence and collaboration brings information from rural third-world farmers to CEOs of Fortune 500 companies. As the price of the smartphone drops, more and more previously unconnected countries will open up vast new opportunities for business to profit from Internet users' access to commercial social media promotions, advertising, and products.

Rural Third World Communities: India and Smartphones

With over 900 million mobile devices, India has 350-400 million Internet connected devices and 150 million smartphones (IDC 2014). In a report from CyberMedia Research, during the first three-months of 2014, 14.5 million Samsung, Micromax, and Nokia brand smartphones were shipped into India. The three mega companies comprised 67% of India's smartphone market.

Naturally, a smartphone boom is anticipated for the rest of 2014 as entry-level smartphones are planned to go for a low 2,500 Indian rupees ($42 US), these early sales numbers may push the smartphone ownership to be available to all in India's market. Many third world citizens without laptops will benefit greatly from connected smartphones (Aulakh, 2014).

Rural Indians and others from developing countries have found the smartphone equipped with social media apps useful for a variety of business and social reasons. Because most people in developing countries lack access to laptop computers, smartphones now fill the gap. Many activities can be accomplished with smartphones where the activities use to demand access to computers. However, in rural communities of the developing world the lack of computers no longer has to limit collaboration and interaction. Smartphones can help farmers by providing weather information and updated market prices for farm related commodities. Indian farmers are able to use Skype to contact agriculture scientists and discuss crop issues. They may also buy and sell needed supplies, such as fertilizers and tools. Smartphones can also assist in promoting rural distance education, by facilitating the use of augmented reality applications to learn collaboration. Connected students can access Massive Open Online Courses (MOOC) classes and engage in online homework groups. They may also play educational games.

Mobile Banking in the Developing World

Several mobile cash options are available in the developing world, such as Airtel money, a common cash alternative service in India. Airtel money may be the future of money and commerce in Asia (Aulakh, 2014). Airtel money is becoming more widely accepted in India and other countries and there are more than 1 million users. Beyond the convenience of using Airtel to transfer money to friends and pay bills, the Airtel's technology app also enables migrant workers, of the many Indian states, to transfer money to their families while providing convenience for Indian workers. Without this service, migrant workers need to wait in bank lines at 6 am in order to transfer money before they start work. The money transferred with Airtel is instantly transferred at any time of day. Another cash option, M-Pesa is the prevailing tech currency in Africa (Aulakh, 2014).

Social Media Apps for Global e-Commerce

An example of the impact of social media benefits to grow and expand business ventures is the potential presented in Facebook's new accelerator, FbStart. Although Facebook has always offered special business benefits to companies featuring apps on Facebook pages, the new FbStart accelerator is designed to help startups create mobile apps at an affordable price. FbStart works by encouraging people to share information from business apps on their Facebook walls. As the sharing network for the business apps expands, the businesses become more reliant on Facebook to grow more business leads, and Facebook becomes harder to replace. Facebook spokesperson Shankar, sees a Facebook partnership with profitable companies seeking to connect with startups, as all benefit from the tools of FbStart apps (Dormehl, 2014). FbStart is the first generation of apps that promote collaboration while breeding more online business ventures.

Technology Trends: Current and Emerging

"A new generation of usercentric, open, dynamic Web, with peer production, sharing, collaboration, collective intelligence, distributed content, and decentralized authority in the background" has become known as Web 2.0 (Sinha, 2012). Web 2.0 includes collaborative components of instant messaging, discussion forums, web conferencing, blogs, wikis, social networking, and social bookmarking.

For emerging trends above and beyond Web 2.0 there are the wearables: Google glasses and smart watches. They provide the impetus for buyers and users of wearables to join the wave that has doubled every month since 2012. The sensory apps are also here. They are specialized apps developed to revolutionize medicine to regulate the user's blood sugar, heartbeat, and measure his or her blood pressure. These specialty apps are incorporated into a smartwatch to assist your fitness goals. Another sensory app, Biorhythm for Sony 2 SmartWatch, monitors a person's health daily. By referring to the user's birth date, the biorhythm program attempts to predict various details of a person's life through simple calculated cycles. This app breaks the cycles into three categories: physical, emotional, and intellectual.

A user may monitor his or her alcohol blood content level. The Sober Up app for Pebble lets him or her count the number of drinks and estimate his or her alcohol blood level. Or perhaps the user would like to try a breakthrough app on Googleglass. With this app, for Photoglassic Memory for Google Glass, the user will never forget. It helps him or her remember and recall people, places, and things by adding them to his or her memory bank with a picture and geo-tag.

The scannables assist business by making a purchase easier to buy with a single swipe of a bar code. The new self-propelled car prototypes are in use providing tours to visitors in Southern and Northern California. Business and personal drones are raising issues about privacy and security rights across the land. In the future, a new era transcending technology convergence will be the ***contextual tech*** era defined by the integration of technology into daily life.

Media Convergence

All the above trending examples are manifestations of ***media convergence***, (when different digital media meet and function on a mobile device). The definition of convergence as it applies to new media was defined by Jenkins (2006) in that

By convergence, I mean the flow of content across multiple media platforms, the cooperation between multiple media industries, and the migratory behavior of media audiences who will go almost anywhere in search of the kinds of entertainment experiences they want.

For businesses in the early 21st century, the most important digital media device is the growing international use of the smartphone. Current discussions revolve around the subject of laptops, which are fading as fast as smartphone users are increasing. The factors that have contributed to this phenomenon are the increasing affordability of the smartphone and ***cloud based Internet connectivity***. A shift is occurring in which the smartphone, a previously inaccessible technology is suddenly affordable and more connected, making them more accessible. This has served the developing economies by bridging a long existing ***digital divide*** between the developed and developing worlds. Presently, Singapore touts the highest level of connectivity per person despite the fact there are few laptops. Kenyans use their phones to transfer money to family across the continent.

Social Media Entrepreneurs in the Developing World

Recent statistics document the proliferation of mobile phone use with the Mobile phone use has reached 80% in the African continent. In 2005, 3% of Kenya's GDP was transferred with mobile money. The number rose to 31% by 2013. African entrepreneurs are benefitting from the promotion social media provides (Slaughter, 2014).

Developing countries' economies are bolstered by micro-business loans from organizations such as Ashoka where craftswomen are encouraged to form collectives around their indigenous craft culture, through weaver, silversmith, textile and mosaic collectives. Western franchises such as Avon, Amway, and Tupperware are creating micro entrepreneur inroads into Asia and Africa. Internet social networks benefit from tight marketing and customer service networks to promote their wares, Mixit of South Africa boasts six and a half million users, while over fifty million Africans access Facebook (Slaughter, 2014).

Section One Summary

Reduction of the digital divide can occur through collaborative social media innovative use. There has been a prevailing concern that those without technical literacy and access to tools would be left in the dust of technological progress. With the availability of smartphones in all corners of the earth, the sky is the limit on what can be achieved through social media collaboration.

Collaboration through social media becomes the new driver of strategic direction in design, production, service, marketing, delivery, and continuous improvement as shown with the MasterCard case. Social media with its exponential growth has become a forerunner in determining strategic direction in companies and personal live. Just as the middle class rose as a result of the Renaissance in Europe, today an emerging new middle class in the developing world is benefiting from an era of disruptive change with global connectivity and collaboration into tomorrow.

Section One: Discussion Review and Small Group Exercises

I. Bridging the International Digital and Economic Divide

Discuss the scenario: How to promote more business profits for a Weavers's Collective in Mumbai, India?
What social media campaign would you use?
How would the collective use the smartphone to manage finances and payments?
How would the collective communicate with prospective buyers from the West?

Figure 1. A weaver's example.

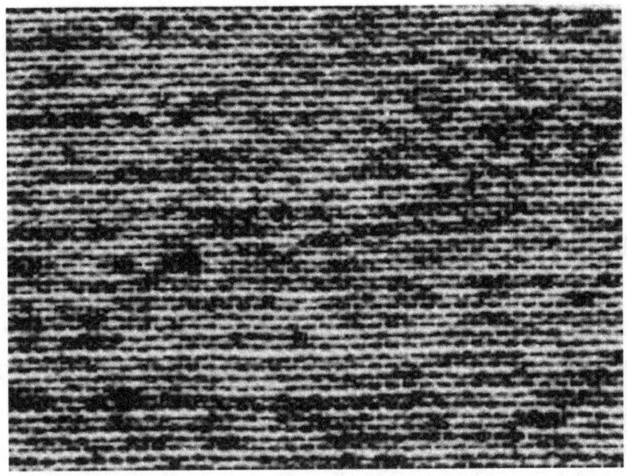

Design a page in Social Media (Facebook, Twitter..) to promote the Collectives weavings
What is special about your product? What audience are you trying to reach?
What are your financial goals?
Explain how social media apps will help grow new business ventures. What dangers exist? Please include as many new concepts introduced in Section One as possible in your answers.

II. Business Orientation Today Requires Strategic Collaboration: The Case of Mastercard

Successful business ventures today require companies to plan using *strategic collaboration* to promote economic sustainability and growth. Many Internet based businesses depend on increasing smartphone use to promote and advertise their products through social media.

Just a few years ago many businesses feared the impact of social media and users who may post negative comments. Although the possibility of negative user comments on social media still exists, savvy businesses have also seized the social media as a chance to build the company brand.

MasterCard developed a social media staff training program to make their employees the company's social media representatives. Each employee is now a MasterCard brand ambassador. Those who are not *digital natives* at MasterCard are trained by *digitally literate* employees so all may participate and reduce the *digital divide.* These changes required new collaboration among the employees. MasterCard's social media training program was initiated with Conversation Suite, a social media listening program created to highlight the multitude of digital interactions about the MasterCard brand (Meister, 2014). Social media has further increased the company's employee collaboration as MasterCard's brand ambassadors, promote the company's success. Conversation Suites is a main piece of a four part strategy to educate, empower and encourage the 8,000 MasterCard employees in this endeavor.

SECTION TWO: COLLABORATION, SOCIAL MEDIA, AND CLOUDS

Six Disciplines Needed for Learning Organizations Ready for 2020 and Beyond

Senge (1990) has traditionally been "anointed" as the organizational leadership guru since publication of his book, *The Fifth Discipline*. The premise of Senge's (1990) work is a learning organization is vital not only for business sustainability but for businesses to be organizations excelling in accomplishing organizational goals. Furthermore, to be a learning organization, five disciplines must be in existence (Senge, 1990). The four initial disciplines Senge (1990) identified were personal mastery, mental models, shared vision building, and team learning. These four are linked together with the advent of the fifth discipline: systems thinking-the ability to make connections clearer and to find the leverage points in a system (Senge, 1990). An effective system is a union of components linked together to accomplish a shared goal. Systems thinking brings together the other four disciplines to guarantee a thriving organization brought about by a fusing of theory into practice. However, without the preliminary four disciplines, systems thinking becomes hollow and devoid of the force required for excelling organizations. Without these five disciplines, organizations cannot continue to grow and to excel in accomplishing strategic goals (Senge, 1990).

A behavior necessary for the five disciplines to be congealed into one piece of art, collaboration, a participatory process, is required (Senge, 2013). Personal mastery cannot be accomplished in isolation but requires association with others for knowledge and wisdom to grow (requiring collaboration). Mental models are the schema by which the world is understood. To understand one's mental models, there is the necessity of holding a mirror pointed inwardly revealing the internal pictures that must be brought to the light and analyzed for goodness of fit in the equations of life. To analyze and determine if mental models are working effectively, they must be exposed and objectively critiqued from various viewpoints (requiring collaboration). By definition, vision building that is shared can only be accomplished through input and consensus with others. Building a shared vision is at the center of leadership and employee success (requiring collaboration). Learning with teams is a fundamental concept that emphasizes the need for an end goal of group cohesion. Team learning is initiated with dialogue in developing the expertise needed to bring the vision into fruition (requiring collaboration). System thinking is nearly impossible with only individuals trying to see whole patterns. Instead related forces are necessary to see the interconnectedness of the parts (Senge, Scharmer, & Winslow, 2013). Every part's action influences the actions of other parts just as when humans are part of the equation, everything and everyone is connected (requiring collaboration).

Therefore, as all five disciplines in order to function require collaboration, collaborative behavior could be perhaps considered a sixth discipline. This sixth discipline is a necessary component linking together all of the other disciplines in moving into the decade of the 2020s and beyond.

In recognizing how the five disciplines are related to social media, one sees the natural ability of social media in bringing people together. In achieving the sixth discipline, a learning organization sees the digital, virtual world of relationships being built effectively for use in developing social media. The potential of globalization is contained within the culture (way we interact), dialogue (technical abilities), and context (technical accessibility) of how we collaborate with social media.

Figure 2. Three legs of a learning organization: Information, Formal Learning, and Collaboration are all needed for the stool to stand supporting a learning organization.

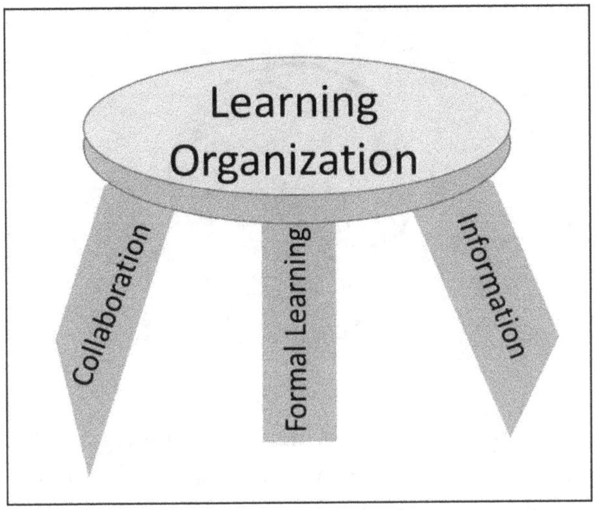

Collaboration has been known to be a key characteristic of innovation since the time of Thomas Edison, who consistently collaborated even into his 8th decade with others in generating ideas for consideration as components of innovation. Edison saw collaboration as a vital component required for continued innovation of a learning organization. A learning organization cannot stand without the three legs of information (job aids, webinars, podcasts, knowledge portals), formal training (self-paced e-learning, virtual classroom learning), and collaboration (blogs, wikis, networking, chat, surveys; see Figure 2).

How can collaboration be achieved in the global, digital world existing today? The answer is in front of us all, through effective use of social media. Social media, therefore, becomes a necessary component in learning organizations, brought to the forefront of importance in acknowledgement of the potential for both synchronous and asynchronous collaboration. The lack of geographic boundaries presents the opportunities for a collaboration infrastructure that is conducive to learning organizations in the generation of knowledge work.

Google, Amazon, Microsoft, and Apple exemplify a corporate culture in a creative collaboration model that stagnate businesses seek to emulate. This paradigm shift promotes and stimulates employee creative collaboration resulting in innovations and greater profitability in a thriving 21st century corporate culture.

Social media, instead of prescribing actions, brings suggestions of actions to users wherein communication is not just direct but collaborative and where consideration of the utility of tangential directions of communication are recognized as necessary for creativity occurring through these interactions that are unpredicted. The resulting tension from wondering off a path of defined dialogue can be loosened by drawing reality toward the vision and snapping the vision back to reality.

What is up in the Clouds? Any Information Needed, Wanted, or Desired

Where is all of the information going from business and social media (see Figure 3)? The trend is up and into the clouds, clouds of computing solutions. Information technology accessed via the Internet can be called *Cloud*. A cloud has four primary categories: Infrastructure, Database, Platform, and Software.

Figure 3. Where is all the information going?

Each of the categories can be differentiated by a) which users interface with the category, b) having technologies that are transparent, and c) no technology can be accessed from the level above (see Figure 4).

For the user, data is accessible from anywhere and the hardware used to access the data is interchangeable, meaning the user is not limited by the specific machine being used to access the database. To the user, data access appears free of downtime and errors because of the technology's infrastructure of elasticity and scalability. Specialized apps will connect tomorrow's media to the data of today due to the broad platform compatibility of the apps.

The Future is Contextual Technology-How will Business Change to Meet the Consumer?

Georgia Tech researchers Anind Dey and Gregory Abowd defined ***contextual tech*** in 2001. They described ubiquitous all-present computers, with abilities to sense the objective and subjective features of a stated scenario. Contextual computers will promote the ability to understand and act in the moment

Figure 4. Four categories make up the cloud computing system.

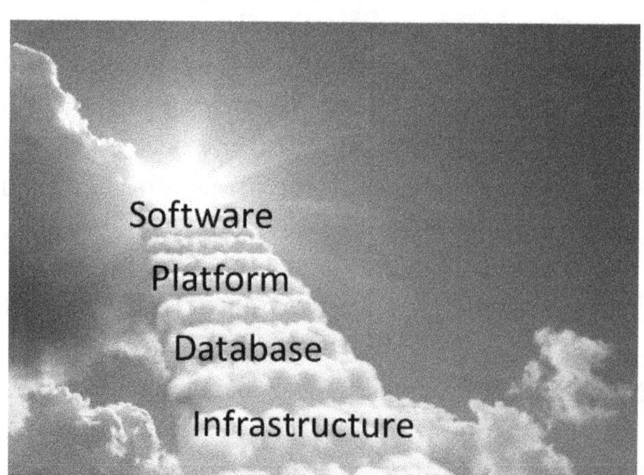

based on where we are, who we're with, and past experiences. These advanced artificial intelligence devices would serve as our sixth, seventh, and eighth senses.

Data graphs make it work. Jump, a business strategy firm, has identified four data graphs essential to the rise of contextual computing: social (connections), interest (likes, curiosities), behavior (what's done), and personal (all beliefs; Mortensen, 2014). These four data graphs are tracked by companies to market products that match the user's profile online. The personal and behavior graphs are not as well developed but social media is finding a way to harness this information for global business development. In a decade, it's predicted *wearable intelligence* will be available.

Section Two Summary

As Senge (2013) reported, collaboration is central to learning organizations and a requisite factor and catalyst of organizational change. Therefore, it is proposed that collaboration be the sixth discipline of a learning organization. The inherent collaborative mechanism of social media makes its use intrinsic to a learning organization. In accessing today's data, cloud computing is becoming more and more the trend. Cloud computing allows for the access of data through machine interchangeability. The future holds limitless possibilities with technological advances today and tomorrow.

Section Two: Discussion Review and Small Group Exercises

I. A New Business Venture: Utilize Social Media for Collaborative Learning

Discuss

Even though Social Media presents many benefits, what disadvantages exist for Businesses? How can product ratings and reviews impact business ventures?

Design your group venture complete with Social media apps.

Explain how the apps will help grow your new business. What dangers exist?

II. Oreo at the Super Bowl: A Sweet Tweet

Ad Age (2013) recognized the digital Oreo ad campaign with an A lister Award for its popular Super Bowl strategy. When the lights went out at the 2013 Super Bowl, Oreo earned the spotlight with the words: "You can still dunk in the dark." It was a "sweet tweet" that grabbed national attention and multiple awards.

Discussion questions after viewing link: https://twitter.com/Oreo/status/298246571718483968

1. Did you see this tweet? Or did someone you know see this tweet around the 2013 Super Bowl?
2. What made this tweet so powerful?
3. How is this ad an example of business use of *media convergence*?
4. Is the ad effective? Why?
5. Would this campaign work in Japan or China or Nigeria? Why, or why not?

Figure 5. You can still dunk in the dark.

Follow up exercise: Small group huddle! Develop a Twitter Campaign!

In small groups discuss how you would develop a Tweet campaign for your favorite food brand, (Cheetos, Doritos, Kashi, etc.).

What would you emphasize in seven to nine words? How would you launch your campaign to ensure success? Share your ideas with your group and then with the class through Twitter.

REFERENCES

Akella, D. (2003). *Unlearning the fifth discipline*. New Delhi, India: Response Books.

Amitabh, A., & Sinha, S. (2012). The learning continuum: Formal and informal learning experiences-enabling learning and creation of new knowledge in an organization. *International Journal of advanced Corporate Learning, 5*(2), 10-14. doi:10.3991/ijac.v5i2.2111

Andriole, S. J. (2012). Seven indisputable technology trends that will define 2015. *Communications of the Associations for Information Systems, 30*(4), 61-72. Retrieved from http://aisel.aisnet.org/cais/

Aulakh, G. (2014). *Philips to re-enter Indian smartphone market.* Retrieved from http://timesofindia.indiatimes.com/tech/tech-news/Philips-to-re-enter-Indian-smartphone- market/articleshow/35614928.cms

Billington, M. G., & Billington, P. J. (2012). Social media tools for leaders and manager. *Journal of Leadership, Accountability and Ethics, 9*(6), 11-18. Retrieved from http://www.na-businesspress.com/jlaeopen.html

Bitcurrent. (2011, February 4). *How to think like a cloud: Rules of the new on-demand architecture.* Retrieved from http://cloud-computing-library.cloudconnectevent.com/download-preview?asset=/1384/how-to-think-like-a-cloud

Caldicott, S. M. (2013). *Midnight lunch: The four phases of team collaboration success from Thomas Edison's lab*. Hoboken, NJ: John Wiley & Sons.

Chun, S. A., & Warner, J. (2010). Finding information in an era of abundance: Towards a collaborative tagging environment in government. *Information Policy, 15*, 89-103. doi:10.3233/IP-2010-0201

Deans, P. C. (2012). Integration of study abroad with social media technologies and decision-Making applications. *Decision Sciences Journal of Innovative Education, 10*(3), 299–336. doi:10.1111/j.1540-4609.2012.00344.x

Dey, A. K., Salber, D., & Abowd, G. D. (2001). A conceptual framework and a toolkit for supporting the rapid prototyping of context-aware applications. *Human-Computer Interaction (HCI) Journal, 16*, 97–166. doi:10.1207/S15327051HCI16234_02

Dixon, B. E. (2010). Towards e-government 2.0: An assessment of where e-government 2.0 is and where it is headed. *Public Administration & Management, 15*(2), 418-454. Retrieved from http://www.spaef.com/pam.php

Dormehl, L. (2014). *Everything you need to know about Facebook's new app accelerator.* Retrieved from http://www.fastcolabs.com/3030615/everything-you-need-to-know-about-facebooks-new-app-accelerator

Goodwin, B. (2013, September). Macmillan drives staff collaboration with cloud-based social media tools. *Computer Weekly.com,* 24-30. Retrieved from http://www.computerweekly.com/

Jenkins, H. (2006). *Convergence culture: Where old and new media collide*. New York, NY: New York University.

Kitchens, R., & Torrence, P. D. (2012). The JOBS Act-crowdfunding and beyond. *Economic Development Journal, 11*(4), 42–47. doi:10.5755/j01.em.18.1.3713

Meeker, M., & Wu, L. (2013, May 19). *Internet trends: D11 conference*. Retrieved from http://www.theguardian.com/technology/blog/2013/may/30/mary-meeker-2013-internet-trends

Meister, J. (2014, March 26). Social media: Moving from danger to brand building opportunity. *Forbes Magazine*. Retrieved from http://www.forbes.com/sites/jeannemeister/2014/03/26/social-media-moving-from-danger-to-brand-building-opportunity/

Mickahail, B., & Andrews, K. (2015). Collaborative innovative leadership and emerging social media technologies. In C. Luntz (Ed.), The Refractive Thinker® – An Anthology of Higher Learning: Effective Business Practices in Leadership and Emerging Technologies (vol. 9). Las Vegas, NV: The Refractor's Press.

Mortensen, P. (2014). *The future of technology is contextual.* Retrieved from http://www.fastcodesign.com/1672531/the-future-of-technology-isnt-mobile-its-contextual

Page, D. (2012). Using electronic portfolios. *Performance Improvement, 51*(8), 34–36. doi:10.1002/pfi.21296

Senge, P. (1990). *The fifth discipline: The art and practice of the learning organization*. New York, NY: Doubleday.

Senge, P. (2013, June 14). *Real collaboration takes more than meetings and powerpoints*. Retrieved from http://nbs.net/real-collaboration-takes-more-than-meetings-and-powerpoints/

Sinha, A. A. S. (2012). *The learning continuum: Formal and informal learning experiences-enabling learning and creation of new knowledge in an organization*. International Journal of Advanced Corporate Learning. doi:10.3991/ijac.v5i2.2111

Slaughter, C. (2014). May the force be with you: Converging innovations will empower entrepreneurs in the developing world. *The Guardian*. Retrieved from http://www.theguardian.com/global-development-professionals-network/partner-zone-path/entrepreneurs-innovation-developing-world

ADDITIONAL READING

Logofalu, M. A. C. (n.d.). The social media impact on small and medium sized businesses. *The Young Economists Journal*, 214-218.

Media convergence-opportunities and challenges. (2011). Retrieved from http://mconvergence.wordpress.com/category/future-of-media-convergence/

Nicholas, D., & Rowlands, I. (2011). Social media use in the research workflow. *Information Services & Use*, *31*, 61–83. doi:10.3233/ISU-2011-0623

Schirr, G. R. (2013). Community-sourcing a new marketing collaboration in social media. *Marketing Education Review*, *23*(3), 225–240. doi:10.2753/MER1052-8008230302

Senge, P., Scharmer, O., & Winslow, D. (2013). 30 years of building learning communicates: A dialogue with Peter Senge, Otto Scharmer and Darcy Winslow, Part I. *Reflections: The SoL Journal*, *12*(4), 40–52.

Spencer, R. W. (2012) A pervasive model for participation in voluntary forums. *Research Technology Management,* May-June, 23-31.

Taranu, A. M., Paun, E., & Florea, N. (2011). The ethics of participation, sharing and collaboration in networked media cultures: New digital technologies and participatory models of knowledge production. *Economics, Management, and Financial Markets, 6*(2), 353-372. Retrieved from http://www.addletonacademicpublishers.com/economics-management-and-financial-markets

Valdez, A. C., Schaar, A. K., & Ziefle, M. (2012). State of the (net)work address developing criteria for applying social networking to the work environment. *Work (Reading, Mass.)*, *41*, 3459–3467. doi:10.3233/WOR-2012-0625-3459 PMID:22317248

Wankel, C. (2009). Management education using social media. *Organizational Management Journal*, *6*(4), 251–262. doi:10.1057/omj.2009.34

KEY TERMS AND DEFINITIONS

Cloud Based Internet Connectivity: Or connectivity-cloud is another word for internet and cloud connectivity that holds the connotation of enterprise networking without businesses having to purchase server hardware and software. It does not have delivery of a product but of a service.

Collaborative Culture: Involves a commitment to and action for the needs of the organization and individual without competitive actions.

Contextual Tech: Involves ubiquitous computers, utilizing situational computerized cookie type technology. Tech that is able to sense the objective and subjective aspects of a given situation. A way computers will extend human ability to perceive and act in the place where they are, who they're with, and their past experiences. It is specialized machine intelligence that will serve as humanity's sixth, seventh, and eighth senses.

Digital Divide: Is a gap of economic and social inequality based on human access to, use of, or knowledge of information and communication technologies. The divide within countries, such as the digital divide in the United States, refers to inequalities between individuals, households, businesses, or geographic areas, usually at different socioeconomic levels or other parts of the country.

Digital Native: Is a concept authored by American writer Marc Prensky in 2001 to describe the global' dominant generation of post millennial. Prensky defines digital natives as those born into an innate "new culture."

Digitally Literate: Describes those with the skills to use information and communication technologies to find, evaluate, create, and communicate information, requiring both critical thinking and technical skills.

Learning Organization: Is an entity composed of individuals adept at generating, gathering, and conveying knowledge in order to create what they want to create.

Media Convergence (MC): Is the merging of computing and information technology companies, telecommunications networks, and content providers from the publishing worlds of newspapers, magazines, music, radio, television films, and entertainment software. MC brings together the "three Cs"-computing, communications, and content.

Tech Convergence: Is the process by which existing technologies merge into new forms that bring together different types of media and applications.

Sixth Discipline: Is collaboration as the next discipline necessary for a learning organization to exist to its fullest capacity.

Wearable Intelligence: Is machine intelligence that can be worn. It is the next step in the evolution of computing platforms (from desktop to mobile to wearable).

ENDNOTE

* First published under Mickahail, B., & Andrews, K. (2015). Collaborative innovative leadership and emerging social media technologies. In C. Luntz (Ed), *The Refractive Thinker® – An Anthology of Higher Learning: Volume IX: Effective Business Practices in Leadership and Emerging Technologies*. Las Vegas, NV: The Refractor's Press.

Chapter 9
The Social Media "Information Explosion" Spectacle:
Perspectives for Documentary Producers

Friedrich H. Kohle
Edinburgh University, UK

ABSTRACT

The semantic web, social media and the amount of user-generated content continues to grow at a staggering rate. Social Media significantly contributed to the information flow during the Arab Spring, the Occupy and Wall Street movement continue to maintain a global online presence using social media technology. But is the social media information explosion really a unique event in media history? How did story telling evolve into social media? In order to place social media in its historical context and anticipate digital native expectations, we explore the origins of narrative and storytelling from the perspective of a documentary producer. How did past media technologies prepare the way for social media? How do digital natives perceive the world via social media and what do they expect from today's documentary producer? What are the viewing habits of digital natives? What do previous 'information explosions' have in common with social and digital media? These are essential questions for media and documentary producers navigating their way through the vast maze of social media technology and the semantic web, in addition to traditional media.

1 THEORETICAL FRAMEWORK

Social media is defined by Kaplan & Haenlein (Kaplan, Andreas M. & Haenlein, Michael 2010). Readers should be familiar with the work of Marshall Mcluhan's (1994) media theory and Briggs and Burke's (2009) history of media.

DOI: 10.4018/978-1-4666-8556-7.ch009

2 BACKGROUND

A physicalist approach was taken to investigate the origins and evolution of media technology and the impact of social media on documentary filmmaking. However, a reductionist explanation as to how social media moves us, remains elusive. A dualist position offers a better explanation into phenomena such as catharsis, but still leaves many questions unanswered. As a result, the author of this paper has come to view social media as an extension of our senses (Mcluhan 1994) and consciousness. What exactly is it that touches our inner world when watching a flash mob play 'Here comes the Sun' in a Spanish unemployment office? (Carne Cruda, 2012), Neuroscientists can measure the activity of our brain during catharsis, yet they cannot determine what *it* is that moves us to tears during those moments. Here we conclude that the 'hard problem' of consciousness (Chalmers, 1995) equally applies to social media *content* as an extension of our consciousness, and as such its deeper purpose remains hidden. Yet throughout history and across all media technologies, storytelling has been the key driving factor. Without compelling content there is no social media.

3 INTRODUCTION

The Internet and social media are frequently referred to as the 'information explosion'. Social Media is ubiquitous and significantly contributed to information dissemination during the Arab spring. It continues to be a major platform used by grass root movements such as Occupy and Wall Street. It is swiftly becoming a tool documentary filmmakers rely on during all production phases, i.e. 'Mon Petit', (Mon Petit, 2012) which tells the story of wheelchair-bound Albert Casal, hitchhiking from his native Barcelona to New Zealand. The producers of 'Mon-Petit' have learned how to apply social and digital media to develop, produce and distribute content. Without social media the production of 'Mon-Petit' would not have been possible in this form. Governments too are learning to use social media, but for different reason: propaganda and surveillance. Authorities find it increasingly difficult to censor the amount of user-generated content online when compared to traditional forms of documentary production. Facebook now exceeds 1 billion members (Shaughnessy, 2012).

But is social media really a unique 'information explosion'? Investigating media history from a physicalist perspective, we draw comparisons to today and put social media into its historical context in order to explore current perspectives for documentary producers. Are we witnessing a paradigm shift or another evolutionary step in the development of media history? Can a reductionist theory fully explain the impact of social media? The documentary "God, Church, Pills & Condoms" offers insight into the methods used to develop, produce and distribute content online and examine online behavior of stakeholders and audiences. A survey conducted among 2nd year undergraduate students at NHTV (NHTV, 2013), University of Applied Sciences, Breda, Netherlands sheds further light onto changing viewing habits of digital natives. Results of this paper provide media and documentary producers with an overview of social media in its historical context, a better understanding of audience behavior and insight into emerging trends.

4 RESEARCH METHODS

We investigate the evolution of social media into its current form using a variety of multi-disciplinary literary sources. We applied and tested our findings using our documentary "God, Church, Pills & Condoms" (Kohle. F, Cuevas. A, 2013), examine online behavior in today's video-verse and compare results with research papers in the field for further discussion, including a survey conducted among 2nd year undergraduate students at NHTV, University of Applied Sciences, Breda Netherlands which investigates changing viewing habits among digital natives.

5 RESULTS

Social Media and the information explosion: Terms such as *the* "information explosion" are often used to describe the impact social and digital media has on society and culture. *This* term is *by no means* a new *term:* it was first coined during the print revolution (Briggs/Burke 2009). *Instead the term 'information explosion' is better defined as a* recurring pattern throughout history. However, social media significantly distinguishes *itself from* previous information revolutions in a number of ways:

1. *The social media revolution is moving forward faster in time than any other previous 'information explosion'.* The rate of growth of social media its and corresponding public spheres (Habermas 1974) is expanding at an exponential rate globally.
2. *Social media no longer requires editorial approval by an elite, nor can censorship effectively be applied by a single state (Kohle 2012), (Bondebjoerg 2009).* Digital natives are used to not only consume content; they are capable of generating an increasing amount of online content without effective censorship.
3. *Social Media is part of a paradigm shift of similar magnitude as the invention of language (Dunbar 2003), (Corbalis 2009).* The invention of writing, printing and broadcast technologies, can be interpreted as an evolutionary process rather than a paradigm shift, whereas social media encompasses all previous forms of media.
4. Viewing habits are changing (Kohle/Cuevas 2010, 2012): *digital natives are quickly adapting to the opportunities the participatory web and social media offers.*
5. *Content development, business models, production and distribution strategies are transforming at an ever-increasing speed (Puttnam 2013).* Producers and directors are challenged to constantly evaluate and develop content and new ways of reaching their audiences.
6. *Digital natives expect instant dialogue with the makers of a programme.* Projects need to feature a number of narratives across various platforms to really meet the growing expectations of their audiences. The narrative among digital natives is rapidly developing and needs to be better understood by producers.
7. *Programme-maker need to carefully consider their target audiences, small or big, and design the programmes relevant transmedia strategy accordingly.* A linear narrative in the form of a film still features prominently in the media vocabulary of a digital native; but digital natives also expect to engage with gamified and transmedia content as well as opportunities for feedback and contribution.

6 DISCUSSION

Origins of Social Media

By nature stories require language, writing and images. Without these elements we lack the means to tell stories and engage with social media. We examine evolution of these elements from prehistory to present day from a physicalist position to provide a contextual background. Is there a repeating pattern in the evolution of the above elements? How does technology influence story telling? What can we learn from media history?

Language

Children have an "Instinct of Inventiveness" (Locke/Bogin 2006) enabling them to invent language even if linguistically deprived. As speech evolved our brains adapted to this new tool. Memory improved as we developed the capacity to speak and refer to events in the past. This ability for 'Mental Time Travel' requires two types of memory: semantic and episodic memory. Episodic memory (Tulving 2002) is more interesting to us as by nature, it lends itself to a narrative structure. The concept of time provides the basic elements needed for story telling: past, present and future. The most basic definition of a story requires a beginning, middle and end with the narrative unfolding over a period of time. Without the invention of language these basic elements needed for a narrative would not exist.

Narrative is needed for all consequent forms of media – and for that reason we submit that language was a paradigm shift, a real game changer in the history of storytelling. *Oral* story telling remained the key instrument of communicating information over distances and time for mankind until writing was invented. Stories passed on verbally over generations maintained their basic plot structure accurately but adapted to the cultural and social circumstances of the time (Cvorovic 2006). The invention of language can be considered the first prehistoric 'information revolution'. The ability to use tools while explaining their use, led to the development of pedagogy (Corbalis 2009) – a key tool regarding the dissemination of information.

Rock Art, Cave Paintings and Early Writing

as speech and language continued to evolve, man began to explore new and more abstract forms of expression and communication, such as rock art and cave paintings. In February 2012 cave paintings were discovered in Spain, which are estimated to be 42000 years old (Pike/Hoffmann 2012). Archaeologists and Anthropologists are debating if these cave paintings were created by the first anatomical modern humans or Neanderthals. The paintings depict hunting scenes, animals, the night sky and silhouettes of hands. Unlike speech, which once spoken leaves no permanent record other than in our own memory, rock art and cave paintings remain visible, forming a permanent external record.

Whichever species left behind some of the earliest records of human creativity, the question remains as to what motived early cave dwellers to decorate their environment in this way. 26 cave symbols (Ravilious 2012) are shared across caves in southern France by early man, implying that this 'creative explosion' took place earlier, perhaps even before the arrival of modern man in Europe. From rock art and cave paintings it is but a small step towards proto-writing (Merola 2007), (Lauinger 2011): a picture system dating back 30-35000 years. (See table below)

Dunbar provides evidence suggesting that the human brain, when compared to other hominids, is capable of a social network encompassing some 200 contacts: a size inadequate for prehistoric permanent settlements exceeding that number. It became necessary for early man to overcome this biological limitation for cognitive evolution by cultural means (Watkins 2010), i.e. monumentalism and *storytelling*. This enabled early men to evolve from a hunter-gatherer group and develop the agricultural skills needed to form permanent settlements.

The formation of tribal settlements undergoing a cultural and spiritual process can be found in early cave paintings: some of the cave paintings in Southern France and Spain are located deep inside cave systems with no natural light available. To venture that deep into the caves, the creators of those paintings deliberately sought out a location isolating them from the outside world. It is reasonable to argue that this is intended because the creators of those paintings wanted to invoke a spiritual experience, an experience not unsimilar to the one a contemporary cinema induces on today's cinema visitor.

With external influences reduced to a minimum the viewers' attention is drawn entirely to the big screen – or cave wall enabling catharsis. A physicalist position reaches the limits of what can be explained at this point. Why is catharsis necessary? How can physicalists explain what it is like to experience catharsis? Media, as an extension of our consciousness, cannot be explained by reductionism alone. Werner Herzog's "The Cave of forgotten Dreams" (Herzog, 2012) investigates the Chauvet cave paintings in Southern France, highlighting just how deep the paintings are located inside the cave. In consequent interviews Herzog also refers to Chauvet as a form of 'proto-cinema'.

The role of the storyteller and painter changed significantly: he or she is no longer master of a narrative forming part of an invisible oral domain. The creators of cave paintings are very much in control of a physical and artful process projected onto a cave wall: a process that repositions the painters and their role as storytellers within the tribe. Karl Marx and Friedrich Engel's Communist Manifesto (Marx & Engels, 1969,p. 98-137) offers an interesting perspective on the evolution of early communes and elites and the consequent class struggle, which persists until today. Critics are quick to point out that Marxism neglects the spiritual dimension in man (Brentlinger 2000) already on display in those early cave paintings.

The Alphabet

documenting history, accounting for resources, trade agreements and accounts of battles lost and won, eventually led to the invention of the Alphabet. (Andis, 2012) This in turn gave rise to an 'information revolution' in antiquity. Who exactly invented the alphabet is still subject to discussion (Goldwasser 2010), but we know that Phoenicians played a major part in spreading the alphabet across the Mediterranean Sea as part of their trade activities.

The alphabet gave rise to a new form of writing: Greek writers repositioned the storyteller with a strong philosophical perspective. Antiquity gave the storyteller a voice, which did not exist before; a form of storytelling that is still found today, i.e. auteur cinema finds its early roots in Greek narrative history. Conflict and war are important *factors in* the evolution of media. As tragic as it seems: without war and conflict, media would not have developed into its current form. Conflict in the Middle-East has been described by many sources (biblical, Egyptian, Phoenician, Greek) for thousands of years. Between

Table 1.

	Approximate Dates Years ago	Event
Lower Paleolithic	2.5 million – 400.000	Invention of Language: 1st 'Information Explosion"
Middle Paleolithic	200.000 – 45.000	Arrival Homo Sapiens: "Creative Explosion"
Upper Paleolithic	45.000 – 10.000	Rock art, Cave paintings: 2nd "Information Explosion"
Neolithic	10.000 – 7.000	Invention of Proto-writing: 3rd "Information Explosion"
Chalcolithic	7.000 – 3.500	Cuneiform writing
Antiquity	3.500 - 600	Invention of the Alphabet: 4th "Information Explosion"
Medieval	600 - 1500	
Early Modern	1500 - 1800	Invention of the Printing Press: 5th "Information Explosion"
Industrialization	1800 - 1900	Invention of Broadcast Technologies: 6th "Information Explosion"
20th Century	1900 - 2000	Invention of the Internet & Social Media
21st Century	2000 - now	Social Media: 7th and current "Information Explosion"

700 and 588 BC nineteen Mediterranean conflicts are accounted for (Conflict History, 2012), with the amount of conflicts steadily increasing in recorded history.

The *below* table provides an overview of the various past media-'revolutions'. There is much debate about the exact timing of the *given dates.* Evidence suggests that Neanderthals were capable of speech including the creation of rock art and cave paintings. Even hominids such as Bonobo apes (Benson, J, Greaves, W., O'Donnell, M, Tagliatlatela, J. 2002) have demonstrated the potential for language, though this remains a hotly debated topic among linguists. Creativity is not a trade exclusively reserved to Homo Sapiens, though some members of the species Homo Sapiens would like to claim otherwise (Pagel 2011).

The Printing Press

It would take a few more millennia before the arrival of the next major 'information explosion': Gutenberg's printing press. Even though printing technology had been developed a millennium before Gutenberg in China, *t*he alphabet combined with Gutenberg's press, suited the literate European elite and middle classes in the 14*th* and 15*th* century Europe best, at a the time when Europe was ravaged once again by conflict.

Culturally and socially Europe was starving for new ideas, which eventually led to European Reformation and Enlightenment. The printing press also gave rise to a new problem: copyright infringement. Briggs and Burke (2009) provide an interesting account of the cultural and political background during enlightenment. To avoid prosecution Amsterdam printers named a non-existent Cologne printer as the publisher of one of the most printed books of all times: the bible. The bible would not remain the most popular book, soon romantic stories and poetry would generate revenues in the form of James Thomson's "The Seasons" published by Alexander Donaldson. Thomson's poetry soon became a copyright landmark case: on February 22, 1774, the House of Lords in the United Kingdom decided in the case of Donaldson v. Becket (Rose 1988) that copyright is not perpetual, but limited in duration, thus leading to the development of today's copyright law.

It can be argued that the media tools of creativity, used by the best writers and artists *of the time* to create prose and art, such as the alphabet and the printing press, are byproducts of thousands of years of Middle-Eastern and European conflict. Unlike language though, print technology was controlled by the elite. As power struggles continued to take their toll across Europe, the next information revolution would take place this time not thousands, but merely a few hundred years later. With each 'information revolution' the arrival of the next revolution occur*r*ed sooner, such as the invention of broadcast technologies.

Europe was beginning to lose grip on her colonial domains and once again she was ravaged by major world wars, as broadcast technologies evolved from Marconi's radio transmitter into *Film and* Television.

The scientific process, which made broadcast technologies possible, spanned some 200 years; a very short time when compared to the time needed to invent speech, writing, the alphabet and printing. During that time the world experienced nearly 900 documented conflicts, double the amount accounted for during Gutenberg's period. History tells us that past information revolutions are not associated with less but more war and conflict. The Internet too is a military invention, with its origins in the cold war: Arpanet (Mowery 2001). Since 1990 the world has experienced 322 documented conflicts, rising in speed and occurrence along with the 'new' information revolution: the Internet and social media. Conflict and media are intrinsically linked, and during times of war new forms of media technologies emerge as a matter of life or death.

As was the case with the invention of language itself, writing and the Alphabet, printing and broadcast technologies, social media requires the audience to develop a new form of media literacy, giving rise to different forms of social, political and cultural awareness. Unlike previous media revolutions, social media encompasses all of the above, providing a platform almost entirely reserved for user-generated content. Content is available instantly, globally and on a variety of screens and devices. 'Mon Petit' and 'God, Church, Pills & Condoms' are excellent examples of how social media provides a platform to guide and shape user generated content during development, production and post-production stage. Social Media was the exclusive tool used to distribute 'God, Church, Pills & Condoms'; consequently it was accepted into Cinemalaya and the Hanoi Film Festival in 2012.

How are Digital Natives Experiencing Social Media?

A total of 26,000 hits were registered on the YouTube site since July 2011 for our documentary "God, Church, Pills & Condoms". Throughout production, post-production and distribution on the festival circuit, we made various short clips available online for stakeholders and the public, steadily building up an audience. Statistical data shows that a great deal of traffic was generated by YouTube Philippines, as the political stage surrounding the documentaries topic, controversial health care legislation, is set there. Filipinos around the world have been watching and commenting on the film on the projects *F*acebook and *Y*ouTube page, at times engaging in passionate online debates. Data collected from both sites show that audiences, especially digital natives, are increasingly accessing content in a "video-verse": 14.6% of viewers access content via smart phones.

42.4% of digital natives enrolled as *2nd* year undergraduate students at NHTV, International University of Applied Sciences, Breda, watch mainstream news and factual programming most of the time. 53% obtain their news from the Internet. Print ranks bottom of the list at 12%. Social media and Internet have become the main source of information for digital natives. 81.1% of participants surveyed claim that mainstream news has more impact on their opinion forming process, with only 18% stating that internet and social media have most impact on their opinion. For 54% of participants the internet and social media has significant impact and combined 72% of all digital natives surveyed consider the internet and social media as a source that has the most if not significant impact: a figure not far away from the 81% still holding on to mainstream news as the main source of information for opinion forming.

Mainstream broadcasters are competing with Social Media, broadcasting user-generated content during unfolding events such as the Arab Spring; they will need to do more to engage and retain digital natives. Experts in a studio interpreting and explaining events to the audience are not enough to satisfy the critical mind of a digital native, which increasingly develops the cognitive tools and ability to generate and react to user-generated content on a regional and global scale. News are no longer solely obtained top-down from a mainstream broadcaster such as CNN, the BBC and others worldwide, who begin to incorporate social media into the existing paradigm, slowly adapting to a new a new form of media literacy.

Digital natives engage and expect to be engaged with programme makers and their content at home, at school and while travelling. Using our documentary "God, Church, Pills & Condoms" we applied social media strategies during development, pre-production, production, post-production and distribution. We examined viewing habits via YouTube, Facebook and the impact of social media combined with real-world events such as the Cinemalaya Festival setting. *"God, Church, Pills & Condoms" is* now in distribution and social media still plays a major part. During development, pre-production, production and production of *"God, Church, Pills & Condoms"*, social media was and still is crucial to engage with stakeholders and audiences on a global scale.

This significantly contributed to its success. Stakeholders for all of the above are able to interact, comment and suggest content at any time. Data from all of these sites confirms that digital natives and many digital immigrants are quickly accustomed to the new video-verse. A real-world setting is nevertheless required. Events in the virtual social media world need to be firmly rooted in events taking place in the real world. The premiere at the Cinemalaya Film Festival of *"God, Church, Pills & Condoms"* was not only sold out, it drove traffic to the projects social media sites. YouTube's and Facebook's statistics section provides an excellent overview of demographic and location data. Data is easily accessed online and suggests that the programme continues to contribute to an ongoing debate in the real as well as virtual world in the Philippines. To underline that point, 'Mon Petit' features two digital natives and accompanies them on their journey to New Zealand. 'Mon Petit' would not have been possible in this way, without social media communication between the production team and the protagonists: content and narrative structure could not have evolved accordingly. Social Media is changing the way content is developed, produced and distributed.

Digital Natives and Revolutionaries in the News: Are the Elite Concerned?

A 16-year-old girl in the Netherlands published her birthday on Facebook and did not mark the event as private. (BBC, Facebook party invite sparks riot in Haren, 2012) As a result some 3000 party visitors descended upon the small village of Haren causing havoc. A 15-year-old Dutch boy was sentenced to a year of juvenile detention for the murder of a 14-year-old girl. (BBC, Teenager convicted of Facebook murder, 2012), The boy was recruited by another teenager as a contract killer over a row on Facebook. These two cases are well documented but are by no means isolated. Cyber bullying and consequent teenage suicides are increasingly making headline news. (BBC, Teenager's death sparks cyber-blackmailing probe, 2013) (Gadkari, 2013), (Arkin, 2013) (Ensor, J., 2013) (Warren, L., Quigley, R., 2013) Schurgin, Clarke-Pearson (2011) make specific recommendations to pediatricians dealing with the negative effects of social media in children.

These examples serve to demonstrate the level of understanding digital natives have of social media and their ability to utilize social media for their purposes. They demonstrate how digital natives have adapted regarding the use of social and digital media to obtain their goals based on some of the strongest motivators in existence: aggression and murder. The above examples highlight how social media transformed life in a quiet small Dutch village into an evening of terror and violence. For a 14-year-old girl, application of social media ended in her death. A 17 year old teenager committed suicide because of cyber-bullying. Digital natives are well aware on the how to use and abuse social and digital media. Parents, teachers, media professionals and the government –all digital immigrants- are lagging behind current developments.

There is no real guidance available to digital natives on the use of social media. Bullying is not new, but cyber-bullying is a practice which has yet to be dealt with by lawmakers, who are struggling to keep up with this new development. Tragic as these events are, they confirm Mcluhan (1994, p.7): the message has not changed: the media is the message. On a global scale, grass root movements such as Occupy, Wall Street and 15M in Spain as well as Wikileaks must present an even scarier scenario to the elite. Over the last few years various Northern African governments were toppled and social media played a crucial role in news reporting as well as providing intelligence to governments, rebels and terrorists alike. Governments and their secret services are taking social media very serious. Recent events surrounding Syria once again demonstrate this reality all too well. While the public was stunned by user generated content coming out of Libya, digital natives and immigrants alike have learned to view the same more critically: the Occupy and Wall Street movements have become permanent platforms providing alternative information to users and mainstream broadcasters.

Will Social Media Replace Traditional, Linear Forms of Documentary Production?

The invention of the broadcast and cinema industry did not eliminate or reduce the popularity of the book. Book sales have increased since the arrival of the Internet. An even greater availability of TV channels, programmes and movies than ever before is available to the consumer. History shows us that broadcast technologies did not make books obsolete, nor did the arrival of the Internet. PWC predicts a healthy 3% growth for the UK Entertainment and broadcast industries and for the global economy a 4.9% growth rate is forecast (PWC, 2012). *Linear documentary narratives will still have a market.*

But there is concern about the future size of that market. More vocal critics of social media and digital technology, such as Prof. Dr. Spitzer (2012) argue that the development of cognitive functions is impaired by inappropriate use of social and digital media. They argue that social and digital media need to be regulated and better understood before it is introduced to children and teenagers as they are still developing their cognitive skills.

*The term 'digital dementia' (*Ryall, J., 2012) (Daily Mail, Digital Dementia on the rise as young people increasingly rely on technology instead of their brain, *2013) has made it into mainstream press, describing a correlation of the loss of cognitive functions among digital natives and digital immigrants alike and their use of social media. This is a development documentary producers should be concerned* *about as they depend on their audiences cognitive abilities to critically view and review programmes presented. Society as a whole needs to be concerned about claims made by Prof. Dr. Spitzer (2013) and his South Korean colleagues Lee, Choi, Lee, Jung and Kwon (2012).*

Internet addiction is comparable to gaming addiction – except unlike gaming, the internet is accessible without any noticeable restrictions. Average consumption of media of a digital native stood at 7 hours and 38 minutes per day in 2010, tendency up. (Rideout, J., Foehr, U., Roberts, F., 2010) Similar results were published by Strasburger, Jordan and Donnerstein (2010) who also made concrete recommendations to schools, the entertainment and advertising industry, as well as researchers and government bodies, on how to deal with the consequences of social media abuse. The scope and extent on how the social media narrative will unfold online is yet to be determined and documentary producers need to understand the evolving digital native language and culture better to maintain their audience ties.

How Does Social Media Differ from Previous 'Information Explosions'?

An elite cannot control the invention of speech, like social media, effectively. The introduction of the alphabet led to a repositioning of the storyteller in antiquity, the print revolution provided the means to develop 'public spheres', followed by the broadcast and cinema industry. But an elite was able to control distribution all of the above effectively, except speech. User generated content on social media networks is free of that limitation; web2 and social media technologies are instant, provide interaction and are nearly impossible to censor and control.

This new freedom requires a change in behavior by content creators and their audiences. During the print revolution the Catholic Church was not enthusiastic about a literate church body. Instead the church preferred the bible to be in Latin to ensure believers receive church dogma administered and interpreted by a priest. Literacy, a result of the print revolution, changed all that and took the class struggle to another level. The print revolution enabled the rise of the middle classes. Today no financial risk is involved in starting up a blog, YouTube channel or Facebook page. There is no editorial office, no government censor who can effectively stem the flood of information published on social media every minute of the day. Social media is taking the class struggle to the next level: without censorship, global and immediate, featuring an increasingly social media literate digital native audience. Julian Assange, Wikileaks and the Dutch village of Haren are the more obvious examples. Digital natives are not merely adapting to social media, they are developing into a generation of social media literate users and content creators capable of challenging the status quo of the existing system.

7 CONCLUSION

What Does All This Mean for Media Producers?

The tools at the disposal of the media producer have developed over thousands of years. Story telling did not just begin with the invention of the printing press or film camera. And it is still the key element all media technologies have in common: all require a form of narrative, the audience is craving for stories. The perspective of today's filmmaker builds on an epistemology established over centuries; from simple pre-historical records to a new perspective in antiquity, mankind always maintained, documented and told stories about their trading ventures and wars. From history we learn that the tools used by storytellers have changed, the stories have not. Mankind is engaged in an ever-increasing number of violent conflicts. And without war and conflict, the media technology we use to criticize violent conflict would not exist. The most talented and most creative writers, film & TV directors are using social media to touch people in new and unexpected ways, for example the flash mob performance in a Spanish unemployment office broadcast on Youtube.

Understanding the Past

documentary film-makers need to better understand the historical origins of the tools they use in order to submit stronger and better arguments, and to better describe the human condition. A physicalist and reductionist position can explain how social media evolved. Today's information explosion is not a new phenomenon and confirms Mcluhan's (1994) assertion that the "medium is the message": whatever it was that drove early cave dwellers deep into their caves to paint and wonder about the resulting painting is equally touching people when they view the flash mob in a Spanish unemployment office. Story telling is a unique phenomenon which can be explained by the needs of a community to pass on traditions, historical events, form part of our collective memory, entertain and more importantly provide catharsis. A unique story telling perspective and a gripping story will do most of the work when combined with the most suitable media production and distribution tools.

Understanding the evolutionary pattern of previous 'information explosions' helps to understand the current situation. With the arrival of each new media technology, users of that technology needed time to acquire the skills needed to develop the necessary media literacy. Only then did it become possible to use the new technology to its full potential. Social media is still in that stage of development. Its full potential has not yet been realized, users and content creators are equally challenged to acquire the social media literacy skills and understand the evolving social media narrative. The task ahead entails to explore the unfolding social media narrative and to actively participate in it.

Copyright infringement too is not new to social media. The *Donaldson vs. Becket* (Rose) landmark ruling in the United Kingdom serves as a reminder of the ongoing legal argument about authorship. Briggs and Burke (2009) provide an excellent account of the history of media. Social and digital media however are not merely another evolutionary step in the history of media. Social media is part of a global paradigm shift, it is changing business models regarding existing and new media production and distribution. Viewing habits of digital natives are changing, abuse of social and digital media has an impact on the way the adolescent brain is developing, digital immigrants are challenged to keep up with developments. Social media is used and abused to spy on entire populations, has a significant political role, i.e. in the Arab Spring and Occupy movement and is transforming how information is disseminated in media and education.

Current events in Syria only demonstrate this point to clearly: social media had a significant role during the Arab Spring. But since the Arab Spring has begun, social media has developed into a permanent platform for users who in a very short time have learned to become 'social media literate': users are increasingly able to distinguish whether or not a government has evidence to invade another country or not. Movements such as Occupy and Wall Street provide a lively platform on which topics such as this are discussed every day. Users of social media platforms no longer just 'follow' government propaganda, but come to their own conclusions based on the evidence they find online. Social media is making the world more transparent – and at the same time provides the elite with the tools needed for mass surveillance. Documentary producers need to remember that and develop a greater awareness of the history of social media, which ironically would not have evolved into what it is today without war and conflict.

Facing this uncomfortable truth is necessary if social media is to develop its full potential as the one and ultimate information revolution empowering the 99% of humanity. Producers now have the opportunity to deliver content to a global audience – directly, and touch people in a profound way. Since the conception of speech mankind has been busy inventing and creating new media forms such as writing, print, broadcast and now social media. History teaches us that the arrival of a new media form does not extinguish previous media technologies; and a reductionist explanation of social media does not explain how social media touches our inner lives. Understanding the past, the technology, changing viewing habits are key aspects of this paradigm shift. Business and funding models too are changing. Crowd-funding sites such as Kickstarter (2013) and IndieGoGo (2013) feature an increasing portfolio of successfully funded documentary projects. Documentary forms are changing too, introducing more interactivity. (Doclab, 2013)

However, the one element that all media technologies have in common though remains in place, whether it be the spoken or written word, radio, TV or film, drama or documentary, Facebook or Youtube: content. Audiences want a good story. Narrative construction and plots may offer more options and interactivity, but technology alone will not make a bad story a huge success. It may help make a mediocre story break even. But a great story will be a success, if the tools needed to tell the story are well understood and applied.

ACKNOWLEDGMENT

I would like to thank the students participating in our survey, Joma Sison, Chairperson, International Coordinating Committee, International League of Peoples' Struggle, and Chairman, Board of Directors, Stichting International Network for Philippine Studies (INPS).

Conflict of interests: the authors of this paper declare that their research has been entirely self-funded.

REFERENCES

Andis, K. (2012). *The Alphabet and the Origins of Writing. epublic GmbH.*

Arkin, D. (2013). Canada charges two in teen cyber-bullying suicide case. *NBC News*. Retrieved 20/8/2013, from http://www.bbc.co.uk/news/world-us-canada-23752923

BBC News. (2013, August 20). Teenager's death sparks cyber-blackmailing probe. *BBC News Scotland*. Retrieved 20/8/2013, from http://www.bbc.co.uk/news/uk-scotland-23723169

Benson, J., Greaves, W., O'Donnell, M., & Taglialatela, J. (2002). Evidence for Symbolic Language Processing in a Bonobo (Pan paniscus). *Journal of Consciousness Studies*, *9*(12), 33–56.

Bondebjoerg. (2009). Behind the headlines: Documentaries, the War on Terror and everyday life. *Studies in Documentary Film*, *3*, 219.

Brentlinger, J. (2000). Revolutionizing spirituality. *Reflections on Marxism and Religion, 64*(2), 171-193.

Briggs, A., & Burke, P. (2009). A social History of Media (3rd ed.). Academic Press.

Carne Cruda 2.0. (2012). Retrieved 28/8/2013 http://youtu.be/kS709ZyZ_YU

Chalmers, D. J. (1995). *Facing up to the problem of consciousness.* Retrieved 20/8/2013, from http://philpapers.org/rec/CHAFUT

Corbalis, M. (2009). *The Evolution of Language.* The Year on Cognitive Neuroscience.

Cvorovic, J. (2006). Gypsies Drown in Shallow Water: Oral Narratives among Macva Gypsies. *Journal of Folklore Research, 43*(2), 129–148. doi:10.2979/JFR.2006.43.2.129

Digital Dementia on the rise as young people increasingly rely on technology instead of their brain. (2013, August 6). *Mail Online.* Retrieved on 6/8/2013, from http://www.dailymail.co.uk/health/article-2347563/Digital-dementia-rise-young-people-increasingly-rely-technology-instead-brain.html

Doclab, I. D. F. A. (n.d.). Retrieved from http://www.doclab.org/

Doclab, T. L. (n.d.). Retrieved 20/8/2013, www.film-and-television.com

Dunbar. (2003). The social brain: Mind, language, and society in evolutionary perspective. *Annual Review of Anthropology, 32*, 163-181.

Ensor, J. (2013). Family of Skype suicide teen calls on David Cameron to tackle cyber bullying. *The Telegraph.* Retrieved 20/8/2013, from http://www.telegraph.co.uk/technology/news/10248058/Family-of-skype-suicide-teen-calls-on-David-Cameron-to-tackle-cyber-bullying.html

Facebook. (n.d.). Retrieved from https://www.facebook.com/tinylittledoclab

Flashmob oficina paro. (n.d.). Retrieved from http://www.npr.org/blogs/thetwo-way/2013/01/14/169345343/flash-mob-brings-some-sunshine-into-spanish-unemployment-office

Gadkari, P. (2013). Ask.fm unveils changes to safety policy. *BBC News.* Retrieved 20/8/2013, from http://www.bbc.co.uk/news/world-us-canada-23752923

Goldwasser, O. (2010). How the Alphabet was born from Hieroglyphs. *Biblical Archealogy Review, 36*(2).

Habermas, J., & Lennox, F. (1974). Article. *New German Critique*, (3), 49-55.

Herzog, W. (2012). *Cave of Forgotten Dreams.* Retrieved 20/8/2013, from http://www.imdb.com/title/tt1664894/

History, C., & Toys, T. (n.d.). Retrieved 20/7/2012, from http://www.conflicthistory.com/#/period/0700BC-0588BC

Indiegogo. (n.d.). Retrieved from http://www.indiegogo.com/

Kaplan, A., & Haenlein, M. (2010). Users of the world, unite! The challenges and opportunities of Social Media. *Business Horizons, 53*(1), 59–68. doi:10.1016/j.bushor.2009.09.003

Kickstarter. (n.d.). Retrieved from http://www.kickstarter.com/

Kohle, F. (2012). *The Arab Spring and the Wall Street Movement: Challenges and Implications for Documentary Filmmakers and Social Media*. Retrieved from http://www.thelondonfilmandmediaconference.com/registered-speakers-2012-a-to-k/

Kohle, F., & Cuevas, A. (2010). A case study in using youtube and facebook as social media tools in enhancing student centered learning and engagement. In *Proceedings of 3rd International Conference of Education, Research and Innovation*. Madrid, Spain: ICERI2010.

Kohle, F., & Cuevas, A. (2012). Social media: Changing the way we teach and changing the way we learn. In *Proceedings of 6th International Technology, Education and Development Conference.* Valencia, Spain: INTED2012.

Lauinger, J. (2011). An Excavated Dossier of Cuneiform Tablets from Level VII Alalah? American Schools of Oriental Research, 362, 21-64.

Lee, H. W., Choi, J.-S., Shin, Y.-C., Lee, J.-Y., Jung, H. Y., & Kwon, J. S. (2012, July). Impulsivity in Internet Addiction: A Comparison with Pathological Gambling. *Cyberpsychology, Behavior, and Social Networking*, *15*(7), 373–377. doi:10.1089/cyber.2012.0063

Locke, J., & Bogin, B. (2006). Language and Life History: A new perspective on the development and evolution of human language. *Behavioral and Brain Sciences*, 267. PMID:17214017

Marx, & Engels. (1969). The Manifesto of the Communist Party. In *Marx/Engels Selected Works* (vol. 1, pp. 98–137). Moscow: Progress Publishers.

McLuhan, M. (1994). *Understanding media: The extensions of man*. MIT Press.

Merola, M. (2007). *Archeaology, Messages of the Dead*. Retrieved 29/8/2013, from www.archive.archaeology.org/0701/abstracts/qatna.html

Mowery, D. C. (2001). Is the Internet a US invention?—An economic and technological history of computer networking. Timothy Simcoe Haas School of Business, University of California at Berkeley.

BBC News. (2012b, October 3). *Teenager convicted of Facebook murder*. Retrieved from http://www.bbc.co.uk/news/world-europe-19467308 retrieved 20/8/2013

News, B. B. C. (2012c, October 22). *Facebook party invite sparks riot in Haren*. Retrieved from http://www.bbc.co.uk/news/world-europe-1968470 retrieved 20/8/2013

NHTV. (n.d.). Retrieved 20/8/2013, from www.nhtv.nl

Pagel, M. (n.d.). *University of Reading*. Retrieved 20/8/2013, from http://www.ted.com/talks/mark_pagel_how_language_transformed_humanity.html

Petit, M. (2012). Retrieved 20/8/2013, from http://www.imdb.com/title/tt1838723/

Pike, H. (2012). U-Series Dating of Paleolithic Art in 11 Caves in Spain. *Science, 336*(6087), 1409-1413. DOI: 10.1126/science.1219957

Puttnam, D. (2013). *In his guest speaker engagement at NHTV*. University of Applied Sciences, Netherlands. Retrieved 20/8/2013, http://youtu.be/HOL14ONdzIU

PWC. (2012). *Media and Entertainment Outlook 2012-2016*. Retrieved 20/2/2013, from http://www.pwc.nl/nl/entertainment-media-outlook/2012site/index.jhtml

Ravilious, K. (2010). The writing on the cave wall. *New Scientist Magazine*, (2748).

Ricciardelli, L. (2009). Documentary Filmmaking in the Postmodern Age: Errol Morris & The Fog of Truth. *Studies in Documentary Film*, *3*, 219.

Rideout, J., Foehr, U., & Roberts, F. (2010). Generation M: Media in the lives of 8-18 year olds. *Kaiser Family Foundation*. Retrieved 25/8/2013, from http://kaiserfamilyfoundation.files.wordpress.com/2013/01/8010.pdf

Rose, M. (1988). The Author as Proprietor: Donaldson v. Becket and the Genealogy of Modern Authorship. *Representations (Berkeley, Calif.)*,(23), 51–85. doi:10.2307/2928566

Ryall, J. (2012). Surge in digital dementia. *The Telegraph*. Retrieved on 6/8/2012, from http://www.telegraph.co.uk/news/worldnews/asia/southkorea/10138403/Surge-in-digital-dementia.html

Schurgin O'Keefe, G., & Clarke-Pearson, K. (2011, April 1). American Academy of Pediatrics. *Pediatrics*, *127*(4), 800–804. PMID:21444588

Shaughnessy, H. (2012, April 10). *Forbes Magazine*. Retrieved 20/8/2013, from http://www.forbes.com/sites/haydnshaughnessy/2012/10/04/facebooks-1-billion-users-why-the-sky-is-still-the-limit/

Spitzer, M. (2012). Digitale Demenz. Droemer Verlag.

Strasburger, Jordan, & Donnerstein. (2010). Article. *Pediatrics*, *125*(756). DOI: 10.1542/peds.2009-2563

Tulving, E. (2002). Episodic Memory: From Mind to Brain. *Annual Review of Psychology*, *53*(1), 1–25. doi:10.1146/annurev.psych.53.100901.135114 PMID:11752477

Warren, L., & Quigley, R. (2013). Police confirm 12 year old girls suicide note said she was being cyber-bullied as her sister reveals she knew about abuse but was sworn to secrecy. *Daily Mail*. Retrieved 20/8/2013, from http://www.dailymail.co.uk/news/article-2331670/Gabrielle-Molina-Police-confirm-12-year-old-girls-suicide-note-said-cyber-bullied-sister-reveals-knew-abuse-sworn-secrecy.html

Watkins, T. (2010). New light on Neolithic revolution in south-west Asia. *Antiquity*, *84*(325), 621–634. doi:10.1017/S0003598X00100122

Youtube. (n.d.). Retrieved 20/8/2013, from https://www.youtube.com/reproductivebill

Chapter 10
Augmenting User Interaction Experience through Embedded Multimodal Media Agents in Social Networks

Maria Matsiola
Aristotle University of Thessaloniki, Greece

George Kalliris
Aristotle University of Thessaloniki, Greece

Charalampos A. Dimoulas
Aristotle University of Thessaloniki, Greece

Andreas A. Veglis
Aristotle University of Thessaloniki, Greece

ABSTRACT

The current chapter proposes media agent and multi-agent models aiming at improving mediated communication and information exchange in social networking. Great progress has been conducted during the last decades in Information and Communication Technologies, which is also reflected in social media. The proposed models exploit the latest media technologies for the augmentation of user-interaction and contribution experience in multiple levels. Features of the suggested agent and multi-agent approaches are discussed and elaborated through the prism of social computing, social media analytics and intelligence, resulting to a sophisticated communication mediator between users and social groups. In addition, enhanced user engagement and collaboration are considered in terms of rich media experience and augmented reality, semantic interaction services, intelligent content processing and management automation over interoperable multiplatform environments. Social media cooperation and integration is envisioned towards the realization of Web 3.0 and beyond, as the main chapter contribution.

INTRODUCTION

Over the past years social media networking became a part of everyday life for a great number of users, being generally accepted that it played, and is still playing, a very unique and important role in the growth of new media and modern global communication services. In general, Social Networking Sites (SNSs), in a broader use of the term incorporating social media and general social networking environments, are

DOI: 10.4018/978-1-4666-8556-7.ch010

defined as Web services that give the ability to construct a public profile, to identify a list of other users with whom a connection is shared and finally the ability to view and track individual connections as well as those made by others (Nadkarni & Hofmann, 2012). In this context, social media platforms promise to create, contribute, share and exchange timely information. Social tagging, feedback and generally commenting are a common habit favoring personalized adaptation services or even hypermedia authoring and non-linear digital storytelling concepts. Most of these services are considered to be part of the so-called Web 2.0 paradigm, while their further exploitation along with future pervasive and ubiquitous computing services are new trends that are expected to change the landscape during the transition to Web 3.0 (Dimoulas, Veglis & Kalliris, 2014b, 2015; Dimoulas, Kalliris, Chatzara, Tsipas, & Papanikolaou, 2014a; Veglis, Dimoulas & Kalliris, 2016). In this journey, enhanced human-machine interaction (HMI) experience is pursued through virtual and augmented reality (AR) interaction environments (Feldon & Kafai, 2008; Zhou, Dun & Billinghurst, 2008), but also semantic understanding and conceptualization modules, that attempt to trail natural, human-like, communication alternatives. Among others, avatars and artificial characters, in general, are proposed as promising solutions to the above pursuance (Avdelidis et al., 2002; Kalliris et al., 2002; Bélisle & Bodur, 2010).

In traditional broadcast media the main advantage is that journalists, newscasters, actors and generally performers are able to communicate with the audience in a direct, immediate, user-friendly and sometimes entertaining way. In order to balance this advantage in new media, multimedia enabled Internet applications were firstly employed using VRML, Java and other programming tools. This kind of "rich media" that are popular today, utilize various techniques to enhance a recipient's experience and can be incorporated in a variety of communication applications, such as email, newsletters, demos and other audio-visual interactive environments. Incorporation of artificial characters and simpler humanoid graphics were attempted in various social media platforms and related users profiles (Ochs, Pelachaud & Sadek, 2008). A common example that can be listed in this category is the use of emoticons that refer to specific emotional state and/or semantic response (Derks, Bos, & Von Grumbkow, 2008; Huang, Yen & Zhang, 2008). Animated digital characters (both 2D and 3D) have been also reported in desktop and web applications but also in Facebook and other SNSs (Avdelidis et al., 2002; Kalliris et al., 2002; Von der Pütten, Krämer, Gratch & Kang, 2010).

On the other hand, there is a variety of social media focusing on different content aspects as their key communication concept and philosophy, so that not all of them favor avatar-based multimedia interaction. For instance, twitter is intended for short text messaging services, while Soundcloud focuses exclusive on audio content contribution and sharing, along with its textual semantic meta-data. Similarly, Instagram and Flickr, but also YouTube and Vimeo, facilitate imaging –based communication by means of experience sharing through pictures and videos (Dimoulas et al., 2014b). In addition, there are social media and their mobile-application front ends that have been seriously accommodated and adapted as to confront limited interfacing capabilities of mobile terminals (Dimoulas et al., 2014b; Veglis et al., 2016). Thereafter, there is a need of various differentiated digital actuators that could adapt to the demands and particularities of different content types (i.e. text, audio, visual), media channels, terminals capabilities and users' preferences in general (i.e. mobile vs desktop computing, content browsing-accessing difficulties and needs that are common in elderly people, technologically illiterate and people with disabilities, etc.). These kinds of multimodal digital characters (or multimodal media agents) that the current chapter focuses on can be very constructive and useful in various types of social media communication, offering context adaptation, intelligent processing and semantic interaction capabilities, integration and collaboration prospects.

BACKGROUND

Underlying Technology and Basic Definitions

Without any doubt today's Information and Communication Technologies (ICT) are employed in all human activities. Users continuously interact with computers and mobile devices in order to conduct a variety of tasks. In the majority of the cases these tasks are accomplished through accessing the Internet. There are four main sub–concepts of interactivity: transmitional (users have the ability to choose information from a continuous stream in a one way communication), consultational (choose by request using a return channel), conversational (produce and contribute new content) and registrational (register users' request and action needs and adapt) (Jensen 1998; Spyridou & Veglis, 2008). In the last decade we have witnessed a tremendous growth in social media and especially in social networking. Social media can be defined as the mechanism for the users to connect, communicate and interact with each other, and their mutual friends, through instant messaging or social networking sites and services (Correa, Hinsley & De Zuniga, 2010). The primary force behind this rapid growth has been the Internet infrastructure, as well as mobile technologies. This is due to the fact that they are providing the technological platform for information dissemination, content generation and interactive exchange of information (Dimoulas et al., 2014b, 2015; Zeng, Chen, Lusch & Li, 2010).

Social media offer the ability of creating and sharing user generated content (UGC), a feature that has caught the attention of many Internet users. Today, social media contain enormous collections of UGC data (Clements, de Vries & Reinders, 2009). This publicly available content is produced by end-users, usually having textual form but also multimodal structures, incorporating multimedia elements (Dimoulas et al., 2014b, 2015; Veglis et al., 2016). Text messages and multimedia posts that are included in forums, blogs, wikis, SNSs, etc. represent common UGC examples (Bothos, Apostolou & Mentzas, 2010). The availability of content in social media gives users the opportunity to explore non-conventional storytelling techniques that combine text, images, videos and sound in a non-linear fashion, through versatile arrangement of information in more complex browsing structures with alternative navigation paths. There are various definitions that classify media content in different models. Two major categories are the Page-Based Model (PBM) and the Time-Based Model (TBM) (Chapman & Chapman, 2009; Veglis et al., 2016). In the PBM case, textual information and images are usually combined and spatially arranged, resembling the way that a page in a book or a magazine looks like. TBM refers to content that is assembled in time-spaced bins, as in the case of audio clips, linear graphic animations and videos. Hybrid models of mixed PBM and TBM content types are also frequently implemented nowadays in forms of Multimodal Media Assets (MMA) (Dimoulas et al., 2014b, 2015; Veglis et al., 2016) and this is the case of user generated content in social media.

The Social Media Networking Experience

The social media networking experience and especially Facebook-use is assumed to be motivated by two primary needs: (1) the need to belong and (2) the need for self-presentation (Nadkarni & Hofmann, 2012). Both needs are covered considerably by Facebook-use, however, many other social media platforms offer sufficient gratification of these two needs as well. Whether a social media platform will be successful and popular or not, at a certain time, place or virtual place, and for a target user group profile or for the general engaged public, is of course affected by many other factors. One important factor was proved to be the marketing and communication strategies combined with quick response of the programming team to the users' needs during the launch period phase but also afterwards during the normal operation times (Kirkpatrick, 2011). Another factor is the ability of social media to easily and timely introduce and test the efficiency and success of new forms and modes of media interaction with the users, in order to accept or decline them according to the users' feedback. One example of a very successful mode of media interaction is the YouTube video hosting service that constitutes a social networking platform on which anyone can upload videos that can be shared, seen and commented by every Internet user or, if so chosen, by closed user groups.

In principle, the first social networking system was the Bulletin Board System (BBS) back in 1978, hosted on personal computers that users dialled in via their modems over phone lines to exchange information with other users (Jones, 2003). Although it was quite slow since only one user could be logged in at a time it was the first system that allowed users to sign in and interact with each other. Social media in the form of networking sites on the Internet was firstly attempted in mid 90s. Among the first, was Geocities in 1994 that allowed users to create their own websites, dividing them into "cities" that were categorized based on content. With the advent of the new millennium, Friendster was the dominant social networking website that acquired 3 million users in its first three months. Consequently, the widely popular MySpace cloned Friendster and was launched after just 10 days of coding (Walker, 2014). In the following years, websites offering social networking of specific interest like Classmates.com, LinkedIn and Tribe.net started to turn up and in 2004 Facebook.com, the most popular social networking website so far, was launched.

Another very promising media interaction form is the artificial characters or the avatars. An avatar is the graphical representation of the user but it may also be the alter ego or the dream character of the user (Cooper, Dibbell & Spaight, 2007). It may take either a three-dimensional form or a two-dimensional form. Avatars are widely used in games, virtual worlds, as icons in Internet forums and in online communities. The use of avatars for "social networking"–like purposes is employed successfully in the online virtual world Second Life (Malaby, 2009). The users in Second Life, called "residents", interact with each other through avatars. Residents explore the world which is known as "the grid", meet other residents, socialize, participate in individual and group activities, create and trade virtual property and services. Second Life is not a real social networking platform; it is a virtual one that most users hide their identity. Avatars offer the advantage of creating the dream character while maintaining anonymity. On the other hand the boundaries between social networking platforms, virtual worlds and role-playing games (RPGs) tend to be very loose and it is a matter of time to see the benefits of using artificial characters for digital storytelling in real world social networking platforms. A typical and promising example of using avatars and Second Life in real world situations and digital storytelling is their deployment for educational purposes and their incorporation in e-learning environments (Warburton, 2009).

The Role of Artificial Characters and Avatars in Digital Storytelling

Contemporary ICTs in conjunction with SNSs give the ability to users to dynamically customize and personalize their applications; those capabilities have a profound impact, especially for young people (Robin, 2008). Digital storytelling, as one of the foretold applications, creates short multimedia stories by making use of low-cost digital equipment combined with various types of media and taking advantage of UGC shared resources (Robin, 2008; Meadows, 2003). Narrative intelligence is a related sector where narrative intersects with intelligence of both artificial and human varieties (Mateas & Sengers, 1999).

Several interactive storytelling-narrative models, author-centric, story-centric and character-centric have been deployed, taking in consideration different parameters. Some of them are using artificial characters either as autonomous virtual actors or in combination with narrative control (Mateas & Sengers, 1999; Riedl & Stern, 2006). Computational systems of narrative intelligence may interact with human users in a natural way since they understand collaborative contexts as emerging narrative and are expressing themselves through storytelling (Riedl & Young, 2010). As narratives across media focus on relationships between characters, researchers are involved in creating computational models of social interaction between characters to the extent of supporting some of them (McCoy et al., 2010).

Avatars are used in virtual worlds and give expressive freedom to an otherwise rather static online presence (Vasalou & Joinson, 2009). They are used in a great variety of online applications like gaming, social media, e-commerce, entertainment and education, being the mediator of the participants in interacting with each other and the environment (Messinger et al., 2008; Qiu, & Benbasat, 2005; Petrakou, 2010). They may be designed through a self-built customization or they may be chosen from already-made ones. Although they were popular in online gaming they did not enter the social networks until later and with a slower pace. Many researches are trying to connect the choice of a self-presentation with an avatar over the personality of the real-user that made the specific choice. There is a tendency to reflect their own appearance (Messinger et al., 2008), lifestyle and preferences if they are created for blogging, however they tend to differentiate their look for other purposes (more attractive for dating and more intellectual for gaming) (Vasalou & Joinson, 2009). In terms of attractiveness, avatars are viable replacements for real people in social behaviors of both adults and children (Principe, Langlois, 2013; Feldon & Kafai, 2008). The entire above are expected since the individuals perform and manage many different roles in their everyday life (Zhang, Yu, Dang & Chen, 2010).

Nowadays, researchers are aiming at acquiring data about avatars' behavior using other avatars as bots. Data, identifiers and behavioral, is automatically and simultaneously collected and in that way an avatar's profile information is matched to his or her behavior (Zhang, Yu, Dang & Chen, 2010). The case that agents would interact with each other and with the user in order to build an autonomous identity is being explored as well (Jacobsson, Rost & Holmquist, 2006). Human-centered computing is a field where computer systems can perceive and understand human behavior by exploiting multimodal inputs (by the users) and respond appropriately by adapting their functionality to their preferences (Cohn, 2007).

Current Challenges and Expectations: Social Media Intelligence and Analytics

While the transition from the Symbiotic (Web 2.5) to the Semantic Web (Web 3.0) is still ongoing and new challenges /web models appear (i.e. Web 4.0 /Web x.0, refer to figure 1), social media analytics and media intelligence are two of the most emerging research fields, which will play a very important role in the near future (Aghaei, Nematbakhsh & Farsani, 2012; Bernal, 2010; Dimoulas et al., 2014b; Jacobsson et al., 2006; Symeonidis & Mitkas, 2005). Social media analytics is a scientific area that deals with developing and evaluating informatics tools and frameworks in order to collect, monitor, analyse, summarize and visualize data that can be found in SNSs (blogs, microblogs, multimedia sharing sites etc.) (Zeng et al., 2010). SNS-shared UGC data may present new challenges to both producers and consumers of information, while social media analytics provide an effective way to track, analyse and manage large quantities of data, thus making explicit the links between users and therefore extract the valuable points of importance (Leskovec, 2011).

Social media networks present a significant source of intelligence since they encode the online activities and inputs of masses of SNS participants, including adaptation and personalization preferences, semantic interaction and feedback during all three content production, consumption and authoring (Dimoulas et al., 2014a, 2014b, 2015; Veglis et al., 2016). The dynamic nature of social media data demands for context, location and time-awareness is also leading to the implementation of AR applications (Feldon & Kafai, 2008; Papadopoulos, Kompatsiaris, Vakali & Spyridonos, 2012; Zhou et al., 2008). Hence, the scientific field of social media intelligence attempts to face such demands by requiring integrated multidisciplinary research. Its particular aims are to extract information from social media, develop corresponding decision-making or decision-aiding frameworks, and provide architectural designs and solution frameworks for existing and new applications (Zeng et al., 2010). This is fuelled by the emerging progress in associated technologies, such as cloud computing and mobile cloud computing paradigms, new software engineering models, natural language processing, artificial digital characters and "experienced" avatars, machine learning algorithms for multimodal content recognition and summarization, emotion analysis, semantic conceptualization and others (Avdelidis et al., 2002; Dimoulas, Avdelidis, Kalliris & Papanikolaou, 2008; Dimoulas Papanikolaou & Petridis, 2011; Drossos, Floros & Kanellopoulos, 2012; Kalal, Mikolajczyk & Matas,2012; Kalliris et al., 2002; Kotsakis, Kalliris & Dimoulas, 2012; Symeonidis & Mitkas, 2005). However, there are still many unresolved issues that are related to the diversity of the UGC sources and the heterogeneity of both the users' preferences and the differences of the access terminals, which need to be properly addressed (Dimoulas et al., 2014a, 2014b, 2015; Kotsakis et al., 2012; Leskovec, 2011; Veglis et al., 2016; Zeng et al., 2010).

MULTIMODAL MEDIA AGENTS AND SOCIAL MEDIA

Enhancing User Interaction Experience through Social Networks: Issues, Controversies, Problems

Looking back at the early days of World Wide Web, it is obvious that a lot of progress has occurred for the enhancement of user interaction and participation experience, regarding Web computing technologies (including mobile and wireless) and their corresponding applications like blogs, wikis and other social media. It all started with Web 1.0 which was static, just for viewing and was used mainly for business (providing information), while there were very few interactive services for users. The launching of social media initiated the transition to the Web 2.0 era, incorporating novel attributes (i.e. portability, adaptation, personalization, etc.) and propelling bi-directional social interaction of large global crowds (Aghaei et al., 2012). Semantic Web or Web 3.0 is meant to provide better functionality to the users of the Web, meaning easier, faster and more precise access to all types of information, including the acts of sharing and combining. Improvement of data management, meaning reconnecting existing data for other uses, accessibility of mobile Internet and collaboration in social Web, are among the provisions of Web 3.0 (Aghaei et al., 2012). It is believed that Semantic Web and ontologies are providing the tools and formalism for representation of the properties of a reflected social reality including individual agents' beliefs, goals, and intentions (Wang, Carley, Zeng & Mao, 2007). In this context, interoperability among SNSs, meaning linking people and objects by using their heterogeneous ties, can be made possible through Semantic Web's representation mechanisms (Breslin & Decker, 2007).

Figure 1. The role of multimodal media agents in collaborative SNSs toward the completion of the semantic web and beyond.

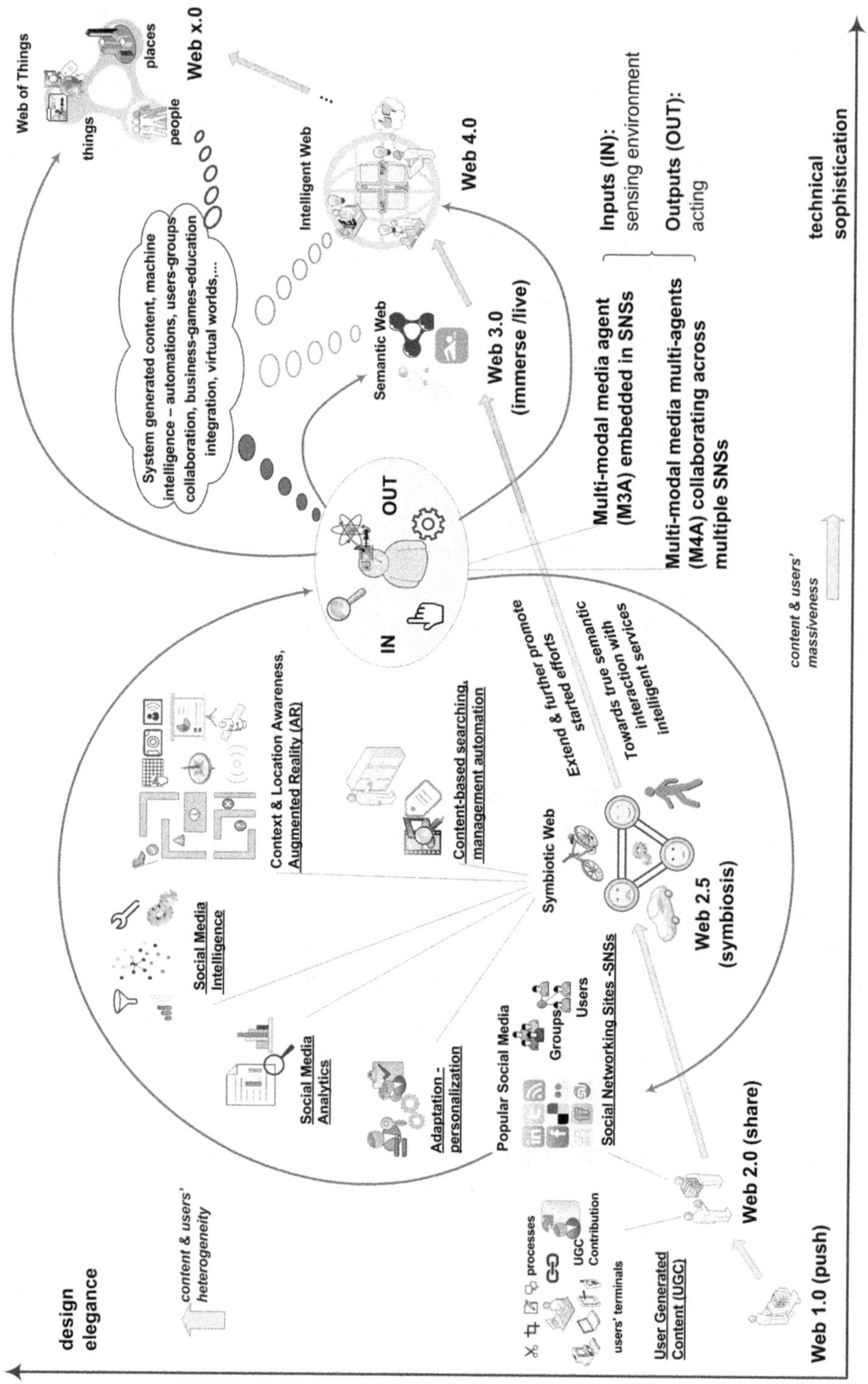

Semantic technologies use machines to interpret and process large amounts of data, in reality they are providing integration of data with services, and make the results available to the users. Lately, data produced by taggers is harvested and used in the Semantic Web. Collaborative tagging describes the process by which anyone may add metadata in the form of keywords to shared content. In that way, information is categorized not only for oneself but for others as well (Golder & Huberman, 2006; Tsirelis & Delopoulos, 2011; Tsipas, Dimoulas, Kalliris & Papanikolaou, 2013). With Twitter being the most popular microblogging platform, where a great variety of recent topics is discussed, researchers use the posts to investigate semantic user modeling, thus allowing personalization and being beneficial for other applications on the social web, as well (Abel, Gao, Houben &Tao, 2011). Likewise, sites such as LinkedIn use information, which is gathered and stored in users' profiles, in order to facilitate connections thus creating a different kind of online communities (Breslin & Decker, 2007). As a consequence, general recommendation systems are implemented, aiming at suggesting interesting or personalized information objects to users, suited to their needs, based on explicit (created by human consent) or implicit (existing in the system) ratings, just like when someone in real life wants to recommend a book or a song to someone else. Those systems require a model that reflects the user's characteristics, preferences and needs. Studies have proved that the use of social tagging (user-generated tags) can be highly beneficial in that area, while semantic processing and meta-data filtering techniques are really necessary (especially when meta-data standardization and formalization is missing) (Golder & Huberman, 2006; Milicevic, Nanopoulos & Ivanovic, 2010; Tsipas et al., 2013).

Social computing has emerged as a cross-disciplinary research and application field, including both computational and social sciences, thus analyzing social behavior and creating artificial social agents. It is the field where Web, database, multimedia, wireless, agent, and software engineering technologies are encompassed. Computational system's designing is based on social theories and agent modeling techniques, aiming at creating artificial societies. Societies' features are described through social information where social knowledge describes agents' cognitive and social states (Wang et al., 2007). Social networking is moving towards simulating real-life's social connections, as people "meet" and interact with other people with common interests (Breslin & Decker, 2007).

As already implied, there are still many difficulties in the road toward Web 3.0 completion that are related to the high level of heterogeneity between content (and their meta-data), users, networks and access /computing terminals (Dimoulas et al., 2014a; Veglis et al., 2016). In fact, the various SNSs have great differences and inhomogeneity. In most cases they are not directly connected with each other, so that they are lacking *Social Networking Portability,* forcing users to create many accounts without being able to automatically transfer around their identities (Breslin & Decker, 2007). Another major issue in all aspects of everyday life, including social networking, which must be preserved, is privacy, which stands as a kind of obstacle towards the completion of transparent social multimedia in collaborative SNSs. The usage of Web 2.0/3.0 services and semantic web applications is exposing a large amount of personally identifiable information to others (Dimoulas et al., 2014b; Weiss, 2009). Hence, certain questions arise regarding the implementation, configuration and use of automated compliance functions, but also the involved levels of individual privacy preferences, especially in data portability use scenarios.

The chapter attempts to provide solutions to the above issues through the use of multimodal media agents (M3A) and SNSs-embedded multimodal media multi-agents (M4A). The term multimodal has a versatile interpretation in the current case, meaning that it deals with many modalities related to various content entities, presentation means, systems involved, diversity of users, SNSs and multiplatform environments in general. In this context, augmentation of HMI is pursued by enhancing presentation of information through the use of multimodal artificial digital actors and avatars, thus offering rich media experience. At the same time, context-, time- and location-aware capabilities are considered in order to serve user adaptation and personalization services, along with semantic interaction and communication mechanisms. Efficient media description and management is also targeted, including cross-platform transparent multimedia networking, with respect to content and users' privacy. Figure 1 provides a graphical representation of the contribution of the current chapter towards the completion of the semantic web (and beyond), by depicting the relationship of the proposed M3A and M4A solutions with the involved emerging technologies that have been presented so far.

Referring to figure 1, the proposed solutions are intended to be utilized in the current landscape of the symbiotic web (Web 2.5), where different users' and systems' behaviors, needs and capabilities are encountered (similarly to the real life's examples of multiple travel ways and emotional statements in fig. 1). In this context, M3A and M4A models attempt to exploit contemporary social computing trends and technologies (i.e media analytics and intelligence, context awareness, adaptation and AR tools, content sharing and intelligent management, etc.) in order to facilitate enhanced semantic interaction and collaboration services. As it is depicted in figure 1, Web 3.0 but also new challenges /web paradigms (i.e. Intelligent Web, Web of Things and beyond, Web 4.0 / Web x.0, etc.) pursue the implementation of such features, although with somewhat different intensity and focusing (as indicated with the different bordering volumes of the annotation callouts in fig. 1). In any case, all these initiatives, including the proposed M3A /M4A models, aim at promoting integration and collaboration of those services in the favor of end-users in all aspects of everyday life (business, information, entertainment, education, etc.).

The Role of Multimodal Media Agents: Wanted Features and Working Definitions

As already implied, the huge amount of UGC data that is produced and shared every day, in combination with the vast heterogeneity between multiple media channels, content formats, access terminals and users' preferences create serious content distribution and management difficulties (Dimoulas et al., 2014a, 2014b, 2015; Kotsakis et al., 2012; Veglis et al., 2016). Hence, sophisticated computing systems are required in order to automatically detect and adapt in a diversity of multimodal behavior from multiple users, information presentation preferences and interaction mechanisms in general (Cohn, 2007). Agents represent such kind of intelligent processing systems, which have become very popular during the last decades, taking advantage from the remarkable progress of the involved technologies (i.e. software engineering tools and practices, cloud computing and mobile cloud computing models, novel machine learning algorithms, knowledge management methods and strategies, etc.). In particular, agents are capable of sensing the contexts of users' surrounding environment that are situated in, so that they can act and/or respond autonomously in order to serve the tasks that have been designed and/or trained for (Symeonidis & Mitkas, 2005).

There are quite a few definitions that are encountered in the bibliography regarding agents' classification and objectives. The common ground is that agents "act on behalf of someone else, after they have been properly authorized" (Symeonidis & Mitkas, 2005). In this context, there is also a differentiation between the terms agent and intelligent agent, with the later representing more complicated situation and behavior, thus requiring more complex descriptions and definitions. While many researchers state that true agent intelligence is not really feasible, still, a kind of artificial (/computational) intelligence (that differs from human intelligence) is developed and exploited as part of most agents' working features

(Symeonidis & Mitkas, 2005; Batzios, Dimou, Symeonidis & Mitkas, 2008; Dimou, Symeonidis & Mitkas, 2008). For this purpose, the terms agent and intelligent agent will be treated as synonyms in the rest of the chapter. Besides autonomy and learning capabilities, other agents' attributes are interactivity (in terms of reactivity or pro-activeness, i.e. perceive the environment and respond to its changes or take initiative), mobility (transition from one environment to other without stop working) and character (incorporation of human-centered attributes, such as personality and emotional state), which are very important in social media storytelling. Other characteristics like adaptability, sociability, cooperativity and competitiveness describe agents' ability to perceive and respond in the presence of other agents within their "sight" or "neighborhood". These features are associated with the so-called multi-agent systems that are presented in the next sub-section (Symeonidis & Mitkas, 2005).

Agents have been massively deployed in media related applications, facilitating content searching, summarization and browsing, information presentation and filtering, context awareness and intelligent multimedia management (Symeonidis & Mitkas, 2005). For instance, media agents have been implemented for information searching and retrieval through web crawling and data mining approaches (Batzios et al., 2008; Dimou et al., 2008), in making media wiser for recommendation and filtering purposes (Jacobsson et al., 2006), for incorporating emotional adaptation tasks (Cohn, 2007), in intelligent story adaptation for interactive storytelling (Riedl & Stern, 2006), for mobile computing resources management and context-awareness in public spaces (Sun, Yang, Zhou & Cruickshank, 2014; Satoh, 2010), for developing social intelligence from social informatics (Wang et al., 2007), for intelligent cross-platform content management (Nadrljanski & Batinica, 2010) and others. All of the above (and further) characteristics are integrated in figure 2 that provides a user-wise graphical representation of the proposed multimodal media agent (M3A).

According to figure 2, M3A is a social media oriented multimodal avatar that acts as a hybrid intelligent agent. The term multimodal refers to the fact that associated information can be processed and presented in a variety of content modalities, such as text, image, audio, video or any multimedia (MMA) structure that can be formed as any combination of all previous entities (Dimoulas et al., 2015; Veglis et al., 2016). This allows the exploitation of contemporary multimodal content processing techniques with hybrid intelligent systems (Dimoulas et al., 2008, 2011). In addition, recent intelligent cross media publishing (iXMP) approaches (Veglis et al., 2016) offer the ability of functional adaptation related to media channel selection and content "appearance" shaping, based on the user's preferences and its surrounding environment. The foretold also incorporate all the necessary processing (i.e. transcoding, editing, authoring etc.), including direct (through explicit content description and semantic annotation tasks) and indirect (through browsing, social tagging, hypermedia authoring and general media analytics outcomes) meta-data acquisition, propagation and meta-processing (Dimoulas et al., 2014a, 2015; Leskovec, 2011; Zeng et al., 2010). In this context, M3A gathers all the necessary information regarding adaptation and personalization purposes, which can be further combined and integrated with ubiquitous and pervasive computing services, such as context-, time- and location-awareness that are quite common in contemporary mobile terminals and AR applications (Dimoulas et al., 2014b; Sun et al., 2014; Satoh, 2010). In the end, M3A acts as a sophisticated communication mediator between the users and the SNSs that: a) allows users to shape their own self-presentation, b) monitors users' client –side configuration regarding personalized browsing and interaction preferences, c) selects, filters and recommends specific media posts that users would be interested in, and d) adapts content presentation and selection of appropriate multichannel media in a context-aware manner, considering also users' access terminal.

Figure 2. The proposed multimodal media agent (M3A) framework

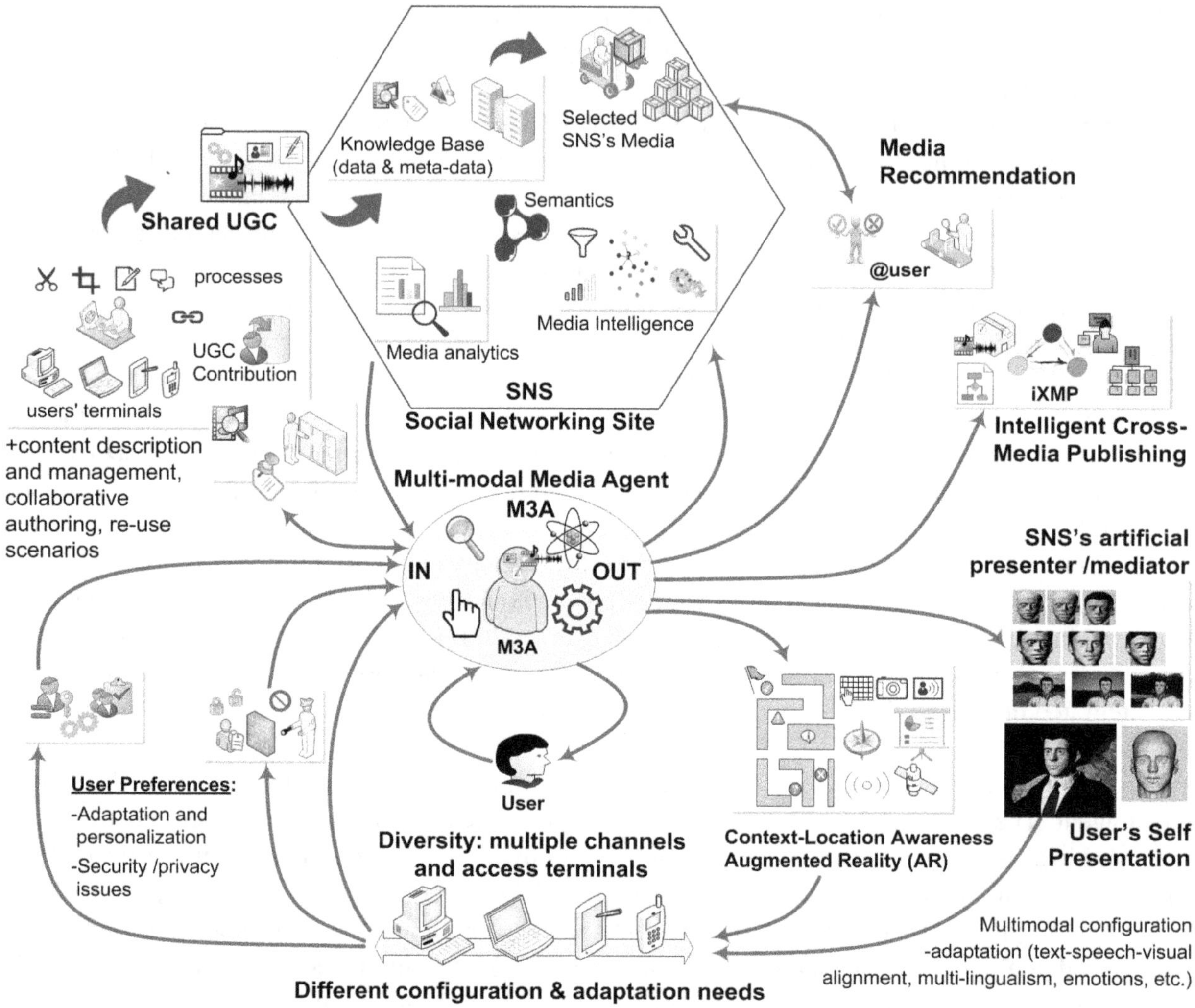

Solutions and Recommendations: Augmenting User Interaction through Social Network Embedded Multimodal Media Multi-Agent Models

The above M3A characteristics can be quite advantageous in the vast majority of infotainment services. For instance, artificial speakers interconnected with multilingual text to speech (TTS) and speech to speech (STS) engines and automated speech translation tools, which could deliver web-gathered textual information, tweets and other SMSs through audio communication (Jones, Berry, & Stevens, 2007; Avdelidis et al., 2002; Kalliris et al., 2002). This approach could combine the advantages of Internet's non-delayed, updated information that is now easily disseminated via flexible web-publishing tools (i.e. RSS feeds, twitter-like websites and other SNSs microblogging services) with the hands-free and undistracted wanted characteristics of radio (and/or web-radio), thus allowing extended use of radio-hybrid services to people with disabilities, as well as to a variety of many other social and professional groups that need timely and instant informing (i.e. at the office, while driving, during physical exercise/

housekeeping activities, etc.).Similarly, speech to text modalities can be used for voice directed message composition especially in interfacing-limited (mobile terminals), taking advantage of the limited bandwidth demands, the immediacy but also the popularity and the spread use of SMS-based communication, which is supported from most social media. Likewise, text-driven digital characters can enhance rich media experience, combining the above advantages (limited bandwidth demands, instant communication with immediate response, etc.), while respecting privacy concerns, offering user-based customization and promoting more-friendly game-like communication. Extending the above, natural language processing (NPL) techniques and text-mining –based topic classification strategies can be utilized along with multimodal content recognition and/or sentiment analysis algorithms, which all could be combined together in order to serve semantic understanding and enhanced interaction concepts (Dimoulas et al., 2014a, 2014b, 2015; Diou et al., 2010; Kotsakis et al., 2012; Symeonidis & Mitkas, 2005; Veglis et al., 2016).

According to figure 2 and the preceding analysis, M3A senses users' environment utilizing as inputs their personalization preferences, the attributes of the corresponding access terminals and their associated context aware parameters. In the actions side, M3A provides as results two main outputs: users' self-presentation to the social media, along with media presentation and adaptation to the users' needs. In terms of agent analysis, design and modeling it might be more simple and advantageous to use different agents for each output. Hence, M3A is altered into a social-network embedded multimodal media multi-agent (M4A) model that is suggested as a solution for augmenting users' interaction in social media (figure 3). This new model allows collaboration and interaction issues between various social media to be taken into consideration and finally be optimally settled in favor of the end-users. With this arrangement, users do not have to create multiple social media profiles (from the scratch) and insert their corresponding customized configuration and personalization preferences multiple times. As already implied, this downside along with the rest social media inhomogeneity difficulties and the lack of their direct linking block SNSs' integration and true semantic web implementation. Given the proposed M4A solution, such kind of obstacles can now be easier and more effectively faced by exploiting multi-agents' adaptability, sociability, cooperativity and competitiveness features. These multi-agents' attributes can significantly propel the construction of robust multi-purpose ground truth data-sets in terms of semantic labelling and social tagging, by successfully deploying (and combining) hierarchical and hybrid classification strategies along with various ensemble-, cascaded- and self-learning approaches (Dimoulas et al., 2008, 2011; Kalal et al., 2012; Kotsakis et al., 2012; Symeonidis & Mitkas, 2005; Tsirelis & Delopoulos, 2011).

For instance, Twitter and similar microblogging SNSs might be preferable for data-mining and textual approaches in semantic recognition, taking advantage of the remarkable progress that has been made in NLP research. These results can be further exploited as training mates in case that the associated media stories (and/or topics) also exist in some of their corresponding multimodal counterparts (i.e. audio –in SoundCloud etc., image –in Instagram, Flicker etc., video –in Youtube, Vimeo etc., non-linear multimedia in Web-docs and hypermedia authoring /storytelling platforms), and vice versa (Dimoulas et al., 2015; Veglis et al., 2016). This explanation also provides an additional interpretation of the "multimodal" term that is used in the M3A and M4A models, meaning that exploitation of multimodal content and their semantic analyses would lead to overall recognition accuracy improvements that are very useful in a variety of desirable semantic services (i.e. content recognition, events detection, summarization and indexing, topic understanding, semantic conceptualization, sentiment analysis, language recognition, multilingual adaptation, etc.) (Dimoulas et al., 2014b, 2015; Kalal et al., 2012; Kotsakis et al., 2012; Symeonidis & Mitkas, 2005; Veglis et al., 2016). In contrast to M3A, the acquisition and propagation of the semantic multimodal ground-truth is not coming from a single SNS and is made available for

Figure 3. Social network embedded multimodal media multi-agent (M4A) model

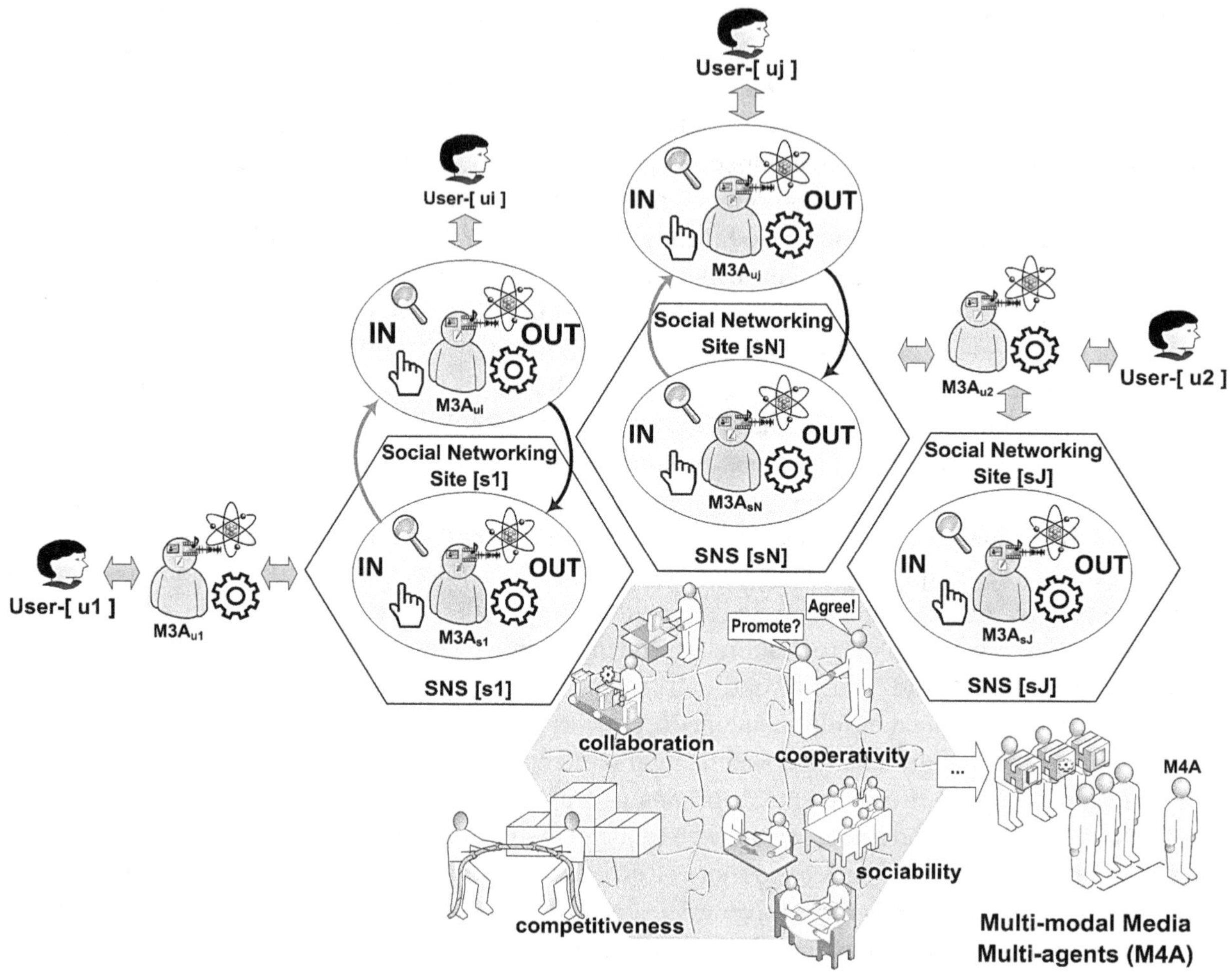

multiple SNSs. Most important of all, this knowledge base can now be rapidly constructed and widely expanded as a combined social media analytics outcome of the every day's content sharing, combining and browsing processes, and not through obligatory and strictly formalized annotation procedures that have to be individually conducted from multiple users (and at different SNSs). Although such an approach seems very difficult, at the moment, due to the extreme massiveness and heterogeneity of all of the different actors (content, SNSs, users' preferences, their terminals, etc.), it also seems that this is rather the only real opportunity for SNSs' integration to be accomplished, towards the realization of the Web 3.0/4.0 paradigms and beyond.

FUTURE RESEARCH DIRECTIONS

The implementation of the proposed framework brings forward a lot of issues that need to be resolved. First of all, the matter of interoperability, where it is quite obvious that large-scale convergence between different media formats, terminals and platforms is difficult to be achieved. Close related with the above issue is security and privacy, which are considered to be crucial for social media users. Security concerns not only users but also content and services. Thus, effective measures have to be taken in order to improve reliability, trust and privacy. In addition to the above, mobile terminals and AR applications that are based on cloud computing and mobile cloud computing appear to have certain implementation limitations and operation difficulties. For instance, most prominent delay and jitter issues are quite common, creating a lot difficulties and inconveniences, especially in demanding rich media application that require high quality of experience (Dimoulas et al., 2014b).

A common ground regarding agents and multi-agents technology, but also semantic web services, context awareness and automation enhanced HMI, is artificial intelligence. It is generally accepted that there is a continuous progress in machine learning algorithms and applications that has a positive impact on the future implementation of intelligent social media processing. Multimedia semantics and multimodal content management mechanisms appear to be very promising. However, sophisticated systems are required for true semantic knowledge extraction and exploitation. Also further enhancement in location-aware pervasive computing services is expected, by taking advantage of the constant context monitoring of the mobile user. One might argue that such kind of surveillance may augment privacy and uncomforting issues, but still there are many location-based AR services that could be positively deployed in the favor of users (Dimoulas et al., 2014b).

The previous mentioned developments can be also combined with advanced speech recognition, natural language understanding and dialog management components, further augmenting HMI potentials, thus overcoming current interfacing limitations (especially in mobile terminals) and offering enhanced multimodal communication (Dimoulas et al., 2014b; Zhou et al., 2008). It is worth noting that recent research reports indicate that content-based searching and retrieval services will also dominate the next web landscapes, offering the change for deploying innovative content sharing and dissemination services in social media. This kind of upcoming technologies are bound to eliminate issues related to presentation and usability from the "mobile environment", offering enhanced user-machine interaction that will copy many attributes of the natural human communication features. As foretold, one must not forget that multimodal intelligent content processing and semantic interaction techniques have not yet reached a maturity for large-scale real world applications, and they also have many prerequisites.

While the above issues require for careful treatment, it seems that multimodal media agents is the right solution to many of the presented difficulties. This can be further fuelled by the continuous progress of contemporary agent technology that has a great expansion and infiltration to many scientific fields (Symeonidis & Mitkas, 2005). However, there is an increased need for close interdisciplinary collaboration between experts of various kinds, such as social media specialists and professionals, journalists and content contributors in general, graphics' designers, mediated communication scientists, machine learning researchers, software engineers and web application developers. Although such large scale collaboration has been initiated, it seems that it will take a while before massive deployment and use of related services will be launched.

CONCLUSION

The current chapter studies social network embedded multimodal media agents and multi-agents as means of augmenting user interaction experience in contemporary networking and mediated communication services. The role of artificial characters and avatars in digital storytelling are initially discussed. Current challenges and expectations with emphasis on social media intelligence and analytics are presented, along with issues, controversies, and problems in enhancing user interaction experience through social networks. The role of multimodal media agents is then enlightened by discussing working definition and desired features. In this context, the sought functionalities of an integrated multimodal media agent (M3A) are thoroughly analyzed, leading to the implementation of an associated M3A model. Based on this, a social-network embedded multimodal media multi-agent (M4A) system is then modeled, offering augmented users' interaction and enhanced collaboration in social media. Both the proposed M3A and M4A models are examined with respect to the upcoming Web 3.0 and 4.0 paradigms, so that they are intended to play an important role in the implementation of these initiatives towards the transition to the Web of Things era (Web x.0). Looking back at the beginning of the social media and the progress that has been achieved so far, it can be foreseen that digital avatars and generally multimodal communication mediators equipped and/or combined with agents' intelligence and multi-agent intelligent systems will overwhelm social networking sites in the near future. The successful deployment of sophisticated social media analytics and media intelligence tools point to the same direction, strengthening and somewhat verifying the above expectation.

ACKNOWLEDGMENT

The authors would like to acknowledge the valuable contribution of Dr. K. A. Avdelidis that shared his expertise and his digital character (N-ico) graphic designs.

REFERENCES

Abel, F., Gao, Q., Houben, G. J., & Tao, K. (2011). Semantic enrichment of twitter posts for user profile construction on the social web. In C. Bussler, J. Davies, D. Fensel, & R. Studer (Eds.), *The Semantic Web: Research and Applications* (pp. 375–389). Springer. doi:10.1007/978-3-642-21064-8_26

Aghaei, S., Nematbakhsh, M. A., & Farsani, H. K. (2012). Evolution of the World Wide Web: From Web 1.0 to Web 4.0. *International Journal of Web & Semantic Technology*, *3*(1), 1–10. doi:10.5121/ijwest.2012.3101

Avdelidis, K., Dimoulas, C., Kalliris, G., Bliatsiou, C., Passias, T., Stoitsis, J., & Papanikolaou, G. (2002). *Multilingual automated digital talking character*. Paper presented at the International Broadcasting Convention, Amsterdam.

Batzios, A., Dimou, C., Symeonidis, A. L., & Mitkas, P. A. (2008). BioCrawler: An intelligent crawler for the semantic web. *Expert Systems with Applications*, *35*(1-2), 524–530. doi:10.1016/j.eswa.2007.07.054

Bélisle, J. F., & Bodur, H. O. (2010). Avatars as information: Perception of consumers based on their avatars in virtual worlds. *Psychology and Marketing*, *27*(8), 741–765. doi:10.1002/mar.20354

Bernal, P. A. (2010). Web 2.5: The symbiotic web. *International Review of Law Computers & Technology*, *24*(1), 25–37. doi:10.1080/13600860903570145

Bothos, E., Apostolou, D., & Mentzas, G. (2010). Using social media to predict future events with agent-based markets. *IEEE Intelligent Systems*, *25*(6), 50–58. doi:10.1109/MIS.2010.152

Breslin, J., & Decker, S. (2007). The future of social networks on the internet: The need for semantics. *IEEE Internet Computing*, *11*(6), 86–90. doi:10.1109/MIC.2007.138

Chapman, N., & Chapman, J. (2009). *Digital multimedia* (3rd ed.). Chichester, UK: John Wiley & Sons, Ltd.

Clements, M., de Vries, A. P., & Reinders, M. J. (2010). The influence of personalization on tag query length in social media search. *Information Processing & Management*, *46*(4), 403–412. doi:10.1016/j.ipm.2009.03.006

Cohn, J. F. (2007). Foundations of human computing: Facial expression and emotion. In J. G. Carbonell & J. Siekmann (Eds.), *Artificial Intelligence for Human Computing* (pp. 1–16). Springer. doi:10.1007/978-3-540-72348-6_1

Cooper, R., Dibbell, J., & Spaight, T. (2007). *Alter ego: avatars and their creators*. London: Chris Boot.

Correa, T., Hinsley, A. W., & De Zuniga, H. G. (2010). Who interacts on the Web?: The intersection of users' personality and social media use. *Computers in Human Behavior*, *26*(2), 247–253. doi:10.1016/j.chb.2009.09.003

Derks, D., Bos, A. E., & Von Grumbkow, J. (2008). Emoticons in computer-mediated communication: Social motives and social context. *Cyberpsychology & Behavior*, *11*(1), 99–101. doi:10.1089/cpb.2007.9926 PMID:18275321

Dimou, C., Symeonidis, A. L., & Mitkas, P. A. (2008). Data Mining and Agent Technology: a Fruitful Symbiosis. In O. Maimon & L. Rokach (Eds.), *Soft Computing for Knowledge Discovery and Data Mining* (pp. 327–362). Springer. doi:10.1007/978-0-387-69935-6_14

Dimoulas, C., Avdelidis, A., Kalliris, G., & Papanikolaou, G. (2008). Joint Wavelet video denoising and motion activity detection in multi-modal human activity analysis: Application to video – Assisted bioacoustic/psycho-physiological monitoring. *EURASIP Journal on Advances in Signal Processing*, *2008*(1), 792028. doi:10.1155/2008/792028

Dimoulas, C., Kalliris, G., Chatzara, E., Tsipas, N., & Papanikolaou, G. (2014a). Audiovisual production, restoration-archiving and content management methods to preserve local tradition and folkloric heritage. *Journal of Cultural Heritage*, *15*(3), 234–241. doi:10.1016/j.culher.2013.05.003

Dimoulas, C., Papanikolaou, G., & Petridis, V. (2011). Pattern classificationand audiovisual content management techniques using hybrid expert systems: A video-assisted bioacoustics application in abdominal sounds pattern analysis. *Expert Systems with Applications*, *38*(10), 13082–13093. doi:10.1016/j.eswa.2011.04.115

Dimoulas, C., Veglis, A., & Kalliris, G. (2014b). Application of Mobile Cloud-Based Technologies in News Reporting: Current Trends and Future Perspectives. In J. Rodrigues, K. Lin, & J. Lloret (Eds.), *Mobile Networks and Cloud Computing Convergence for Progressive Services and Applications* (pp. 320–343). Hershey, PA: Information Science Reference; doi:10.4018/978-1-4666-4781-7.ch017

Dimoulas, C., Veglis, A., & Kalliris, G. (2015). Audiovisual hypermedia in the semantic Web. In M. Khosrow-Pour (Ed.), *Encyclopedia of Information Science and Technology* (3rd ed.; pp. 7594–7604). IGI global. doi:10.4018/978-1-4666-5888-2.ch748

Diou, C., Stephanopoulos, G., Panagiotopoulos, P., Papachristou, C., Dimitriou, N., & Delopoulos, A. (2010). Large-scale concept detection in multimedia data using small training sets and cross-domain concept fusion. *IEEE Transactions on Circuits and Systems for Video Technology*, *20*(12), 1808–1821. doi:10.1109/TCSVT.2010.2087814

Drossos, K., Floros, A., & Kanellopoulos, N. G. (2012). Affective acoustic ecology: towards emotionally enhanced sound events. In *Proceedings of the 7th Audio Mostly Conference: A Conference on Interaction with Sound* (pp. 109-116). ACM. doi:10.1145/2371456.2371474

Feldon, D. F., & Kafai, Y. B. (2008). Mixed methods for mixed reality: Understanding users' avatar activities in virtual worlds. *Educational Technology Research and Development*, *56*(5-6), 575–593. doi:10.1007/s11423-007-9081-2

Golder, S. A., & Huberman, B. A. (2006). Usage patterns of collaborative tagging systems. *Journal of Information Science*, *32*(2), 198–208. doi:10.1177/0165551506062337

Huang, A. H., Yen, D. C., & Zhang, X. (2008). Exploring the potential effects of emoticons. *Information & Management*, *45*(7), 466–473. doi:10.1016/j.im.2008.07.001

Jacobsson, M., Rost, M., & Holmquist, L. E. (2006). When Media Gets Wise: collaborative filtering with mobile media agents. In *Proceedings of the 11th international conference on Intelligent user interfaces* (pp. 291-293). ACM. doi:10.1145/1111449.1111514

Jensen, J. (1998). Interactivity – Tracking a new concept in media and communication studies. *Nordicom Review*, *19*(1), 185–204.

Jones, C., Berry, L., & Stevens, C. (2007). Synthesized speech intelligibility and persuasion: Speech rate and non-native listeners. *Computer Speech & Language*, *21*(4), 641–651. doi:10.1016/j.csl.2007.03.001

Jones, S. (2003). *Encyclopedia of new media: an essential reference to communication and technology*. Thousand Oaks, CA: Sage Publications. doi:10.4135/9781412950657

Kalal, Z., Mikolajczyk, K., & Matas, J. (2012). Tracking-learning-detection. *Pattern Analysis and Machine Intelligence. IEEE Transactions on*, *34*(7), 1409–1422.

Kalliris, G. M., Dimoulas, C. A., Papanikolaou, G. V., Avdelidis, K., Passias, T. N., & Stoitsis, J. S. (2002). Phoneme recognition for 3d modeled digital character talking emulation. In Audio Engineering Society Convention 112 (paper no. 5626).

Kirkpatrick, D. (2011). The Facebook effect. New York: Simon & Schuster.

Kotsakis, R., Kalliris, G., & Dimoulas, C. (2012). Investigation of broadcast-audio semantic analysis scenarios employing radio-programme-adaptive pattern classification. *Speech Communication*, *54*(6), 743–762. doi:10.1016/j.specom.2012.01.004

Leskovec, J. (2011). Social media analytics: tracking, modeling and predicting the flow of information through networks. In *Proceedings of the 20th international conference companion on World wide web* (pp. 277-278). ACM. doi:10.1145/1963192.1963309

Malaby, T. (2009). *Making virtual worlds: Linden Lab and Second Life*. Ithaca, NY: Cornell University Press.

Mateas, M., & Sengers, P. (1999). Narrative intelligence. In *Proceedings AAAI Fall Symposium on Narrative Intelligence* (pp. 1-10). AAAI.

McCoy, J., Treanor, M., Samuel, B., Tearse, B., Mateas, M., & Wardrip-Fruin, N. (2010). Authoring game-based interactive narrative using social games and comme il faut. In *Proceedings of the 4th International Conference & Festival of the Electronic Literature Organization: Archive & Innovate*. Academic Press.

Meadows, D. (2003). Digital storytelling: Research-based practice in new media. *Visual Communication*, *2*(2), 189–193. doi:10.1177/1470357203002002004

Messinger, P. R., Ge, X., Stroulia, E., Lyons, K., Smirnov, K., & Bone, M. (2008). On the relationship between my avatar and myself. *Journal of Virtual Worlds Research*, *1*(2).

Milicevic, A. K., Nanopoulos, A., & Ivanovic, M. (2010). Social tagging in recommender systems: A survey of the state-of-the-art and possible extensions. *Artificial Intelligence Review*, *33*(3), 187–209. doi:10.1007/s10462-009-9153-2

Nadkarni, A., & Hofmann, S. (2012). Why do people use Facebook? *Personality and Individual Differences*, *52*(3), 243–249. doi:10.1016/j.paid.2011.11.007 PMID:22544987

Nadrljanski, M., & Batinica, V. (2010). Intelligent media agents. *International Journal of Intelligent Defence Support Systems*, *3*(1), 128–138. doi:10.1504/IJIDSS.2010.033682

Ochs, M., Pelachaud, C., & Sadek, D. (2008). An empathic virtual dialog agent to improve human-machine interaction. In *Proceedings of the 7th international joint conference on Autonomous agents and multiagent systems-Volume 1* (pp. 89-96). International Foundation for Autonomous Agents and Multiagent Systems.

Papadopoulos, S., Kompatsiaris, Y., Vakali, A., & Spyridonos, P. (2012). Community detection in social media. *Data Mining and Knowledge Discovery*, *24*(3), 515–554. doi:10.1007/s10618-011-0224-z

Petrakou, A. (2010). Interacting through avatars: Virtual worlds as a context for online education. *Computers & Education*, *54*(4), 1020–1027. doi:10.1016/j.compedu.2009.10.007

Qiu, L., & Benbasat, I. (2005). An investigation into the effects of Text-To-Speech voice and 3D avatars on the perception of presence and flow of live help in electronic commerce. *ACM Transactions on Computer-Human Interaction*, *12*(4), 329–355. doi:10.1145/1121112.1121113

Riedl, M. O., & Stern, A. (2006). Believable agents and intelligent story adaptation for interactive storytelling. In *Technologies for Interactive Digital Storytelling and Entertainment* (pp. 1–12). Springer. doi:10.1007/11944577_1

Riedl, M. O., & Young, R. M. (2010). Narrative planning: Balancing plot and character. *Journal of Artificial Intelligence Research*, *39*(1), 217–268.

Robin, B. R. (2008). Digital storytelling: A powerful technology tool for the 21st century classroom. *Theory into Practice*, *47*(3), 220–228. doi:10.1080/00405840802153916

Satoh, I. (2010). *Context-aware Media Agent for Public Spaces*. ECAI.

Spyridou, P.-L., & Veglis, A. (2008). Exploring Structural Interactivity in Online Newspapers: A Look at the Greek Web Landscape. *First Monday*, *13*(5). doi:10.5210/fm.v13i5.2164

Sun, Z., Yang, Y., Zhou, Y., & Cruickshank, H. (2014). Agent-Based Resource Management for Mobile Cloud. In J. Rodrigues, K. Lin, & J. Lloret (Eds.), *Mobile Networks and Cloud Computing Convergence for Progressive Services and Applications* (pp. 118–134). Hershey, PA: Information Science Reference; doi:10.4018/978-1-4666-4781-7.ch007

Symeonidis, A. L., & Mitkas, P. A. (2005). Agent intelligence through data mining. Springer.

Tsipas, N., Dimoulas, C., Kalliris, G., & Papanikolaou, G. (2013). Collaborative annotation platform for audio semantics. In *134th Audio Engineering Society Convention 2013*, (pp. 218-222).

Tsirelis, T., & Delopoulos, A. (2011). Automatic ground-truth image generation from user tags. In *12th International Workshop on Image Analysis for Multimedia Interactive Services (WIAMIS 2011)*. Academic Press.

Vasalou, A., & Joinson, A. N. (2009). Me, myself and I: The role of interactional context on self-presentation through avatars. *Computers in Human Behavior*, *25*(2), 510–520. doi:10.1016/j.chb.2008.11.007

Veglis, A., Dimoulas, C., & Kalliris, G. (2016). Towards intelligent cross-media publishing: media practices and technology convergence perspectives. In A. Lugmayr & C. Dal Zotto (Eds.), Media convergence Handbook (Vol. 2). Germany: Springer-Verlag..

Von der Puetten, A. M., Krämer, N. C., Gratch, J., & Kang, S. H. (2010). "It doesn't matter what you are!" explaining social effects of agents and avatars. *Computers in Human Behavior*, *26*(6), 1641–1650. doi:10.1016/j.chb.2010.06.012

Walker, M. (2014). *The History of Social Networking - WebMaster View*. Available at: http://www.webmasterview.com/2011/08/social-networking-history/

Wang, F. Y., Carley, K. M., Zeng, D., & Mao, W. (2007). Social computing: From social informatics to social intelligence. *IEEE Intelligent Systems*, *22*(2), 79–83. doi:10.1109/MIS.2007.41

Warburton, S. (2009). Second Life in higher education: Assessing the potential for and the barriers to deploying virtual worlds in learning and teaching. *British Journal of Educational Technology*, *40*(3), 414–426. doi:10.1111/j.1467-8535.2009.00952.x

Weiss, S. (2009). Privacy threat model for data portability in social network applications. *International Journal of Information Management*, *29*(4), 249–254. doi:10.1016/j.ijinfomgt.2009.03.007

Zeng, D., Chen, H., Lusch, R., & Li, S. H. (2010). Social media analytics and intelligence. *IEEE Intelligent Systems*, *25*(6), 13–16. doi:10.1109/MIS.2010.151

Zhang, Y., Yu, X., Dang, Y., & Chen, H. (2010). An integrated framework for avatar data collection from the virtual world. *IEEE Intelligent Systems*, *25*(6), 17–23. doi:10.1109/MIS.2010.138

Zhou, F., Dun, H. B.-L., & Billinghurst, M. (2008). Trends in augmented reality tracking, interaction and display: A review of ten years of ISMAR. In *Proceedings - 7th IEEE International Symposium on Mixed and Augmented Reality 2008, ISMAR 2008,* (pp. 193-202). IEEE.

ADDITIONAL READING

Atzori, L., Delgado, J., & Giusto, D. D. (2012). Special issue on pervasive mobile multimedia. *Signal Processing Image Communication*, *27*(8), 785–934. doi:10.1016/j.image.2012.05.002

Avdelidis, K., Dimoulas, C., Kalliris, G., & Papanikolaou, G. (2010). Adaptive phoneme alignment based on rough set theory. In *Rough Sets and Current Trends in Computing* (pp. 100–109). Springer. doi:10.1007/978-3-642-13529-3_12

Avdelidis, K. A. (2012). Synchronous voice and speech analysis and synthesis techniques using visual and auditory information. (Unpublished Ph.D.). Dept. of Electrical and Computer Engineering, Aristotle University of Thessaloniki, Thessaloniki Greece. (129567), http://invenio.lib.auth.gr/record/129567/

Benz, D., Körner, C., Hotho, A., Stumme, G., & Strohmaier, M. (2011). One tag to bind them all: Measuring term abstractness in social metadata. In The Semanic Web: Research and Applications (pp. 360-374). Springer.

Borrino, R., Furini, M., & Roccetti, M. (2009). Augmenting social media accessibility. In *Proceedings of the 2009 International Cross-Disciplinary Conference on Web Accessibililty (W4A)* (pp. 54-57). ACM. doi:10.1145/1535654.1535666

Boulos, M., Hetherington, L., & Wheeler, S. (2007). Second Life: An overview of the potential of 3-D virtual worlds in medical and health education. Health Information \&. *Library Journal*, *24*(4), 233–245.

Cavazza, M., Charles, F., & Mead, S. J. (2002). Planning characters' behavior in interactive storytelling. *The Journal of Visualization and Computer Animation*, *13*(2), 121–131. doi:10.1002/vis.285

Cool, J. (2008). Communities of Innovation: Cyborganic and the Birth of Networked Social Media. PhD dissertation, University of Southern California.

Cordara, G., Bober, M., & Reznik, Y. (2013). Special issue on visual search and augmented reality. *Signal Processing Image Communication*, *28*(4), 309–402. doi:10.1016/j.image.2012.11.001

Diemert, B., Abel, M. H., & Moulin, C. (2011). Semantic audiovisual asset model - the case of TV production. *Multimedia Tools and Applications*, 1–28.

Dijck, J. (2013). *The culture of connectivity* (1st ed.). Oxford: Oxford University Press. doi:10.1093/acprof:oso/9780199970773.001.0001

Dimoulas, C., & Kalliris, C. (2013). Investigation of wavelet approaches for joint temporal, spectral and cepstral features in audio semantics. In *134th Audio Engineering Society Convention 2013*, 509-518.

Domingue, J., Fensel, D., & Hendler, J. (Eds.). (2011). *Handbook of Semantic Web Technologies*. Springer. doi:10.1007/978-3-540-92913-0

Drossos, K., Kotsakis, R., Kalliris, G., & Floros, A. (2013). Sound events and emotions: Investigating the relation of rhythmic characteristics and arousal. In *Information, Intelligence, Systems and Applications (IISA), 2013 Fourth International Conference on* (pp. 1-6). IEEE.

Karpouzis, K., Caridakis, G., Kessous, L., Amir, N., Raouzaiou, A., Malatesta, L., & Kollias, S. (2007). Modeling naturalistic affective states via facial, vocal, and bodily expressions recognition. In *Artifical Intelligence for Human Computing* (pp. 91–112). Springer Berlin Heidelberg. doi:10.1007/978-3-540-72348-6_5

Kim, H. N., Alkhaldi, A., El Saddik, A., & Jo, G. S. (2011). Collaborative user modeling with user-generated tags for social recommender systems. *Expert Systems with Applications*, *38*(7), 8488–8496. doi:10.1016/j.eswa.2011.01.048

Kim, Y. J., Baker, J., & Song, J. (2007). An exploratory study of social factors influencing virtual community members' satisfaction with avatars. *Communications of the Association for Information Systems*, *20*(1), 36.

Kompatsiaris, Y., & Hobson, P. (2008). *Semantic Multimedia and Ontologies*. London, UK: Springer–Verlag Limited. doi:10.1007/978-1-84800-076-6

Kompatsiaris, Y., Merialdo, B., & Lian, S. (Eds.). (2012). *TV Content Analysis: Techniques and Applications*. CRC Press.

Liao, S., Chu, P., & Hsiao, P. (2012). Data mining techniques and applications - A decade review from 2000 to 2011. *Expert Systems with Applications*, *39*(12), 11303–11311. doi:10.1016/j.eswa.2012.02.063

Monaghan, F., Handschuh, S., & O'Sullivan, D. (2011). ACRONYM: Context metrics for linking people to user-generated media content. *International Journal on Semantic Web and Information Systems*, *7*(4), 1–35. doi:10.4018/jswis.2011100101

Pagani, M. (2009). Encyclopedia of Multimedia Technology and Networking (2nd ed.). (Vols. 1–3). pp. 1–1756)., doi:10.4018/978-1-60566-014-1.ch001

Psarras, V., Floros, A., Drosos, K., & Strapatsakis, M. (2011). Emotional control and visual representation using advanced audiovisual interaction. *International Journal of Arts and Technology*, *4*(4), 480–498. doi:10.1504/IJART.2011.043446

Quénot, G., Benois-Pineau, J., & André-Obrecht, R. (Eds.). (. (2012). Content-based Multimedia Indexing, Special issue of MTAP following CBMI 2010. *Multimedia Tools and Applications*, *60*(2), 257–482. doi:10.1007/s11042-010-0648-2

Schimratzki, O., Bakalov, F., Knoth, A., & König-Ries, B. (2010). Semantic enrichment of social media resources for adaptation. In *Proceedings of International Workshop on Adaptation in Social and Semantic Web (SAS-WEB 2010), Big Island of Hawaii* (pp. 31-41).

Spyridou, L. P., Matsiola, M., Veglis, A., Kalliris, G., & Dimoulas, C. (2013). Journalism in a state of flux: Journalists as agents of technology innovation and emerging news practices. *International Communication Gazette*, *75*(1), 76–98. doi:10.1177/1748048512461763

Tsikrika, T., Diou, C., de Vries, A. P., & Delopoulos, A. (2011). Reliability and effectiveness of clickthrough data for automatic image annotation. *Multimedia Tools and Applications*, *55*(1), 27–52. doi:10.1007/s11042-010-0584-1

Vasalou, A., Joinson, A., Bänziger, T., Goldie, P., & Pitt, J. (2008). Avatars in social media: Balancing accuracy, playfulness and embodied messages. *International Journal of Human-Computer Studies*, *66*(11), 801–811. doi:10.1016/j.ijhcs.2008.08.002

Vasalou, A., Joinson, A. N., & Pitt, J. (2007, April). Constructing my online self: avatars that increase self-focused attention. In *Proceedings of the SIGCHI conference on Human factors in computing systems* (pp. 445-448). ACM. doi:10.1145/1240624.1240696

Vrysis, L., Dimoulas, C., Kalliris, C., & Papanikolaou, G. (2013). Mobile audio measurements platform: bringing audio semantic intelligence into ubiquitous computing environments. In *134th Audio Engineering Society Convention 2013*, 183-189.

KEY TERMS AND DEFINITIONS

Augmented Interaction: Exploitation of augmented reality concepts where physical and virtual worlds are merged, offering augmented information views and interaction in the favor of mobile users.

Avatar: Denotes a graphical object corresponding to the user's virtual body in on-line virtual worlds, usually encompassing kinds of autonomy, experience and/or intelligence (the term originates from the word "avatara", which means "incarnation" in Sanskrit).

Digital Storytelling: A technology application that creates short multimedia stories by making use of low-cost digital equipment combined with various types of media and taking advantage of user generated content (UGC) shared resources.

Multimodal Media Agent: A media-oriented software entity that has (or uses) multiples or many semiotic modes and communication channels and that is able to sense its surrounding environment and autonomously adapt on its changes, responding on behalf of other entities (humans or software systems).

Media Analytics: The scientific field that deals with developing and evaluating informatics tools and frameworks in order to collect, monitor, analyze, summarize and visualize data that can be found in SNSs.

Media Intelligence: The scientific field that aims at extracting information from social media, developing corresponding decision-making or decision-aiding frameworks, and providing architectural designs and solution frameworks for existing and new applications.

Multimedia Semantics: Intelligent content analysis usually implementing machine learning with text and multimedia features, purposing to provide semantic interpretation and conceptualization outcomes.

Social Networking Sites (SNS): A group of Internet-based applications that were built on the ideological and technological foundations of Web 2.0, which allow the creation and exchange of user-generated content but also pursue the vision of large scale semantic interaction, collaboration and integration.

Chapter 11
Digital Paranoia:
Unfriendly Social Media Climate Affecting Social Networking Activities

Ramona Sue McNeal
University of Northern Iowa, USA

Mary Schmeida
Kent State University, USA

ABSTRACT

Participation in social networks, forums, and other discussion groups is a growing trend in the United States. Aside from the benefits of online social media, there is a growing concern about privacy and safety from the devolvement of personal information online. As a result of this unfriendly social media climate, Americans are taking measures to protect personal identity and to avoid surveillance by others. The purpose of this chapter is to analyze factors predicting which groups are most concerned about Internet privacy. In addition, this chapter explores how concerns regarding Information privacy are impacting usage of social network sites. We explore these questions using multivariate regression analysis and individual level data from the Pew Internet and American Life Project. Our findings suggest that those with the greatest fears regarding online privacy are not staying offline but are taking necessary precautions to address concerns.

INTRODUCTION

The social networking climate in the U.S. has become unfriendly for some users, and possibly affecting their emotional health. This is a concern particularly with social media interaction becoming a popular activity with 67% reporting using some form of social networking site and most appealing to those under age 50 years, women, and urban residents (Pew Internet & American Life Project, 2013a, p. 2). With social media use popular among Americans both on and off the job, some employers now require employees to disclose user names and passwords in effort to comply with financial obligations to government, to protect proprietary information, and to minimize company liabilities. As of October

DOI: 10.4018/978-1-4666-8556-7.ch011

2013, 36 states have introduced legislation to prohibit employers from requesting social media personal account information from users (National Conference of State Legislatures, 2014, p. 2). Currently, 32 states have passed data disposal laws requiring businesses and government to make personal identifying information that is collected and stored by these institutions destroyed and/or altered so that it is no longer comprehendible (National Conference of State Legislatures, 2015). As a result of this unfriendly social media climate, growing numbers of Americans are concerned about their personal identity and cybercrime, and are taking measures to avoid surveillance by others. A 2013 Pew Internet & American Life Project survey, found as many as 55% reported avoiding observation by other entities such as employers, government, organizations, people and groups; while 86% of adult Internet users have taken measures to promote anonymity, privacy, and security online. Cybercrime continues to grow with 11% reporting stolen bank account or credit care information, and/or Social Security Number. In an attempt to avoid surveillance, online user behaviors are diverse and range from masking personal information, clearing search histories, to using public computer instead of personal home computer (Pew Internet & American Life Project, 2013b, pp. 1-2, 5, 9).

This change in the social media climate in the U.S. raises a number of questions regarding its impact on individual trust of both government and other organizations. In this chapter we will explore this phenomenon "digital paranoia" by empirically analyzing 1) what demographic groups are most concerned about Internet privacy and 2) how are these concerns impacting usage of social network sites (SNS). This chapter will report research analysis from 2013 using multivariate regression analysis and Pew Internet & American Life Project social media user opinion data.

BACKGROUND

Socialization online is becoming a popular activity in the United States. Accessibility to social media interaction is easy, with new generation technological devices such as tablet computers and the multifunctional cell phones (for example Smartphones), becoming more available. The benefits to social networking and online forums range from sharing personal opinion(s) to discussion of political events and political activism. Social media is also becoming a common meeting place for the online health community for sharing caregiver experiences and provide emotional support. However, the devolvement of personal information during socialization has led to a heightened awareness of privacy and security matters, not only for adults, but for children. The issue of Internet privacy has moved beyond concerns of individual consumer to the government agenda. In 2007, the Federal Trade Commission (2007) identified social networking and mobile Internet as children's online privacy issues. As a response, individuals, organizations, and government are taking measures to protect information that may jeopardize personal and professional integrity.

Social media users are growing in number and are a broad demographic group. According to the Pew Internet & American Life Project (2013c) since May 2013, 72% of U.S. adults have visited social networking sites (SNS), an increase of 5% from 2012. Adults age 65 and over are also visiting these sites growing from 13% in early 2009 to 43% mid 2012 (Pew Internet & American Life Project, 2013c, pp. 2, 4). Although youth are more likely to take part in social network activities, this activity continues to grow with six out of ten Internet users visiting social sites for age group 50 to 64 (Pew Internet & American Life Project, 2013d, p. 2). In 2012, social networking was most common for women, individuals under age 50, urban residents, with those in age group 18 to 29 years leading other groups in socialization using Facebook, Twitter, Instagram, and Tumblr. In addition, women, African-Americans, and Latinos also show high interest in sites like Pinterest (Pew Internet & American Life Project, 2013e, p. 2).

Youth under the age of 13 are participating in social networking where website operators in compliance with federal protective laws, may not be aware that information gathering on their sites may be from children falsifying their age to participate in networking activities (Federal Trade Commission, 2007). Today's teens are strong users of digital media participating in various activities, such as texting, e-mailing, Web browsing and online purchases. Their digital activities are more intense than adult users and they may not be as informed about the drawbacks to online information sharing (Federal Trade Commission, 2010). Personal data can be used by outsiders for surveillance including Internet service providers tracking online behavior for advertizing (Pew Internet & American Life Project, 2012a, p. 6), by employers, and criminals.

Other groups participating in social media are the home healthcare givers in the U.S. The digital environment has brought support to this population providing health assistance in their home setting. Mobile technology allows this group to interface with highly interactive social health platforms, download health applications, and visit government medical websites. Healthcare social networking forums are becoming a meeting place to exchange ideas, personal healthcare experiences and act as a support system to home caregivers. These home health consumers are more likely to use multifunctional cell phones, such as Smartphones and download health applications led by females, those under the age of 50, those highly educated and with higher incomes (Pew Internet & American Life Project, 2012b). Personal information disclosed during social media activities, such as personal experiences may be used by non-intended individuals in ways that may jeopardize personal and professional integrity.

Aside from socialization, the Internet has become a convenient medium for banking online, applying for licenses, submitting taxes, ordering prescription medications, purchasing vehicles, etc. However, the social media climate in the U.S. has become unfriendly because of cybercrime and is precipitating protective-type behaviors. Concerned about privacy matters, many are taking measures to protect their identity and personal information and to avoid surveillance by others (Pew Internet & American Life Project, 2013e, p. 2). Some individuals go so far as to stay offline. Initially, researchers (Belanger & Carter, 2009) argued that citizens who did not go online either lacked access to the Internet or did not have the necessary skills to navigate the World Wide Web. Van Dijk (1999) however argued that there was a third reason for individual choice to stay offline. This barrier, the psychological divide, could be described as limited Internet use because of mental barriers including lack of interest and fear of computers.

Protecting Personal Online Integrity

Individuals who avoid the Internet entirely because of security and privacy concerns represent a small percentage of the U.S. population: approximately 1% U.S. households (National Telecommunications and Information Administration [NTIA], 2014, p. 26). Nevertheless, citizens are becoming more concerned about Internet security and privacy issues and some are altering their online behavior as a result. As online socialization grows, so do concerns regarding personal information that may be devolved in social networking and forums, information such as location, age, race, occupation, and political affiliation. Although a small number, about 9% of teens identify being "seriously" concerned about third party access to their personal information shared on social media sites; they are increasingly taking action to protect personal information. Actions taken by teenagers to managing their identity includes changing content, masking information from others, and deleting other people from their network or list (Pew Internet &

American Life Project, 2013e). According to the Pew Internet & American Life Project (2013c), 58% of 12-17 year olds report having downloaded a software application (app) to their cell phone or computer (p. 3). However, 26% of teen app users have uninstalled an application that they found was collecting personal information on them (p. 6). Approximately 50% of teen app users turned off location tracking on their cell phone or app (p. 7) so as to protect location data, especially girls (who downloaded apps) at 59% as compared to 37% boys (p. 8).

Other protective activities are reported by social media users and people interfacing with institutional entities online. In a Pew Internet & American Life Project survey, as many as 55% reported avoiding observation or surveillance by other entities such as employers, government, organizations, people and groups; while 86% of adult Internet users have taken measures to promote anonymity, privacy, and security online. Cybercrime continues to grow with 11% reporting stolen bank account or credit card information, and/or Social Security number. In an attempt to avoid these crimes and/ or surveillance by outsiders, people are masking personal information, changing privacy settings, clearing search histories, using different profiles, and using a public computer versus personal home computer (Pew Internet & American Life Project, 2013f, pp. 1-2, 5, 9). False names and email addresses are typical evasive behavior, even though this may be in violation of service terms by some social networking websites. Software that masks Internet profile addresses and deleting cookies after social networking is another way some individuals maintain anonymity (Privacy Rights Clearinghouse, 2014a).

Depending on employer policy, some employers may monitor social media activities of employees or hire third-party companies to monitor for them. Surveillance of social networking sites is of interest for those employers concerned about employee behavior that may unfavorably effect company reputation (Privacy Rights Clearing House, 2014b). Employers may require employees to disclose user names and passwords to learn about employee online activities in attempt to protect proprietary information, to minimize company liabilities, to comply with financial obligations to government, and to secure or keep one's employment (National Conference of State Legislatures, 2014). As a response to employer surveillance, employees can inquire about employer policy. Other steps that are taken by employees include searching database websites listing social media policies for listed companies that can inform the employee about employer surveillance (Privacy Rights Clearing House, 2014b).

Government Response to an Unfriendly Social Media Climate

Citizen concern regarding the collection of personal information in computer files began before the advent of the Internet. A 1983 poll by Harris found that 60% of those surveyed did not feel that information on computers was adequately protected. Furthermore, a 1998 Harris Poll found that 88% of Internet users worried that others could be reading their email without their knowledge (Robbin, 2001, p.499). These fears translated to demand by citizens for government action to regulate how information could be collected from the Internet and used. A 1998 Harris Poll found that 50% of respondents felt that the government should pass laws regarding the collection and use of information online. The number that supported such laws rose to 57.4% in a 2000 Harris Poll (Robbin, 2001, p.500). More recently, a Pew Internet & American Life Survey (2013b) found that 66% of those surveyed did not feel that current laws were sufficient to protect online privacy.

Despite public concerns, federal legislation that targets online privacy violations and risks has been piecemeal in nature with states taking the lead on these laws. Federal laws protecting mobile privacy have not been progressively made, because of opposition from industry, such as advertizing, and because of the laggard nature of passing federal laws in light of a fast growing mobile industry. Those that have passed are not uniform. The cornerstone federal law passed during the crisis of 9/11 in short title called the USA Patriot Act of 2001, known as "Uniting and Strengthening America by Providing Appropriate Tools Required to Intercept and Obstruct Terrorism" (USA PATRIOT) Act of 2001, aims to improve U.S. counterterrorism efforts. Among its many provisions are those that bring federal laws up to date on new technology threats, such as computer hacking. Victims can now seek law enforcement assistance to monitor hackers on their computers (Public Law 107-56, 2001). Passed before the Internet became a public communication tool was the Electronic Communications Privacy Act (ECPA) of 1986. This law sets restrictions on government entities who wire tap telephone calls, computer electronic data transmission, and who track or engage in surveillance or other digital communication means. It protects the general public and upholds citizen constitutional right to privacy if they are under government surveillance. It protects the citizen from any aggressive behavior that a governmental entity may implement in effort to gain their information. It sets requirements for law enforcement to obtain warrants related to observed communication behavior (Public Law 99-508, 1986).

Although the USA PATRIOT Act and ECPA should provide citizens some peace of mind regarding Internet privacy, they are actually raising greater fear regarding the protection of online privacy. The ECPA along with the Right to Financial Privacy Act (RFPA) of 1988 were the first to introduce the use of National Security Letters (NSL). The NSLs give the FBI the ability to obtain access to individual financial institution records, telephone records and other communication service provider information without going through the normal procedures of obtaining a warrant. National Security Letters were meant to help the FBI obtain information on foreign powers and agents of foreign powers. However, section 505 of the USA PATRIOT Act loosened restrictions for issuance of a NSL, raising fear regarding information privacy of ordinary citizens (Gorham-Oscilowski, & Jaeger, 2008).

In addition to federal laws that target government behavior, there are those that regulate the actions of financial institutions. By law, financial institutions can consolidate. The Gramm-Leach-Bliley Act on financial privacy allowed financial affiliates to operate as one. As an effect of their lawful consolidation, they can merge customer data collected for their services and sell or rent it to non-affiliates or third parties. This transaction can affect customer privacy since personal data has moved to a third party. However, several provisions of the Gramm-Leach-Bliley Act do provide protection of consumer's nonpublic personal financial information. It outlines the obligation of financial institutions to protect the customer, sets limitations on the sharing of information, such as account number for marketing purposes, and identifies the regulatory agencies and enforcement rules (Public Law 106-102, 1999).

Other federal legislation targets the Internet, social media activities and consumer privacy concerns. For example, the Children's Online Privacy Protection Act of 1998 (COPPA) is a protective-type of regulatory law enforced by the Federal Trade Commission (FTC). COPPA was passed to protect the privacy of children age 13 and under who use the World Wide Web. It targets commercial Websites. Among its provisions, COPPA permits the FTC to bring civil penalties against violators of the act. Parental consent must be obtained from the Website (website operators) prior to collecting information from children. In addition, Websites must provide a link to a notice disclosing their information gathering policies and at each point where data is gathered. Parental consent is not required for all data gathering information, and there are varying types of methods for gaining parental consent. Schools as "parental agents" may provide consent for students (Federal Trade Commission, 1999).

Recently, "knowledge-based authentication" has been approved as a method to verify the consenter's identity, for a child's use of a particular website. In this new method, atypical questions are asked that measure beyond identifying data of a driver's license, to verify the actual consenter's identity. This method promises to overcome falsification of information by users seeking website access without true parental permission (Federal Trade Commission, 2013). The FTC has acknowledged that an extension to COPPA to protect teen privacy may prove to be not as effective for this age group as compared to those under age 13 which COPPA targets. Adolescents are considered to be more impulsive in behavior compared to their under age 13 counterparts and may be more likely to alter parental consent information and enter into online activity despite parental disapproval. As a result, the FTC has taken on an educational initiative aimed at adolescents and their parents to promote more responsible online activity geared toward safe social media use by adolescents (Federal Trade Commission, 2010). The educational consumer guide *Net Cetera: Chatting with Kids about Being Online* has recently been updated as a parental tool in guiding responsible online teen behavior, and targets current forms of online teen activity, such as downloading mobile applications (Federal Trade Commission, 2014).

States have pick up where federal effort leaves off by passing regulatory laws that protect social networkers and other online users from threats to privacy and security. As a result of employer surveillance of employees online, states have introduced legislation prohibiting employers from requesting social media personal account information from users. Some states have passed laws requiring employers to give employees notice of monitoring of employee online activity. For example, Tennessee and Colorado now require any institution that operates an email system and monitoring employee activity to have a written policy giving employees written notice (National Conference of State Legislatures, 2011a). In 2012, state-level laws were introduced prohibiting employers from asking employees to disclose their private Internet account passwords, and similarly protecting students in higher public education institutions with their social networking accounts (National Conference of State Legislatures, 2014). States have also picked up where federal legislation such as, the Gramm-Leach-Bliley Act on financial privacy (Public Law 106-102, 1999) has left off in protecting consumer's personal financial information. States have introduced a variety of regulatory laws to protect those using the Internet for activities requiring one to disclose personal financial information, such as banking online information. According to the Pew Internet & American Life Project (2013f), about 61% of Internet users are banking online with 32% of adults banking by mobile phone (p. 2).

More than half of the states have passed data disposal laws requiring government and private businesses who collect and store personal information to dispose, destroy, or make unreadable this information. California has been an innovator state in passing protective type of regulatory legislation on cybercrime. For example, California passed laws regulating businesses in the management of customer personal records and disclosure of breached data (National Conference of State Legislatures, 2015) and is one of many innovator states passing laws protecting consumers against computer spyware or adware software that can track online activities, collect personal data, and alter the computer system (National Conference of State Legislatures, 2013). California has also passed the California Online Privacy Act requiring any website operator collecting personal information from California residents on their website, to post an online privacy policy. They are required to inform residents on "what" is collected and the third-parties who will view the information (California Business and Professions Code, 2003; Privacy Rights Clearinghouse, 2014a).

Government agencies use information technology for improving functionality of government, to provide better service quality, improve accessibility of resident information, and to improve the cost-effectiveness of government services to citizens, particularly important since most states face a budget crisis year-to-year. This information collected on citizens is also subject to the privacy laws, particularly with the sharing of information among government agencies. As a result, some states have passed legislation requiring government Web sites or state portals to develop privacy laws, and post them on their Web sites. In Delaware for example, each state agency must post a direct link to its privacy policy and provide a paper copy to institutions upon request (National Conference of State Legislatures, 2011b).

Although both the federal and state governments have taken action to protect Internet privacy and insure online security, news headlines have given citizens reason to be worried. Headlines over the last year raising red flags with regard to Internet privacy and online security have included Edward Snowden's allegations regarding the National Security Agency (NSA); a security breach of customer data at Target and most recently the heartbleed virus. These events raise a number of questions. Which individuals are most likely to be concerned about Internet privacy and security? What actions are individuals taking to insure information privacy? Are online fears causing individuals to avoid certain Internet activities altogether?

EMPIRICAL MODEL: DATA AND METHODS

This study utilizes the *2013 Pew Anonymity (omnibus) Survey* conducted for the Pew Internet & American Life Project, by the Princeton Survey Research Associates. The Pew Internet & American Life Project is part of the Pew Research Center, a nonpartisan, nonprofit group that conducts studies and provides information regarding factors that shape American society. Their surveys are random digit dial national telephone surveys limited to individuals 18 years or older and live in the continental United States. Participants of this survey were asked questions regarding anonymity online. This study is based on a survey with a sample size (n) of 1,002 adults living in the United States collected from July 11 to July 14, 2013.

Digital Privacy and Security Concerns

This study begins with an examination of the demographic and geographical factors associated with an individual's worries regarding information privacy and security. Independent variables were selected based on findings from prior research. Two dependent variables were utilized. The first is as a dummy variable constructed using the following question: "Do you ever worry about how much information is available about you on the Internet, or is that not something you worry about?" The second dependent variable is a dummy variable constructed using the following question: "Considering everything you know & have heard about the Internet, do you think it is possible to use the Internet completely anonymously-so that none of their online activities can be easily traced backed to them?" Both variables are coded 1 if an individual responded yes and 0 otherwise.

Explanatory or independent variables includes an *Income* measure on a 9-point scale where 1 indicated that family incomes ranges from \$0 to \$10,000 and 9 signifies a family income of \$150,000 or more. *Education* is measured using a 7-point scale, ranging from eighth-grade education or less to post-graduate training/ professional training, which includes Masters/ Ph.D., law and medical school. *Age* is measured in years. Research (Selwyn, 2004; Mossberger, Tolbert, & McNeal, 2007) has found that greater usage can lead to improved technology skills. This includes technical competencies which includes knowing how to use hardware and software. It is expected that those individuals who are more likely to use the Internet regularly (younger, more affluent and better educated) would have greater technical competency as well as a greater knowledge of and concerns of online threats to security and privacy. *Gender* is measured using a binary variable coded 1 for male and 0 for female. The literature (Pew Internet & American Life Project, 2013c) suggests that females are more likely to be concerned about online security and privacy.

To control for race and ethnicity, dummy variables were included for *African Americans, Asian Americans*, and *Latinos* with *non-Hispanic Whites* as the reference group. Earlier research (Mossberger, Tolbert, & Stansbury, 2003) found that African American and Latino populations are more likely than whites to express positive attitudes toward telecommunication technology. Based on this research, it is expected that whites will be more concerned about online privacy and security. Dummy variables for the *Midwest, Northeast and West* regions were included to control for regional differences. The South was designated as the reference group because states in the South have been slower to adopt legislation prohibiting employers from requesting social media personal account information from users or pass data disposal laws (National Conference of State Legislatures, 2011a, 2015). Since research (Pew Internet & American Life Project, 2012a) finds that individuals living in suburban areas are more likely to say that current laws do not sufficiently protect online privacy, two dummy variables were included for the respondent's community type (*Suburban and Urban*) with rural as the reference group. Finally, dummy variables for *Employed Fulltime* and *Employed Part-Time* were added to control for employment status. Because some employers have requested social media personal account information from users, individuals who are employed may be more sensitive to privacy issues.

Social Network Avoidance

Because this study also explores how fears regarding Internet privacy and security alter online behavior, additional models were run to determine if these fears limited social network activities. Control variables were selected based on findings from prior research on barriers to Internet usage. Two dependent variables were used. The first is as a dummy variable constructed using the following question: "Do you ever use the Internet to use a social networking site like Facebook, LinkedIn or Google Plus?" The second dependent variable is a dummy variable constructed using the following question: "Do you ever use the Internet to use Twitter?" Both variables are coded 1 if the respondent answered yes and 0 otherwise. The two main independent variables in these models are the dependent variables from Table 1.

As with the previous model (Table 1), demographic and geographical measures will be used as control variables but their inclusion is based on a different literature. More specifically, control variables were chosen based on the "digital divide" or inequalities in Internet access and usage. There are three main reasons why individuals do not go online; the first two are lack of Internet access and technical skills. Mossberger et al. (2003) describe technology skills in terms of two broad categories. The first category is technical competencies that include the necessary skills to use hardware and software such as typing and using a mouse. The second category (information literacy) concerns the ability to determine which information to obtain from the Internet for specific tasks (Mossberger et al., 2003, p. 38). Their research found that individuals lacking technical skills tended to be older, less educated, Latino, African American and less affluent. Early research on Internet access in the United States (NTIA, 2000) found a similar divide based on access. Those least likely to have Internet access were female, less affluent, older, less

Table 1. Digital Privacy and Security Concerns, 2013

	Worried About Internet Privacy		Feels They can use the Internet Anonymously	
Variables	β (se)	p>lzl	β (se)	p>lzl
Environmental Variables				
Suburban	-.079(.170)	.643	- .282(.183)	.122
Urban	**-.370(.215)**	**.084**	**-.491(.228)**	**.031**
Midwest	-.130(.203)	.520	-.186(.214)	.385
Northeast	.068(.223)	.760	**-.420(.246)**	**.088**
West	**.453(.196)**	**.021**	- .109(.211)	.606
Individual Level Variables				
Age	**-.020(.005)**	**.000**	**- .020(.005)**	**.000**
Male	**-.390(.152)**	**.010**	**.301(.163)**	**.064**
Latino	.006(.277)	.983	.213(.282)	.451
Black	-.330(.256)	.196	-.139(.262)	.595
Asian	.340(.427)	.426	.654(.426)	.125
Education	**.136(.047)**	**.004**	**-.119(.051)**	**.022**
Income	**.064(.036)**	**.074**	**.067(.039)**	**.026**
Work Part Time	.294(.247)	.233	.029(.276)	.916
Work Full Time	.221(.187)	.236	**.349(.201)**	**.082**
Constant	-.367(.380)	.334	.324(.406)	.425
Pseudo R2	.0711		.0592	
LR Chi² (14)	80.67	.000	58.96	.000
N	853		823	

Sources: *2013 Pew Anonymity (omnibus) Survey.* Logistic regression estimates with standard errors in parentheses. Reported probabilities are based on two-tailed tests. Statistically significant coefficients at .10 or less in bold.

educated, African American or Latino, had mental or physical disabilities and resided in rural areas. In addition, Van Dijk (1999) argued that there is a third divide or psychological divide which could be described as limited Internet use because of mental barriers including lack of interest and fear of computers.

While much of the literature focuses on individual level factors to explain inequalities in Internet usage, there are also environmental level factors related to usage. The first issue is a rural/urban divide. Urban areas are more likely to attract Internet providers because of advantages related to technological, economics, infrastructure and geography (Gabel, 2007; Mack & Grubesic, 2009). The rural/urban divide has proven to be a difficult gap to close. The NTIA (2014) found that in 2012 that were considerable variation in Internet adoption by state. The differences by state fell along the rural/urban divide with more urban states having greater broadband adoption. Finally, West and Miller (2006) found regional differences with certain regions having higher levels of Internet connectivity.

A recent report by the NTIA (2014) finds that Americans are going online in greater numbers than ever. Nevertheless, some groups were less likely to go online than others. Those least likely to use the Internet were less affluent, older, less educated, African American or Latino and resided in a rural area.

Based on this literature, the measures from Table 1 included age, income, education, race/ethnicity, and region and community type are included as controls. Although gender is not considered a determinate of Internet usage, it is included as a control because women are found to be more likely to engage in social networking (Pew Internet & American Life Project, 2013e). Finally, a dummy variable for *parent* was included coded 1 if the individual is a parent or guardian of a child under 18 and 0 otherwise because households with children were found to be more likely to have Internet access (NTIA, 2014).

FINDINGS AND DISCUSSION

Digital Privacy and Security Concerns

In Table 1, the dependent variables were coded so that higher scores are associated with increased likelihood of either worrying about information privacy or believing an individual can use the Internet completely anonymously. Since the dependent variables are binary, the models are estimated using logistic regression. Although many of the findings correspond to prior research on concerns regarding Internet privacy and security, a few were counterintuitive. Specifically the findings indicate that individuals who were most likely to worry about how much information is available about them on the Internet were more affluent, younger, female, better educated and lived in the West. Individuals who lived in urban areas were among those least worried about information available about them on the Internet while race/ethnicity and work status were unrelated to information privacy. Individuals found to be more likely to believe that it is possible to use the Internet completely anonymously were males, younger, less educated, work full time and are more affluent. Among those least likely to believe that you could use the Internet completely anonymously were those that lived in urban areas and the Northeast. Race/ethnicity was found to be unrelated to beliefs regarding the ability to use the Internet anonymously.

Among these findings that were counterintuitive is the fact that while those individuals who were more affluent or younger were among those worried about the information about them online, they were also among the demographic groups who believed that a person could go online anonymously. One possible explanation is technical competency or knowledge of how to use hardware and software. Younger individuals and the more affluent are among those who are more likely to use the Internet and have a greater level of technical competency. While they may be more familiar with threats to online security and privacy, they may also be more familiar with steps to protect privacy and security. These steps may include such activities as being careful about what information to give out online, cleaning browser and cookie history and using public computers for browsing.

The final outcome that was initially puzzling was the fact that even though states in the South were among some of the slowest to adopt laws prohibiting employers from requesting social media personal account information from users or pass data disposal laws, residence of this region were not among those found to be most worried about information privacy. Residents in the West were among those most concerned about online privacy and these states were among the first to adopt these policies (National Conference of State Legislatures, 2011a, 2015). A possible explanation follows from Mooney & Lee (1995) who found that politics, resources and public demand are important factors in explaining policy adoption and implementation. It is possible that one reason why states in the West are innovators in this policy area is that they have a greater public demand for them.

Social Network Avoidance

In Table 2, the dependent variables are coded so that higher scores are associated with increased likelihood of taking part in social networking activities. Since the dependent variables are binary, models are estimated using logistic regression. As with Table 1, some of the findings were consistent with prior research while others were contradictory. Specifically the findings indicate that individuals most likely to use social network sites lived in rural areas, younger, female, more affluent, better educated, and worried about online privacy and believed that a person could go online anonymously. Region, parental status and race/ethnicity were found to be unrelated to social network site visits. Individuals who were more likely to use Twitter were younger, more affluent, better educated, parents and Asian America. None of the environmental variables or those measuring belief about online privacy were found related to Twitter usage.

Table 2. Social Network Activities, 2013

	Uses Social Network Sites		Uses Twitter	
Variables	β (se)	p>|z|	β (se)	p>|z|
Environmental Variables				
Suburban	**-.245(.204)**	**.016**	- .233(.249)	.350
Urban	**-.704(.241)**	**.004**	-.346(.328)	.292
Midwest	-.024(.226)	.915	.362(.306)	.236
Northeast	.023(.258)	.929	.160(.343)	.641
West	-.161(.232)	.486	.352(.290)	.225
Individual Level Variables				
Worry About Privacy	**.834(.181)**	**.000**	.146(.226)	.518
Can be Anonymous	**.847(.194)**	**.000**	.124(.237)	.600
Age	**-.054(.005)**	**.000**	**- .061(.008)**	**.000**
Male	**-.583(.175)**	**.001**	.192(.227)	.398
Latino	.345(.336)	.305	.072(.376)	.848
Black	-.250(.288)	.386	-.027(.356)	.940
Asian	1.099(.683)	.108	**.989(.465)**	**.034**
Education	**.158(.054)**	**.003**	**.202(.071)**	**.005**
Income	**.137(.041)**	**.001**	**.108(.052)**	**.037**
Parent	.222(.217)	.305	**.397(.232)**	**.087**
Constant	1.646(.468)	.000	-1.14(.551)	.038
Pseudo R2	.2614		.1891	
LR Chi² (15)	296.02	.000	125.67	.000
N	820		819	

Sources: *2013 Pew Anonymity (omnibus) Survey.* Logistic regression estimates with standard errors in parentheses. Reported probabilities are based on two-tailed tests. Statistically significant coefficients at .10 or less in bold.

Although many of the findings were consistent with earlier research, a few were unexpected. The first was that residents of rural areas were more likely to visit social networking sites. Rural areas are less likely to attract Internet providers, leaving their residents on the wrong side of the digital divide. There are two factors that might help to explain this finding. The first is that the smartphones are helping to bring individuals online, particularly for the social networking activities (Pew Internet & American Life Project, 2012c). Smartphones are helping to alleviate some of the problems facing individuals living in rural areas. Adams et al. (2005) provide additional help in explaining this outcome. They found that motivation was important in predicting both Internet usage and developing technological skills. For those living in isolated areas, social interaction may be more difficult, hence there could be greater motivation to use the Internet (at least for social interaction). The fact that individuals who are most concerned about their Internet privacy are the ones most likely to take part in social networking is also counterintuitive. This can again be explained by technical competency. Those most concerned about online privacy may also be the ones most knowledgeable about steps to safeguard their personal information.

The final outcome that was initially unanticipated is that a small set of demographic variables were predictive of Twitter usage. Specifically those most likely to use Twitter were individuals who were younger, more affluent, better educated, parents and Asian American. In addition, none of the environmental variables or those measuring belief about online privacy were found related to Twitter usage. This is not surprising when you consider that approximately only 8% of the adult online population was using Twitter in 2012 (Pew Internet & American Life Project, 2012c). The group of individuals who have historically been associated with greater Internet access and usage are the same ones predicted by this model to use Twitter (NTIA, 2014). It is not surprising that these specific demographic groups would be among the first to adopt Twitter.

FUTURE RESEARCH DIRECTIONS

In examining the impact of concerns regarding online security and privacy on social network activities, this chapter relied on secondary data analysis or data collected by another entity. There are advantages and disadvantages associated with all forms of data analysis. One disadvantage associated with secondary data analysis is that the data was collected for a specific purpose and may not be ideally suited for other research. In general, this dataset has been well suited for this study. One concern regarding this dataset is its relatively smaller size (n=1002). Future research would benefit from larger surveys.

Another issue to consider for future research is that this study relies on a cross-sectional design that depicts individual action at one point in time. Both the dependent and independent variables are being measured at roughly the same time. This makes it more difficult to establish time order, therefore weakening internal validity. While this study suggests that certain demographic factors are important in determining who has concerns about information privacy and its impact on social networking activities, it is more difficult to make the argument that there is a causal effect. Finally, the use of a cross-sectional design suggests factors, which may only be relevant during this time period. The results of a longitudinal study, on the other hand, might find that these variables are only significantly related to this topic during a specific time period. Such a study could find that, in the long run, other factors play a more important role. This study was done in 2013 where there were a number of news headlines regarding both privacy and Internet security. It is very possible that media coverage of these topics impacted respondent answers.

Finally, this study found for a number of demographic groups, those most worried about threats to their online information privacy were the same individuals who felt that it was possible to be completely anonymous online. This is counterintuitive but suggests that these individuals may have a greater understanding of online threats and how to avoid them. Future studies should explore this further by examining what steps individuals are taking to protect their online safety and who are taking these steps.

CONCLUSION

This chapter began with two questions, "what demographic groups are most concerned about Internet privacy?" and "how are these concerns impacting usage of social network sites (SNS)?" There are reasons to be concerned about the impact that concerns regarding Internet security and threats to online privacy are having on Internet usage. Citizens are regularly exposed to media coverage regarding such threats. Over the last year, news headline included Edward Snowden's allegations regarding the National Security Agency (NSA); a security breach of customer data at Target and most recently the heartbleed virus.

In addition, surveys continue to show that citizens do not feel that the government is doing enough to protect their online privacy. Concerns about privacy began even before the Internet. Early surveys showed that citizens worried that the government were using computers to create databases on individual citizens without adequate protection. For example, a 1983 poll by Harris found that 60% of those surveyed did not feel that information on computers was adequately protected. Familiarity with computers and the Internet has not lessened fears and has increased public demand for laws protecting online privacy. For example, a Pew Internet & American Life Survey (2013b) found that 66% of those surveyed did not feel that current laws were sufficient to protect online privacy. One thing that has change over time is public response to these fears. Early research (Stanley, 2003) found a number of psychological barriers were keeping certain individuals offline including relevance, fear and self-concept. The findings from this study suggest that the Internet has become too ingrained in society for fear to act as a barrier to Internet usage. Instead, the findings indicate that those with the greatest understanding of the threats to privacy and security may be taking necessary precautions to address these issues.

REFERENCES

Adams, N., Stubbs, V., & Woods, V. (2005). Psychological barriers to internet usage among older adults in the U.K. *Medical Informatics and the Internet in Medicine*, *30*(1), 3–17. doi:10.1080/14639230500066876 PMID:16036626

Belanger, F., & Carter, L. (2009). The impact of the digital divide on e-government use. *Communications of the ACM*, *54*(4), 132–135. doi:10.1145/1498765.1498801

California Business and Professions Code. (2003). *California Online Privacy Protection Act. Sections 22575-22579*. Author.

Federal Trade Commission. (1999). *New rule will protect privacy of children online. Effective April 2000 certain web sites must obtain parental consent before collecting personal information from children.* Retrieved from http://www.ftc.gov/news-events/press-releases/1999/10/new-rule-will-protect-privacy-children-online

Federal Trade Commission. (2007). *COPPA protects children but challenges lie ahead.* Retrieved from http://www.ftc.gov/news-events/press-releases/2007/02/coppa-protects-children-challenges-lie-ahead

Federal Trade Commission. (2010). *FTC testifies on protecting teen privacy.* Retrieved from http://www.ftc.gov/news-events/press-releases/2010/07/ftc-testifies-protecting-teen-privacy

Federal Trade Commission. (2013). *FTC grants approval for new COPPA verifiable parental consent method.* Retrieved from http://www.ftc.gov/news-events/press-releases/2013/12/ftc-grants-approval-new-coppa-verifiable-parental-consent-method

Federal Trade Commission. (2014). *FTC's 'net cetera' advises parents on how to talk to their kids about internet use newly released update includes advice on mobile apps, Wi-Fi, text message spam and more.* Retrieved from http://www.ftc.gov/news-events/press-releases/2014/01/ftcs-net-cetera-advises-parents-how-talk-their-kids-about-0

Gabel, D. (2007). Broadband and universal service. *Telecommunications Policy, 31*(6/7), 347–358.

Gorham-Oscilowski, U., & Jaeger, P. (2008). National Security Letters, the USA PATRIOT Act, and the Constitution: The tensions between national security and civil rights. *Government Information Quarterly, 25*(4), 625–644. doi:10.1016/j.giq.2008.02.001

Mack, E., & Grubesic, T. (2009). Forecasting broadband provision. *Information Economics and Policy, 21*(1), 57–67.

Mooney, C., & Lee, M. (1995). Legislating morality in the American states: The case of pre-Roe abortion regulation reform. *American Journal of Political Science, 39*(3), 599–627. doi:10.2307/2111646

Mossberger, K., Tolbert, C., & McNeal, R. (2007). *Digital citizenship: the internet, society and participation.* Cambridge, MA: MIT Press.

Mossberger, K., Tolbert, C., & Stansbury, M. (2003). *Virtual inequality: beyond the digital divide.* Washington, DC: Georgetown Press.

National Conference of State Legislatures. (2011a). *Privacy laws of government websites. Notice of monitoring of employee e-mail communications and internet access.* Retrieved from http://www.ncsl.org/issues-research/telecommunications-information-technology/state-laws-related-to-internet-privacy.aspx#govpolicies

National Conference of State Legislatures. (2011b). *Privacy policies: Government websites.* Retrieved from http://www.delcode.delaware.gov/title29/c090c/sc02/index.shtml

National Conference of State Legislatures. (2013). *State spyware laws.* Retrieved from http://www.ncsl.org/research/telecommunications-and-information-technology/state-spyware-laws.aspx

National Conference of State Legislatures. (2014). *Employer access to social media usernames and passwords 2013.* Retrieved from http://www.ncsl.org/research/telecommunications-and-information-technology/employer-access-to-social-media-passwords-2013.aspx

National Conference of State Legislatures. (2015). *Data disposal laws.* Retrieved from http://www.ncsl.org/research/telecommunications-and-information-technology/data-disposal-laws.aspx

National Telecommunications and Information Administration. (2000). *Falling through the net: Toward digital inclusion. A report on Americans access to technology tools.* Retrieved from http://www.ntia.doc.gov/files/ntia/publications/fttn00.pdf

National Telecommunications and Information Administration. (2014). *Exploring the digital nation: Embracing the mobile internet.* Retrieved from http://www.ntia.doc.gov/files/ntia/publications/exploring_the_digital_nation_embracing_the_mobile_internet_10162014.pdf

Pew Internet & American Life Project. (2012a). *Parents, teens, and online privacy.* Retrieved from http://www.pewinternet.org/files/oldmedia/Files/Reports/2012/PIP_ParentsTeensAndPrivacy.pdf

Pew Internet & American Life Project. (2012b). *Mobile health 2012.* Retrieved from http://www.pewinternet.org/Reports/2012/Mobile-Health.aspx

Pew Internet & American Life Project. (2012c). *Twitter users 2012.* Retrieved from http://www.pewinternet.org/files/old-media//Files/Reports/2012/PIP_Twitter_Use_2012.pdf

Pew Internet & American Life Project. (2013a). *The demographics of social media users – 2012.* Retrieved from http://pewinternet.org/Reports/2013/Social-media-users.aspx

Pew Internet & American Life Project. (2013b). *Anonymity, privacy, and security online.* Retrieved from http://pewinternet.org/Reports/2013/Anonymity-online.aspx

Pew Internet & American Life Project. (2013c). *Teens and mobile apps privacy.* Retrieved from http://www.pewinternet.org/Reports/2013/Teens-and-Mobile-Apps-Privacy.aspx

Pew Internet & American Life Project. (2013d). *72% of online adults are social networking site users.* Retrieved from http://www.pewinternet.org/files/oldmedia/Files/Reports/2013/PIP_Social_networking_sites_update_PDF.pdf

Pew Internet & American Life Project. (2013e). *Teens, social media, and privacy.* Retrieved from http://www.pewinternet.org/2013/05/21/teens-social-media-and-privacy

Pew Internet & American Life Project. (2013f). *51% of U.S. adults bank online 32% of adults transact bank business on their mobile phones.* Retrieved from http://www.pewinternet.org/2013/08/07/51-of-u-s-adults-bank-online

Privacy Rights Clearing House. (2014b). *Fact sheet 7: Workplace privacy and employee monitoring.* Retrieved from https://www.privacyrights.org/workplace-privacy-and-employee-monitoring#socialmedia

Privacy Rights Clearinghouse. (2014a). *Fact sheet 35: Social networking privacy: How to be safe, secure and social.* Retrieved from https://www.privacyrights.org/social-networking-privacy-how-be-safe-secure-and-social#user-shares

Public Law 106-102. (1999). Gramm-Leach-Bliley Act. 106th Congress.

Public Law 107-56. (2001). Uniting and Strengthening America by Providing Appropriate Tools Required to Intercept and Obstruct Terrorism (USA PATRIOT ACT) of 2001. 107th Congress.

Public Law 99-508. (1986). The Electronic Communications Privacy Act of 1986. 99th Congress.

Robbins, A. (2001). The loss of personal privacy and the consequences for social research. *Journal of Government Information*, *28*(4), 493–527. doi:10.1016/S1352-0237(02)00343-X

Selwyn, N. (2004). Reconsidering political and popular understandings of the digital divide. *New Media & Society*, *6*(3), 341–362. doi:10.1177/1461444804042519

Stanley, L. D. (2003). Beyond access: Psychosocial barrier to computer literacy. *The Information Society*, *19*(5), 407–416. doi:10.1080/715720560

U.S Department of Homeland Security. (2013). *Fact sheet: Executive order on cybersecurity / presidential policy directive on critical infrastructure security and resilience*. Retrieved from http://www.dhs.gov/news/2013/02/13/fact-sheet-executive-order-cybersecurity-presidential-policy-directive-critical

Van Dijk, J. (1999). *The network society, social aspects of new media*. London: Sage.

West, D., & Miller, E. (2006). The digital divide in public e-health: Barriers to accessibility and privacy in state health department websites. *Journal of Health Care for the Poor and Underserved*, *17*(3), 652–666. doi:10.1353/hpu.2006.0115 PMID:16960328

ADDITIONAL READING

Bailey, D. (2004). *The open society paradox: Why the 21st century calls for more openness--not less*. Washington, DC: Brassey's.

Best, S., & Krueger, B. (2008). Political conflict and public perceptions of government surveillance on the internet: An experiment of online search terms. *Technology & Politics*, *5*(2), 191–212. doi:10.1080/19331680802294479

Hargittai, E., & Hinnant, A. (2008). Digital inequalities: Differences in young adults' use of the internet. *Communication Research*, *35*(5), 602–621. doi:10.1177/0093650208321782

Hernon, P., Relyea, H., Dugan, R., & Cheverie, J. (2002). *United States government information: Policies and sources*. Westport, CT: Libraries Unlimited.

Jaeger, P., & Bertot, J. (2010). Transparency and technological change: Ensuring equal and sustained public access to government information. *Government Information Quarterly*, *27*(4), 371–376. doi:10.1016/j.giq.2010.05.003

Jaeger, P., McClure, C., & Fraser, B. (2002). The structures of centralized governmental privacy protection: Approaches, models, and analysis. *Government Information Quarterly*, *19*(4), 317–336. doi:10.1016/S0740-624X(02)00111-9

Jho, W. (2005). Challenges of e-governance: Protests from civil society on the protection of privacy in e-governance in Korea. *International Review of Administrative Sciences*, *71*(1), 151–166. doi:10.1177/0020852305051690

Kapp, K., Denney, G., & Barner, M. (2011). Key issues in data center security: An investigation of government audit reports. *Government Information Quarterly*, *28*(4), 533–541. doi:10.1016/j.giq.2010.10.008

Kimball, M. (2011). Mandated state-level open government training programs. *Government Information Quarterly*, *28*(4), 474–483. doi:10.1016/j.giq.2011.04.003

Lawson, S. (2013). Beyond cyber-doom: Assessing the limits of hypothetical scenarios in the framing of cyber threats. *Journal of Information Technology & Politics*, *10*(1), 86–103. doi:10.1080/19331681.2012.759059

Littler Workplace Policy Institute. (2013). Social media password protection and privacy — the patchwork of state laws and how it affects employers. Retrieved from http://www.littler.com/files/press/pdf/WPI-Social-Media-Password-Protection-Privacy-May-2013.pdf

McDermott, P. (2010). Building open government. *Government Information Quarterly*, *27*(4), 401–413. doi:10.1016/j.giq.2010.07.002

McNeal, R., Hale, K., & Dotterweich, L. (2008). Citizen-government interaction and the internet: Expectations and accomplishments in contact, quality and trust. *Journal of Information Technology & Politics*, *5*(2), 213–229. doi:10.1080/19331680802298298

Moynihan, D. (1998). *Secrecy*. New Haven, CT: Yale University Press.

National Alliance for Caregiving. (2009). Caregiving in the U.S. 2009. Retrieved from http://www.caregiving.org/data/Caregiving_in_the_US_2009_full_report.pdf

National Telecommunications and Information Administration. (2011). Digital nation: Expanding internet usage. Retrieved from http://www.ntia.doc.gov/files/ntia/publications/ntia_internet_use_report_february_2011.pdf

Paquette, S., Jaeger, P., & Wilson, S. (2010). Identifying the security risk associated with governmental use of cloud computing. *Government Information Quarterly*, *27*(4), 545–253.

Privacy Rights Clearing House. (2013). Fact sheet 24: Protecting financial privacy: The burden is on you. Retrieved April 11, 2012, from https://www.privacyrights.org/protecting-financial-privacy-burden-you

Reddick, C. G. (2012). *Public administration and information technology*. Burlington, MA: Jones & Bartlett Learning.

Regan, P. (2004). Old issues, new context: Privacy, information collection and Homeland Security. *Government Information Quarterly*, *21*(4), 481–497. doi:10.1016/j.giq.2004.08.003

Sawyer, S., & Tapia, A. (2005). The sociotechnical nature of mobile computing work: Evidence from a study of policing in the United States. *International Journal of Technology and Human Interaction*, *1*(3), 1–14. doi:10.4018/jthi.2005070101

Sharp, E. B. (1986). *Citizen demand-making in the urban context.* Tuscaloosa: University of Alabama Press.

Van Dijk, J., & Hacker, K. (2003). The digital divide as a complex and dynamic phenomenon. *The Information Society, 19*(4), 315–326. doi:10.1080/01972240309487

Wagner, B. (2013). Governing internet expression: How public and private regulation shape expression governance. *Journal of Information Technology & Politics, 10*(4), 389–403. doi:10.1080/19331681.2013.799051

KEY TERMS AND DEFINITIONS

Facebook: Online social network service.

Federal Trade Commission: The consumer protection agency for the U.S.

Instagram: Online photo and video sharing service and social network service.

LinkedIn: Business or professional social networking website.

Mobile Application: Software program that can be downloaded and used from a mobile device.

Social Media Personal Information: Information that is shared online during socialization considered identifying data, for example: name, address, age, gender, race, occupation, social security number, driver's license number, bank account number, credit card account number, and medical history.

Spyware or Adware: Software that can track or collect online user information and activities.

Tumblr: Online social network website and microblogging platform (smaller broadcasting medium).

Twitter: Online social network and microblogging service for brief text messages.

Related References

To continue our tradition of advancing academic research, we have compiled a list of recommended IGI Global readings. These references will provide additional information and guidance to further enrich your knowledge and assist you with your own research and future publications.

Adeyemo, O. (2013). The nationwide health information network: A biometric approach to prevent medical identity theft. In *User-driven healthcare: Concepts, methodologies, tools, and applications* (pp. 1636–1649). Hershey, PA: Medical Information Science Reference; doi:10.4018/978-1-4666-2770-3.ch081

Adler, M., & Henman, P. (2009). Justice beyond the courts: The implications of computerisation for procedural justice in social security. In A. Martínez & P. Abat (Eds.), *E-justice: Using information communication technologies in the court system* (pp. 65–86). Hershey, PA: Information Science Reference; doi:10.4018/978-1-59904-998-4.ch005

Aflalo, E., & Gabay, E. (2013). An information system for coping with student dropout. In L. Tomei (Ed.), *Learning tools and teaching approaches through ICT advancements* (pp. 176–187). Hershey, PA: Information Science Reference; doi:10.4018/978-1-4666-2017-9.ch016

Ahmed, M. A., Janssen, M., & van den Hoven, J. (2012). Value sensitive transfer (VST) of systems among countries: Towards a framework. *International Journal of Electronic Government Research*, *8*(1), 26–42. doi:10.4018/jegr.2012010102

Aikins, S. K. (2008). Issues and trends in internet-based citizen participation. In G. Garson & M. Khosrow-Pour (Eds.), *Handbook of research on public information technology* (pp. 31–40). Hershey, PA: Information Science Reference; doi:10.4018/978-1-59904-857-4.ch004

Aikins, S. K. (2009). A comparative study of municipal adoption of internet-based citizen participation. In C. Reddick (Ed.), *Handbook of research on strategies for local e-government adoption and implementation: Comparative studies* (pp. 206–230). Hershey, PA: Information Science Reference; doi:10.4018/978-1-60566-282-4.ch011

Aikins, S. K. (2012). Improving e-government project management: Best practices and critical success factors. In *Digital democracy: Concepts, methodologies, tools, and applications* (pp. 1314–1332). Hershey, PA: Information Science Reference; doi:10.4018/978-1-4666-1740-7.ch065

Akabawi, M. S. (2011). Ghabbour group ERP deployment: Learning from past technology failures. In E. Business Research and Case Center (Ed.), Cases on business and management in the MENA region: New trends and opportunities (pp. 177-203). Hershey, PA: Business Science Reference. doi:10.4018/978-1-60960-583-4.ch012

Akabawi, M. S. (2013). Ghabbour group ERP deployment: Learning from past technology failures. In *Industrial engineering: Concepts, methodologies, tools, and applications* (pp. 933–958). Hershey, PA: Engineering Science Reference; doi:10.4018/978-1-4666-1945-6.ch051

Akbulut, A. Y., & Motwani, J. (2008). Integration and information sharing in e-government. In G. Putnik & M. Cruz-Cunha (Eds.), *Encyclopedia of networked and virtual organizations* (pp. 729–734). Hershey, PA: Information Science Reference; doi:10.4018/978-1-59904-885-7.ch096

Akers, E. J. (2008). Technology diffusion in public administration. In G. Garson & M. Khosrow-Pour (Eds.), *Handbook of research on public information technology* (pp. 339–348). Hershey, PA: Information Science Reference; doi:10.4018/978-1-59904-857-4.ch033

Al-Shafi, S. (2008). Free wireless internet park services: An investigation of technology adoption in Qatar from a citizens' perspective. *Journal of Cases on Information Technology*, *10*(3), 21–34. doi:10.4018/jcit.2008070103

Al-Shafi, S., & Weerakkody, V. (2009). Implementing free wi-fi in public parks: An empirical study in Qatar. *International Journal of Electronic Government Research*, *5*(3), 21–35. doi:10.4018/jegr.2009070102

Aladwani, A. M. (2002). Organizational actions, computer attitudes and end-user satisfaction in public organizations: An empirical study. In C. Snodgrass & E. Szewczak (Eds.), *Human factors in information systems* (pp. 153–168). Hershey, PA: IRM Press; doi:10.4018/978-1-931777-10-0.ch012

Aladwani, A. M. (2002). Organizational actions, computer attitudes, and end-user satisfaction in public organizations: An empirical study. *Journal of Organizational and End User Computing*, *14*(1), 42–49. doi:10.4018/joeuc.2002010104

Allen, B., Juillet, L., Paquet, G., & Roy, J. (2005). E-government and private-public partnerships: Relational challenges and strategic directions. In M. Khosrow-Pour (Ed.), *Practicing e-government: A global perspective* (pp. 364–382). Hershey, PA: Idea Group Publishing; doi:10.4018/978-1-59140-637-2.ch016

Alshawaf, A., & Knalil, O. E. (2008). IS success factors and IS organizational impact: Does ownership type matter in Kuwait? *International Journal of Enterprise Information Systems*, *4*(2), 13–33. doi:10.4018/jeis.2008040102

Ambali, A. R. (2009). Digital divide and its implication on Malaysian e-government: Policy initiatives. In H. Rahman (Ed.), *Social and political implications of data mining: Knowledge management in e-government* (pp. 267–287). Hershey, PA: Information Science Reference; doi:10.4018/978-1-60566-230-5.ch016

Amoretti, F. (2007). Digital international governance. In A. Anttiroiko & M. Malkia (Eds.), *Encyclopedia of digital government* (pp. 365–370). Hershey, PA: Information Science Reference; doi:10.4018/978-1-59140-789-8.ch056

Amoretti, F. (2008). Digital international governance. In A. Anttiroiko (Ed.), *Electronic government: Concepts, methodologies, tools, and applications* (pp. 688–696). Hershey, PA: Information Science Reference; doi:10.4018/978-1-59904-947-2.ch058

Amoretti, F. (2008). E-government at supranational level in the European Union. In A. Anttiroiko (Ed.), *Electronic government: Concepts, methodologies, tools, and applications* (pp. 1047–1055). Hershey, PA: Information Science Reference; doi:10.4018/978-1-59904-947-2.ch079

Amoretti, F. (2008). E-government regimes. In A. Anttiroiko (Ed.), *Electronic government: Concepts, methodologies, tools, and applications* (pp. 3846–3856). Hershey, PA: Information Science Reference; doi:10.4018/978-1-59904-947-2.ch280

Amoretti, F. (2009). Electronic constitution: A Braudelian perspective. In F. Amoretti (Ed.), *Electronic constitution: Social, cultural, and political implications* (pp. 1–19). Hershey, PA: Information Science Reference; doi:10.4018/978-1-60566-254-1.ch001

Amoretti, F., & Musella, F. (2009). Institutional isomorphism and new technologies. In M. Khosrow-Pour (Ed.), *Encyclopedia of information science and technology* (2nd ed., pp. 2066–2071). Hershey, PA: Information Science Reference; doi:10.4018/978-1-60566-026-4.ch325

Andersen, K. V., & Henriksen, H. Z. (2007). E-government research: Capabilities, interaction, orientation, and values. In D. Norris (Ed.), *Current issues and trends in e-government research* (pp. 269–288). Hershey, PA: CyberTech Publishing; doi:10.4018/978-1-59904-283-1.ch013

Anderson, K. V., & Henriksen, H. Z. (2005). The first leg of e-government research: Domains and application areas 1998-2003. *International Journal of Electronic Government Research*, *1*(4), 26–44. doi:10.4018/jegr.2005100102

Anttiroiko, A. (2009). Democratic e-governance. In M. Khosrow-Pour (Ed.), *Encyclopedia of information science and technology* (2nd ed., pp. 990–995). Hershey, PA: Information Science Reference; doi:10.4018/978-1-60566-026-4.ch158

Association, I. R. (2010). Networking and telecommunications: Concepts, methodologies, tools and applications (Vols. 1–3). Hershey, PA: IGI Global; doi:10.4018/978-1-60566-986-1

Association, I. R. (2010). Web-based education: Concepts, methodologies, tools and applications (Vols. 1–3). Hershey, PA: IGI Global; doi:10.4018/978-1-61520-963-7

Baker, P. M., Bell, A., & Moon, N. W. (2009). Accessibility issues in municipal wireless networks. In C. Reddick (Ed.), *Handbook of research on strategies for local e-government adoption and implementation: Comparative studies* (pp. 569–588). Hershey, PA: Information Science Reference; doi:10.4018/978-1-60566-282-4.ch030

Becker, S. A., Keimer, R., & Muth, T. (2010). A case on university and community collaboration: The sci-tech entrepreneurial training services (ETS) program. In S. Becker & R. Niebuhr (Eds.), *Cases on technology innovation: Entrepreneurial successes and pitfalls* (pp. 68–90). Hershey, PA: Business Science Reference; doi:10.4018/978-1-61520-609-4.ch003

Becker, S. A., Keimer, R., & Muth, T. (2012). A case on university and community collaboration: The sci-tech entrepreneurial training services (ETS) program. In Regional development: Concepts, methodologies, tools, and applications (pp. 947-969). Hershey, PA: Information Science Reference. doi:10.4018/978-1-4666-0882-5.ch507

Bernardi, R. (2012). Information technology and resistance to public sector reforms: A case study in Kenya. In T. Papadopoulos & P. Kanellis (Eds.), *Public sector reform using information technologies: Transforming policy into practice* (pp. 59–78). Hershey, PA: Information Science Reference; doi:10.4018/978-1-60960-839-2.ch004

Bernardi, R. (2013). Information technology and resistance to public sector reforms: A case study in Kenya. In *User-driven healthcare: Concepts, methodologies, tools, and applications* (pp. 14–33). Hershey, PA: Medical Information Science Reference; doi:10.4018/978-1-4666-2770-3.ch002

Bolívar, M. P., Pérez, M. D., & Hernández, A. M. (2012). Municipal e-government services in emerging economies: The Latin-American and Caribbean experiences. In Y. Chen & P. Chu (Eds.), *Electronic governance and cross-boundary collaboration: Innovations and advancing tools* (pp. 198–226). Hershey, PA: Information Science Reference; doi:10.4018/978-1-60960-753-1.ch011

Borycki, E. M., & Kushniruk, A. W. (2010). Use of clinical simulations to evaluate the impact of health information systems and ubiquitous computing devices upon health professional work. In S. Mohammed & J. Fiaidhi (Eds.), *Ubiquitous health and medical informatics: The ubiquity 2.0 trend and beyond* (pp. 552–573). Hershey, PA: Medical Information Science Reference; doi:10.4018/978-1-61520-777-0.ch026

Borycki, E. M., & Kushniruk, A. W. (2011). Use of clinical simulations to evaluate the impact of health information systems and ubiquitous computing devices upon health professional work. In *Clinical technologies: Concepts, methodologies, tools and applications* (pp. 532–553). Hershey, PA: Medical Information Science Reference; doi:10.4018/978-1-60960-561-2.ch220

Buchan, J. (2011). Developing a dynamic and responsive online learning environment: A case study of a large Australian university. In B. Czerkawski (Ed.), *Free and open source software for e-learning: Issues, successes and challenges* (pp. 92–109). Hershey, PA: Information Science Reference; doi:10.4018/978-1-61520-917-0.ch006

Buenger, A. W. (2008). Digital convergence and cybersecurity policy. In G. Garson & M. Khosrow-Pour (Eds.), *Handbook of research on public information technology* (pp. 395–405). Hershey, PA: Information Science Reference; doi:10.4018/978-1-59904-857-4.ch038

Burn, J. M., & Loch, K. D. (2002). The societal impact of world wide web - Key challenges for the 21st century. In A. Salehnia (Ed.), *Ethical issues of information systems* (pp. 88–106). Hershey, PA: IRM Press; doi:10.4018/978-1-931777-15-5.ch007

Burn, J. M., & Loch, K. D. (2003). The societal impact of the world wide web-Key challenges for the 21st century. In M. Khosrow-Pour (Ed.), *Advanced topics in information resources management* (Vol. 2, pp. 32–51). Hershey, PA: Idea Group Publishing; doi:10.4018/978-1-59140-062-2.ch002

Bwalya, K. J., Du Plessis, T., & Rensleigh, C. (2012). The "quicksilver initiatives" as a framework for e-government strategy design in developing economies. In K. Bwalya & S. Zulu (Eds.), *Handbook of research on e-government in emerging economies: Adoption, e-participation, and legal frameworks* (pp. 605–623). Hershey, PA: Information Science Reference; doi:10.4018/978-1-4666-0324-0.ch031

Cabotaje, C. E., & Alampay, E. A. (2013). Social media and citizen engagement: Two cases from the Philippines. In S. Saeed & C. Reddick (Eds.), *Human-centered system design for electronic governance* (pp. 225–238). Hershey, PA: Information Science Reference; doi:10.4018/978-1-4666-3640-8.ch013

Camillo, A., Di Pietro, L., Di Virgilio, F., & Franco, M. (2013). Work-groups conflict at PetroTech-Italy, S.R.L.: The influence of culture on conflict dynamics. In B. Christiansen, E. Turkina, & N. Williams (Eds.), *Cultural and technological influences on global business* (pp. 272–289). Hershey, PA: Business Science Reference; doi:10.4018/978-1-4666-3966-9.ch015

Capra, E., Francalanci, C., & Marinoni, C. (2008). Soft success factors for m-government. In A. Anttiroiko (Ed.), *Electronic government: Concepts, methodologies, tools, and applications* (pp. 1213–1233). Hershey, PA: Information Science Reference; doi:10.4018/978-1-59904-947-2.ch089

Cartelli, A. (2009). The implementation of practices with ICT as a new teaching-learning paradigm. In A. Cartelli & M. Palma (Eds.), *Encyclopedia of information communication technology* (pp. 413–417). Hershey, PA: Information Science Reference; doi:10.4018/978-1-59904-845-1.ch055

Charalabidis, Y., Lampathaki, F., & Askounis, D. (2010). Investigating the landscape in national interoperability frameworks. *International Journal of E-Services and Mobile Applications*, *2*(4), 28–41. doi:10.4018/jesma.2010100103

Charalabidis, Y., Lampathaki, F., & Askounis, D. (2012). Investigating the landscape in national interoperability frameworks. In A. Scupola (Ed.), *Innovative mobile platform developments for electronic services design and delivery* (pp. 218–231). Hershey, PA: Business Science Reference; doi:10.4018/978-1-4666-1568-7.ch013

Chen, I. (2005). Distance education associations. In C. Howard, J. Boettcher, L. Justice, K. Schenk, P. Rogers, & G. Berg (Eds.), *Encyclopedia of distance learning* (pp. 599–612). Hershey, PA: Information Science Reference; doi:10.4018/978-1-59140-555-9.ch087

Chen, I. (2008). Distance education associations. In L. Tomei (Ed.), *Online and distance learning: Concepts, methodologies, tools, and applications* (pp. 562–579). Hershey, PA: Information Science Reference; doi:10.4018/978-1-59904-935-9.ch048

Chen, Y. (2008). Managing IT outsourcing for digital government. In A. Anttiroiko (Ed.), *Electronic government: Concepts, methodologies, tools, and applications* (pp. 3107–3114). Hershey, PA: Information Science Reference; doi:10.4018/978-1-59904-947-2.ch229

Chen, Y., & Dimitrova, D. V. (2006). Electronic government and online engagement: Citizen interaction with government via web portals. *International Journal of Electronic Government Research*, *2*(1), 54–76. doi:10.4018/jegr.2006010104

Chen, Y., & Knepper, R. (2005). Digital government development strategies: Lessons for policy makers from a comparative perspective. In W. Huang, K. Siau, & K. Wei (Eds.), *Electronic government strategies and implementation* (pp. 394–420). Hershey, PA: Idea Group Publishing; doi:10.4018/978-1-59140-348-7.ch017

Chen, Y., & Knepper, R. (2008). Digital government development strategies: Lessons for policy makers from a comparative perspective. In H. Rahman (Ed.), *Developing successful ICT strategies: Competitive advantages in a global knowledge-driven society* (pp. 334–356). Hershey, PA: Information Science Reference; doi:10.4018/978-1-59904-654-9.ch017

Cherian, E. J., & Ryan, T. W. (2014). Incongruent needs: Why differences in the iron-triangle of priorities make health information technology adoption and use difficult. In C. El Morr (Ed.), *Research perspectives on the role of informatics in health policy and management* (pp. 209–221). Hershey, PA: Medical Information Science Reference; doi:10.4018/978-1-4666-4321-5.ch012

Cho, H. J., & Hwang, S. (2010). Government 2.0 in Korea: Focusing on e-participation services. In C. Reddick (Ed.), *Politics, democracy and e-government: Participation and service delivery* (pp. 94–114). Hershey, PA: Information Science Reference; doi:10.4018/978-1-61520-933-0.ch006

Chorus, C., & Timmermans, H. (2010). Ubiquitous travel environments and travel control strategies: Prospects and challenges. In M. Wachowicz (Ed.), *Movement-aware applications for sustainable mobility: Technologies and approaches* (pp. 30–51). Hershey, PA: Information Science Reference; doi:10.4018/978-1-61520-769-5.ch003

Chuanshen, R. (2007). E-government construction and China's administrative litigation act. In A. Anttiroiko & M. Malkia (Eds.), *Encyclopedia of digital government* (pp. 507–510). Hershey, PA: Information Science Reference; doi:10.4018/978-1-59140-789-8.ch077

Ciaghi, A., & Villafiorita, A. (2012). Law modeling and BPR for public administration improvement. In K. Bwalya & S. Zulu (Eds.), *Handbook of research on e-government in emerging economies: Adoption, e-participation, and legal frameworks* (pp. 391–410). Hershey, PA: Information Science Reference; doi:10.4018/978-1-4666-0324-0.ch019

Ciaramitaro, B. L., & Skrocki, M. (2012). mHealth: Mobile healthcare. In B. Ciaramitaro (Ed.), Mobile technology consumption: Opportunities and challenges (pp. 99-109). Hershey, PA: Information Science Reference. doi:10.4018/978-1-61350-150-4.ch007

Comite, U. (2012). Innovative processes and managerial effectiveness of e-procurement in healthcare. In A. Manoharan & M. Holzer (Eds.), *Active citizen participation in e-government: A global perspective* (pp. 206–229). Hershey, PA: Information Science Reference; doi:10.4018/978-1-4666-0116-1.ch011

Cordella, A. (2013). E-government success: How to account for ICT, administrative rationalization, and institutional change. In J. Gil-Garcia (Ed.), *E-government success factors and measures: Theories, concepts, and methodologies* (pp. 40–51). Hershey, PA: Information Science Reference; doi:10.4018/978-1-4666-4058-0.ch003

Cropf, R. A. (2009). ICT and e-democracy. In M. Khosrow-Pour (Ed.), *Encyclopedia of information science and technology* (2nd ed., pp. 1789–1793). Hershey, PA: Information Science Reference; doi:10.4018/978-1-60566-026-4.ch281

Cropf, R. A. (2009). The virtual public sphere. In M. Pagani (Ed.), *Encyclopedia of multimedia technology and networking* (2nd ed., pp. 1525–1530). Hershey, PA: Information Science Reference; doi:10.4018/978-1-60566-014-1.ch206

D'Abundo, M. L. (2013). Electronic health record implementation in the United States healthcare industry: Making the process of change manageable. In V. Wang (Ed.), *Handbook of research on technologies for improving the 21st century workforce: Tools for lifelong learning* (pp. 272–286). Hershey, PA: Information Science Publishing; doi:10.4018/978-1-4666-2181-7.ch018

Damurski, L. (2012). E-participation in urban planning: Online tools for citizen engagement in Poland and in Germany. *International Journal of E-Planning Research, 1*(3), 40–67. doi:10.4018/ijepr.2012070103

de Almeida, M. O. (2007). E-government strategy in Brazil: Increasing transparency and efficiency through e-government procurement. In M. Gascó-Hernandez (Ed.), *Latin America online: Cases, successes and pitfalls* (pp. 34–82). Hershey, PA: IRM Press; doi:10.4018/978-1-59140-974-8.ch002

de Juana Espinosa, S. (2008). Empirical study of the municipalitites' motivations for adopting online presence. In A. Anttiroiko (Ed.), *Electronic government: Concepts, methodologies, tools, and applications* (pp. 3593–3608). Hershey, PA: Information Science Reference; doi:10.4018/978-1-59904-947-2.ch262

de Souza Dias, D. (2002). Motivation for using information technology. In C. Snodgrass & E. Szewczak (Eds.), *Human factors in information systems* (pp. 55–60). Hershey, PA: IRM Press; doi:10.4018/978-1-931777-10-0.ch005

Demediuk, P. (2006). Government procurement ICT's impact on the sustainability of SMEs and regional communities. In S. Marshall, W. Taylor, & X. Yu (Eds.), *Encyclopedia of developing regional communities with information and communication technology* (pp. 321–324). Hershey, PA: Information Science Reference; doi:10.4018/978-1-59140-575-7.ch056

Devonshire, E., Forsyth, H., Reid, S., & Simpson, J. M. (2013). The challenges and opportunities of online postgraduate coursework programs in a traditional university context. In B. Tynan, J. Willems, & R. James (Eds.), *Outlooks and opportunities in blended and distance learning* (pp. 353–368). Hershey, PA: Information Science Reference; doi:10.4018/978-1-4666-4205-8.ch026

Di Cerbo, F., Scotto, M., Sillitti, A., Succi, G., & Vernazza, T. (2007). Toward a GNU/Linux distribution for corporate environments. In S. Sowe, I. Stamelos, & I. Samoladas (Eds.), *Emerging free and open source software practices* (pp. 215–236). Hershey, PA: Idea Group Publishing; doi:10.4018/978-1-59904-210-7.ch010

Diesner, J., & Carley, K. M. (2005). Revealing social structure from texts: Meta-matrix text analysis as a novel method for network text analysis. In V. Narayanan & D. Armstrong (Eds.), *Causal mapping for research in information technology* (pp. 81–108). Hershey, PA: Idea Group Publishing; doi:10.4018/978-1-59140-396-8.ch004

Dologite, D. G., Mockler, R. J., Bai, Q., & Viszhanyo, P. F. (2006). IS change agents in practice in a US-Chinese joint venture. In M. Hunter & F. Tan (Eds.), *Advanced topics in global information management* (Vol. 5, pp. 331–352). Hershey, PA: Idea Group Publishing; doi:10.4018/978-1-59140-923-6.ch015

Drnevich, P., Brush, T. H., & Luckock, G. T. (2011). Process and structural implications for IT-enabled outsourcing. *International Journal of Strategic Information Technology and Applications*, *2*(4), 30–43. doi:10.4018/jsita.2011100103

Dwivedi, A. N. (2009). Handbook of research on information technology management and clinical data administration in healthcare (Vols. 1–2). Hershey, PA: IGI Global; doi:10.4018/978-1-60566-356-2

Elbeltagi, I., McBride, N., & Hardaker, G. (2006). Evaluating the factors affecting DSS usage by senior managers in local authorities in Egypt. In M. Hunter & F. Tan (Eds.), *Advanced topics in global information management* (Vol. 5, pp. 283–307). Hershey, PA: Idea Group Publishing; doi:10.4018/978-1-59140-923-6.ch013

Eom, S., & Fountain, J. E. (2013). Enhancing information services through public-private partnerships: Information technology knowledge transfer underlying structures to develop shared services in the U.S. and Korea. In J. Gil-Garcia (Ed.), *E-government success around the world: Cases, empirical studies, and practical recommendations* (pp. 15–40). Hershey, PA: Information Science Reference; doi:10.4018/978-1-4666-4173-0.ch002

Esteves, T., Leuenberger, D., & Van Leuven, N. (2012). Reaching citizen 2.0: How government uses social media to send public messages during times of calm and times of crisis. In K. Kloby & M. D'Agostino (Eds.), *Citizen 2.0: Public and governmental interaction through web 2.0 technologies* (pp. 250–268). Hershey, PA: Information Science Reference; doi:10.4018/978-1-4666-0318-9.ch013

Estevez, E., Fillottrani, P., Janowski, T., & Ojo, A. (2012). Government information sharing: A framework for policy formulation. In Y. Chen & P. Chu (Eds.), *Electronic governance and cross-boundary collaboration: Innovations and advancing tools* (pp. 23–55). Hershey, PA: Information Science Reference; doi:10.4018/978-1-60960-753-1.ch002

Ezz, I. E. (2008). E-governement emerging trends: Organizational challenges. In A. Anttiroiko (Ed.), *Electronic government: Concepts, methodologies, tools, and applications* (pp. 3721–3737). Hershey, PA: Information Science Reference; doi:10.4018/978-1-59904-947-2.ch269

Fabri, M. (2009). The Italian style of e-justice in a comparative perspective. In A. Martínez & P. Abat (Eds.), *E-justice: Using information communication technologies in the court system* (pp. 1–19). Hershey, PA: Information Science Reference; doi:10.4018/978-1-59904-998-4.ch001

Fagbe, T., & Adekola, O. D. (2010). Workplace safety and personnel well-being: The impact of information technology. *International Journal of Green Computing*, *1*(1), 28–33. doi:10.4018/jgc.2010010103

Fagbe, T., & Adekola, O. D. (2011). Workplace safety and personnel well-being: The impact of information technology. In *Global business: Concepts, methodologies, tools and applications* (pp. 1438–1444). Hershey, PA: Business Science Reference; doi:10.4018/978-1-60960-587-2.ch509

Farmer, L. (2008). Affective collaborative instruction with librarians. In S. Kelsey & K. St.Amant (Eds.), *Handbook of research on computer mediated communication* (pp. 15–24). Hershey, PA: Information Science Reference; doi:10.4018/978-1-59904-863-5.ch002

Favier, L., & Mekhantar, J. (2007). Use of OSS by local e-administration: The French situation. In K. St.Amant & B. Still (Eds.), *Handbook of research on open source software: Technological, economic, and social perspectives* (pp. 428–444). Hershey, PA: Information Science Reference; doi:10.4018/978-1-59140-999-1.ch033

Fernando, S. (2009). Issues of e-learning in third world countries. In M. Khosrow-Pour (Ed.), *Encyclopedia of information science and technology* (2nd ed., pp. 2273–2277). Hershey, PA: Information Science Reference; doi:10.4018/978-1-60566-026-4.ch360

Filho, J. R., & dos Santos Junior, J. R. (2009). Local e-government in Brazil: Poor interaction and local politics as usual. In C. Reddick (Ed.), *Handbook of research on strategies for local e-government adoption and implementation: Comparative studies* (pp. 863–878). Hershey, PA: Information Science Reference; doi:10.4018/978-1-60566-282-4.ch045

Fletcher, P. D. (2004). Portals and policy: Implications of electronic access to U.S. federal government information services. In A. Pavlichev & G. Garson (Eds.), *Digital government: Principles and best practices* (pp. 52–62). Hershey, PA: Idea Group Publishing; doi:10.4018/978-1-59140-122-3.ch004

Fletcher, P. D. (2008). Portals and policy: Implications of electronic access to U.S. federal government information services. In A. Anttiroiko (Ed.), *Electronic government: Concepts, methodologies, tools, and applications* (pp. 3970–3979). Hershey, PA: Information Science Reference; doi:10.4018/978-1-59904-947-2.ch289

Forlano, L. (2004). The emergence of digital government: International perspectives. In A. Pavlichev & G. Garson (Eds.), *Digital government: Principles and best practices* (pp. 34–51). Hershey, PA: Idea Group Publishing; doi:10.4018/978-1-59140-122-3.ch003

Franzel, J. M., & Coursey, D. H. (2004). Government web portals: Management issues and the approaches of five states. In A. Pavlichev & G. Garson (Eds.), *Digital government: Principles and best practices* (pp. 63–77). Hershey, PA: Idea Group Publishing; doi:10.4018/978-1-59140-122-3.ch005

Gaivéo, J. M. (2013). Security of ICTs supporting healthcare activities. In M. Cruz-Cunha, I. Miranda, & P. Gonçalves (Eds.), *Handbook of research on ICTs for human-centered healthcare and social care services* (pp. 208–228). Hershey, PA: Medical Information Science Reference; doi:10.4018/978-1-4666-3986-7.ch011

Garson, G. D. (1999). *Information technology and computer applications in public administration: Issues and trends*. Hershey, PA: IGI Global; doi:10.4018/978-1-87828-952-0

Garson, G. D. (2003). Toward an information technology research agenda for public administration. In G. Garson (Ed.), *Public information technology: Policy and management issues* (pp. 331–357). Hershey, PA: Idea Group Publishing; doi:10.4018/978-1-59140-060-8.ch014

Garson, G. D. (2004). The promise of digital government. In A. Pavlichev & G. Garson (Eds.), *Digital government: Principles and best practices* (pp. 2–15). Hershey, PA: Idea Group Publishing; doi:10.4018/978-1-59140-122-3.ch001

Garson, G. D. (2007). An information technology research agenda for public administration. In G. Garson (Ed.), *Modern public information technology systems: Issues and challenges* (pp. 365–392). Hershey, PA: Idea Group Publishing; doi:10.4018/978-1-59904-051-6.ch018

Gasco, M. (2007). Civil servants' resistance towards e-government development. In A. Anttiroiko & M. Malkia (Eds.), *Encyclopedia of digital government* (pp. 190–195). Hershey, PA: Information Science Reference; doi:10.4018/978-1-59140-789-8.ch028

Gasco, M. (2008). Civil servants' resistance towards e-government development. In A. Anttiroiko (Ed.), *Electronic government: Concepts, methodologies, tools, and applications* (pp. 2580–2588). Hershey, PA: Information Science Reference; doi:10.4018/978-1-59904-947-2.ch190

Ghere, R. K. (2010). Accountability and information technology enactment: Implications for social empowerment. In E. Ferro, Y. Dwivedi, J. Gil-Garcia, & M. Williams (Eds.), *Handbook of research on overcoming digital divides: Constructing an equitable and competitive information society* (pp. 515–532). Hershey, PA: Information Science Reference; doi:10.4018/978-1-60566-699-0.ch028

Gibson, I. W. (2012). Simulation modeling of healthcare delivery. In A. Kolker & P. Story (Eds.), *Management engineering for effective healthcare delivery: Principles and applications* (pp. 69–89). Hershey, PA: Medical Information Science Reference; doi:10.4018/978-1-60960-872-9.ch003

Gil-Garcia, J. R. (2007). Exploring e-government benefits and success factors. In A. Anttiroiko & M. Malkia (Eds.), *Encyclopedia of digital government* (pp. 803–811). Hershey, PA: Information Science Reference; doi:10.4018/978-1-59140-789-8.ch122

Gil-Garcia, J. R., & González Miranda, F. (2010). E-government and opportunities for participation: The case of the Mexican state web portals. In C. Reddick (Ed.), *Politics, democracy and e-government: Participation and service delivery* (pp. 56–74). Hershey, PA: Information Science Reference; doi:10.4018/978-1-61520-933-0.ch004

Goldfinch, S. (2012). Public trust in government, trust in e-government, and use of e-government. In Z. Yan (Ed.), *Encyclopedia of cyber behavior* (pp. 987–995). Hershey, PA: Information Science Reference; doi:10.4018/978-1-4666-0315-8.ch081

Goodyear, M. (2012). Organizational change contributions to e-government project transitions. In S. Aikins (Ed.), *Managing e-government projects: Concepts, issues, and best practices* (pp. 1–21). Hershey, PA: Information Science Reference; doi:10.4018/978-1-4666-0086-7.ch001

Gordon, S., & Mulligan, P. (2003). Strategic models for the delivery of personal financial services: The role of infocracy. In S. Gordon (Ed.), *Computing information technology: The human side* (pp. 220–232). Hershey, PA: IRM Press; doi:10.4018/978-1-93177-752-0.ch014

Gordon, T. F. (2007). Legal knowledge systems. In A. Anttiroiko & M. Malkia (Eds.), *Encyclopedia of digital government* (pp. 1161–1166). Hershey, PA: Information Science Reference; doi:10.4018/978-1-59140-789-8.ch175

Graham, J. E., & Semich, G. W. (2008). Integrating technology to transform pedagogy: Revisiting the progress of the three phase TUI model for faculty development. In L. Tomei (Ed.), *Adapting information and communication technologies for effective education* (pp. 1–12). Hershey, PA: Information Science Reference; doi:10.4018/978-1-59904-922-9.ch001

Grandinetti, L., & Pisacane, O. (2012). Web services for healthcare management. In D. Prakash Vidyarthi (Ed.), *Technologies and protocols for the future of internet design: Reinventing the web* (pp. 60–94). Hershey, PA: Information Science Reference; doi:10.4018/978-1-4666-0203-8.ch004

Groenewegen, P., & Wagenaar, F. P. (2008). VO as an alternative to hierarchy in the Dutch police sector. In G. Putnik & M. Cruz-Cunha (Eds.), *Encyclopedia of networked and virtual organizations* (pp. 1851–1857). Hershey, PA: Information Science Reference; doi:10.4018/978-1-59904-885-7.ch245

Gronlund, A. (2001). Building an infrastructure to manage electronic services. In S. Dasgupta (Ed.), *Managing internet and intranet technologies in organizations: Challenges and opportunities* (pp. 71–103). Hershey, PA: Idea Group Publishing; doi:10.4018/978-1-878289-95-7.ch006

Gronlund, A. (2002). Introduction to electronic government: Design, applications and management. In Å. Grönlund (Ed.), *Electronic government: Design, applications and management* (pp. 1–21). Hershey, PA: Idea Group Publishing; doi:10.4018/978-1-930708-19-8.ch001

Gupta, A., Woosley, R., Crk, I., & Sarnikar, S. (2009). An information technology architecture for drug effectiveness reporting and post-marketing surveillance. In J. Tan (Ed.), *Medical informatics: Concepts, methodologies, tools, and applications* (pp. 631–646). Hershey, PA: Medical Information Science Reference; doi:10.4018/978-1-60566-050-9.ch047

Hallin, A., & Lundevall, K. (2007). mCity: User focused development of mobile services within the city of Stockholm. In I. Kushchu (Ed.), Mobile government: An emerging direction in e-government (pp. 12-29). Hershey, PA: Idea Group Publishing. doi:10.4018/978-1-59140-884-0.ch002

Hallin, A., & Lundevall, K. (2009). mCity: User focused development of mobile services within the city of Stockholm. In S. Clarke (Ed.), Evolutionary concepts in end user productivity and performance: Applications for organizational progress (pp. 268-280). Hershey, PA: Information Science Reference. doi:10.4018/978-1-60566-136-0.ch017

Hallin, A., & Lundevall, K. (2009). mCity: User focused development of mobile services within the city of Stockholm. In D. Taniar (Ed.), Mobile computing: Concepts, methodologies, tools, and applications (pp. 3455-3467). Hershey, PA: Information Science Reference. doi:10.4018/978-1-60566-054-7.ch253

Hanson, A. (2005). Overcoming barriers in the planning of a virtual library. In M. Khosrow-Pour (Ed.), *Encyclopedia of information science and technology* (pp. 2255–2259). Hershey, PA: Information Science Reference; doi:10.4018/978-1-59140-553-5.ch397

Haque, A. (2008). Information technology and surveillance: Implications for public administration in a new word order. In T. Loendorf & G. Garson (Eds.), *Patriotic information systems* (pp. 177–185). Hershey, PA: IGI Publishing; doi:10.4018/978-1-59904-594-8.ch008

Hauck, R. V., Thatcher, S. M., & Weisband, S. P. (2012). Temporal aspects of information technology use: Increasing shift work effectiveness. In J. Wang (Ed.), *Advancing the service sector with evolving technologies: Techniques and principles* (pp. 87–104). Hershey, PA: Business Science Reference; doi:10.4018/978-1-4666-0044-7.ch006

Hawk, S., & Witt, T. (2006). Telecommunications courses in information systems programs. *International Journal of Information and Communication Technology Education*, *2*(1), 79–92. doi:10.4018/jicte.2006010107

Helms, M. M., Moore, R., & Ahmadi, M. (2009). Information technology (IT) and the healthcare industry: A SWOT analysis. In J. Tan (Ed.), *Medical informatics: Concepts, methodologies, tools, and applications* (pp. 134–152). Hershey, PA: Medical Information Science Reference; doi:10.4018/978-1-60566-050-9.ch012

Hendrickson, S. M., & Young, M. E. (2014). Electronic records management at a federally funded research and development center. In J. Krueger (Ed.), *Cases on electronic records and resource management implementation in diverse environments* (pp. 334–350). Hershey, PA: Information Science Reference; doi:10.4018/978-1-4666-4466-3.ch020

Henman, P. (2010). Social policy and information communication technologies. In J. Martin & L. Hawkins (Eds.), *Information communication technologies for human services education and delivery: Concepts and cases* (pp. 215–229). Hershey, PA: Information Science Reference; doi:10.4018/978-1-60566-735-5.ch014

Hismanoglu, M. (2011). Important issues in online education: E-pedagogy and marketing. In U. Demiray & S. Sever (Eds.), *Marketing online education programs: Frameworks for promotion and communication* (pp. 184–209). Hershey, PA: Information Science Reference; doi:10.4018/978-1-60960-074-7.ch012

Ho, K. K. (2008). The e-government development, IT strategies, and portals of the Hong Kong SAR government. In A. Anttiroiko (Ed.), *Electronic government: Concepts, methodologies, tools, and applications* (pp. 715–733). Hershey, PA: Information Science Reference; doi:10.4018/978-1-59904-947-2.ch060

Holden, S. H. (2003). The evolution of information technology management at the federal level: Implications for public administration. In G. Garson (Ed.), *Public information technology: Policy and management issues* (pp. 53–73). Hershey, PA: Idea Group Publishing; doi:10.4018/978-1-59140-060-8.ch003

Holden, S. H. (2007). The evolution of federal information technology management literature: Does IT finally matter? In G. Garson (Ed.), *Modern public information technology systems: Issues and challenges* (pp. 17–34). Hershey, PA: Idea Group Publishing; doi:10.4018/978-1-59904-051-6.ch002

Holland, J. W. (2009). Automation of American criminal justice. In M. Khosrow-Pour (Ed.), *Encyclopedia of information science and technology* (2nd ed., pp. 300–302). Hershey, PA: Information Science Reference; doi:10.4018/978-1-60566-026-4.ch051

Holloway, K. (2013). Fair use, copyright, and academic integrity in an online academic environment. In *Digital rights management: Concepts, methodologies, tools, and applications* (pp. 917–928). Hershey, PA: Information Science Reference; doi:10.4018/978-1-4666-2136-7.ch044

Horiuchi, C. (2005). E-government databases. In L. Rivero, J. Doorn, & V. Ferraggine (Eds.), *Encyclopedia of database technologies and applications* (pp. 206–210). Hershey, PA: Information Science Reference; doi:10.4018/978-1-59140-560-3.ch035

Horiuchi, C. (2006). Creating IS quality in government settings. In E. Duggan & J. Reichgelt (Eds.), *Measuring information systems delivery quality* (pp. 311–327). Hershey, PA: Idea Group Publishing; doi:10.4018/978-1-59140-857-4.ch014

Hsiao, N., Chu, P., & Lee, C. (2012). Impact of e-governance on businesses: Model development and case study. In *Digital democracy: Concepts, methodologies, tools, and applications* (pp. 1407–1425). Hershey, PA: Information Science Reference; doi:10.4018/978-1-4666-1740-7.ch070

Huang, T., & Lee, C. (2010). Evaluating the impact of e-government on citizens: Cost-benefit analysis. In C. Reddick (Ed.), *Citizens and e-government: Evaluating policy and management* (pp. 37–52). Hershey, PA: Information Science Reference; doi:10.4018/978-1-61520-931-6.ch003

Hunter, M. G., Diochon, M., Pugsley, D., & Wright, B. (2002). Unique challenges for small business adoption of information technology: The case of the Nova Scotia ten. In S. Burgess (Ed.), *Managing information technology in small business: Challenges and solutions* (pp. 98–117). Hershey, PA: Idea Group Publishing; doi:10.4018/978-1-930708-35-8.ch006

Hurskainen, J. (2003). Integration of business systems and applications in merger and alliance: Case metso automation. In T. Reponen (Ed.), *Information technology enabled global customer service* (pp. 207–225). Hershey, PA: Idea Group Publishing; doi:10.4018/978-1-59140-048-6.ch012

Iazzolino, G., & Pietrantonio, R. (2011). The soveria.it project: A best practice of e-government in southern Italy. In D. Piaggesi, K. Sund, & W. Castelnovo (Eds.), *Global strategy and practice of e-governance: Examples from around the world* (pp. 34–56). Hershey, PA: Information Science Reference; doi:10.4018/978-1-60960-489-9.ch003

Imran, A., & Gregor, S. (2012). A process model for successful e-government adoption in the least developed countries: A case of Bangladesh. In F. Tan (Ed.), *International comparisons of information communication technologies: Advancing applications* (pp. 321–350). Hershey, PA: Information Science Reference; doi:10.4018/978-1-61350-480-2.ch014

Inoue, Y., & Bell, S. T. (2005). Electronic/digital government innovation, and publishing trends with IT. In M. Khosrow-Pour (Ed.), *Encyclopedia of information science and technology* (pp. 1018–1023). Hershey, PA: Information Science Reference; doi:10.4018/978-1-59140-553-5.ch180

Islam, M. M., & Ehsan, M. (2013). Understanding e-governance: A theoretical approach. In M. Islam & M. Ehsan (Eds.), *From government to e-governance: Public administration in the digital age* (pp. 38–49). Hershey, PA: Information Science Reference; doi:10.4018/978-1-4666-1909-8.ch003

Jaeger, B. (2009). E-government and e-democracy in the making. In M. Khosrow-Pour (Ed.), *Encyclopedia of information science and technology* (2nd ed., pp. 1318–1322). Hershey, PA: Information Science Reference; doi:10.4018/978-1-60566-026-4.ch208

Jain, R. B. (2007). Revamping the administrative structure and processes in India for online diplomacy. In A. Anttiroiko & M. Malkia (Eds.), *Encyclopedia of digital government* (pp. 1418–1423). Hershey, PA: Information Science Reference; doi:10.4018/978-1-59140-789-8.ch217

Jain, R. B. (2008). Revamping the administrative structure and processes in India for online diplomacy. In A. Anttiroiko (Ed.), *Electronic government: Concepts, methodologies, tools, and applications* (pp. 3142–3149). Hershey, PA: Information Science Reference; doi:10.4018/978-1-59904-947-2.ch233

Jauhiainen, J. S., & Inkinen, T. (2009). E-governance and the information society in periphery. In C. Reddick (Ed.), *Handbook of research on strategies for local e-government adoption and implementation: Comparative studies* (pp. 497–514). Hershey, PA: Information Science Reference; doi:10.4018/978-1-60566-282-4.ch026

Jensen, M. J. (2009). Electronic democracy and citizen influence in government. In C. Reddick (Ed.), *Handbook of research on strategies for local e-government adoption and implementation: Comparative studies* (pp. 288–305). Hershey, PA: Information Science Reference; doi:10.4018/978-1-60566-282-4.ch015

Jiao, Y., Hurson, A. R., Potok, T. E., & Beckerman, B. G. (2009). Integrating mobile-based systems with healthcare databases. In J. Erickson (Ed.), *Database technologies: Concepts, methodologies, tools, and applications* (pp. 484–504). Hershey, PA: Information Science Reference; doi:10.4018/978-1-60566-058-5.ch031

Joia, L. A. (2002). A systematic model to integrate information technology into metabusinesses: A case study in the engineering realms. In F. Tan (Ed.), *Advanced topics in global information management* (Vol. 1, pp. 250–267). Hershey, PA: Idea Group Publishing; doi:10.4018/978-1-930708-43-3.ch016

Jones, T. H., & Song, I. (2000). Binary equivalents of ternary relationships in entity-relationship modeling: A logical decomposition approach. *Journal of Database Management*, *11*(2), 12–19. doi:10.4018/jdm.2000040102

Juana-Espinosa, S. D. (2007). Empirical study of the municipalitites' motivations for adopting online presence. In L. Al-Hakim (Ed.), *Global e-government: Theory, applications and benchmarking* (pp. 261–279). Hershey, PA: Idea Group Publishing; doi:10.4018/978-1-59904-027-1.ch015

Jun, K., & Weare, C. (2012). Bridging from e-government practice to e-government research: Past trends and future directions. In K. Bwalya & S. Zulu (Eds.), *Handbook of research on e-government in emerging economies: Adoption, e-participation, and legal frameworks* (pp. 263–289). Hershey, PA: Information Science Reference; doi:10.4018/978-1-4666-0324-0.ch013

Junqueira, A., Diniz, E. H., & Fernandez, M. (2010). Electronic government implementation projects with multiple agencies: Analysis of the electronic invoice project under PMBOK framework. In J. Cordoba-Pachon & A. Ochoa-Arias (Eds.), *Systems thinking and e-participation: ICT in the governance of society* (pp. 135–153). Hershey, PA: Information Science Reference; doi:10.4018/978-1-60566-860-4.ch009

Juntunen, A. (2009). Joint service development with the local authorities. In C. Reddick (Ed.), *Handbook of research on strategies for local e-government adoption and implementation: Comparative studies* (pp. 902–920). Hershey, PA: Information Science Reference; doi:10.4018/978-1-60566-282-4.ch047

Kamel, S. (2001). *Using DSS for crisis management.* Hershey, PA: IGI Global; doi:10.4018/978-1-87828-961-2.ch020

Kamel, S. (2006). DSS for strategic decision making. In M. Khosrow-Pour (Ed.), *Cases on information technology and organizational politics & culture* (pp. 230–246). Hershey, PA: Idea Group Publishing; doi:10.4018/978-1-59904-411-8.ch015

Kamel, S. (2009). The software industry in Egypt as a potential contributor to economic growth. In M. Khosrow-Pour (Ed.), *Encyclopedia of information science and technology* (2nd ed., pp. 3531–3537). Hershey, PA: Information Science Reference; doi:10.4018/978-1-60566-026-4.ch562

Kamel, S., & Hussein, M. (2008). Xceed: Pioneering the contact center industry in Egypt. *Journal of Cases on Information Technology, 10*(1), 67–91. doi:10.4018/jcit.2008010105

Kamel, S., & Wahba, K. (2003). The use of a hybrid model in web-based education: "The Global campus project. In A. Aggarwal (Ed.), *Web-based education: Learning from experience* (pp. 331–346). Hershey, PA: Information Science Publishing; doi:10.4018/978-1-59140-102-5.ch020

Kardaras, D. K., & Papathanassiou, E. A. (2008). An exploratory study of the e-government services in Greece. In G. Garson & M. Khosrow-Pour (Eds.), *Handbook of research on public information technology* (pp. 162–174). Hershey, PA: Information Science Reference; doi:10.4018/978-1-59904-857-4.ch016

Kassahun, A. E., Molla, A., & Sarkar, P. (2012). Government process reengineering: What we know and what we need to know. In *Digital democracy: Concepts, methodologies, tools, and applications* (pp. 1730–1752). Hershey, PA: Information Science Reference; doi:10.4018/978-1-4666-1740-7.ch086

Khan, B. (2005). Technological issues. In B. Khan (Ed.), *Managing e-learning strategies: Design, delivery, implementation and evaluation* (pp. 154–180). Hershey, PA: Information Science Publishing; doi:10.4018/978-1-59140-634-1.ch004

Khasawneh, A., Bsoul, M., Obeidat, I., & Al Azzam, I. (2012). Technology fears: A study of e-commerce loyalty perception by Jordanian customers. In J. Wang (Ed.), *Advancing the service sector with evolving technologies: Techniques and principles* (pp. 158–165). Hershey, PA: Business Science Reference; doi:10.4018/978-1-4666-0044-7.ch010

Khatibi, V., & Montazer, G. A. (2012). E-research methodology. In A. Juan, T. Daradoumis, M. Roca, S. Grasman, & J. Faulin (Eds.), *Collaborative and distributed e-research: Innovations in technologies, strategies and applications* (pp. 62–81). Hershey, PA: Information Science Reference; doi:10.4018/978-1-4666-0125-3.ch003

Kidd, T. (2011). The dragon in the school's backyard: A review of literature on the uses of technology in urban schools. In L. Tomei (Ed.), *Online courses and ICT in education: Emerging practices and applications* (pp. 242–257). Hershey, PA: Information Science Reference; doi:10.4018/978-1-60960-150-8.ch019

Kidd, T. T. (2010). My experience tells the story: Exploring technology adoption from a qualitative perspective - A pilot study. In H. Song & T. Kidd (Eds.), *Handbook of research on human performance and instructional technology* (pp. 247–262). Hershey, PA: Information Science Reference; doi:10.4018/978-1-60566-782-9.ch015

Kieley, B., Lane, G., Paquet, G., & Roy, J. (2002). e-Government in Canada: Services online or public service renewal? In Å. Grönlund (Ed.), Electronic government: Design, applications and management (pp. 340-355). Hershey, PA: Idea Group Publishing. doi:10.4018/978-1-930708-19-8.ch016

Kim, P. (2012). "Stay out of the way! My kid is video blogging through a phone!": A lesson learned from math tutoring social media for children in underserved communities. In *Wireless technologies: Concepts, methodologies, tools and applications* (pp. 1415–1428). Hershey, PA: Information Science Reference; doi:10.4018/978-1-61350-101-6.ch517

Kirlidog, M. (2010). Financial aspects of national ICT strategies. In S. Kamel (Ed.), *E-strategies for technological diffusion and adoption: National ICT approaches for socioeconomic development* (pp. 277–292). Hershey, PA: Information Science Reference; doi:10.4018/978-1-60566-388-3.ch016

Kisielnicki, J. (2006). Transfer of information and knowledge in the project management. In E. Coakes & S. Clarke (Eds.), *Encyclopedia of communities of practice in information and knowledge management* (pp. 544–551). Hershey, PA: Information Science Reference; doi:10.4018/978-1-59140-556-6.ch091

Kittner, M., & Van Slyke, C. (2006). Reorganizing information technology services in an academic environment. In M. Khosrow-Pour (Ed.), *Cases on the human side of information technology* (pp. 49–66). Hershey, PA: Idea Group Publishing; doi:10.4018/978-1-59904-405-7.ch004

Knoell, H. D. (2008). Semi virtual workplaces in German financial service enterprises. In P. Zemliansky & K. St.Amant (Eds.), *Handbook of research on virtual workplaces and the new nature of business practices* (pp. 570–581). Hershey, PA: Information Science Reference; doi:10.4018/978-1-59904-893-2.ch041

Koh, S. L., & Maguire, S. (2009). Competing in the age of information technology in a developing economy: Experiences of an Indian bank. In S. Koh & S. Maguire (Eds.), *Information and communication technologies management in turbulent business environments* (pp. 326–350). Hershey, PA: Information Science Reference; doi:10.4018/978-1-60566-424-8.ch018

Kollmann, T., & Häsel, M. (2009). Competence of information technology professionals in internet-based ventures. In I. Lee (Ed.), *Electronic business: Concepts, methodologies, tools, and applications* (pp. 1905–1919). Hershey, PA: Information Science Reference; doi:10.4018/978-1-60566-056-1.ch118

Kollmann, T., & Häsel, M. (2009). Competence of information technology professionals in internet-based ventures. In A. Cater-Steel (Ed.), *Information technology governance and service management: Frameworks and adaptations* (pp. 239–253). Hershey, PA: Information Science Reference; doi:10.4018/978-1-60566-008-0.ch013

Kollmann, T., & Häsel, M. (2010). Competence of information technology professionals in internet-based ventures. In *Electronic services: Concepts, methodologies, tools and applications* (pp. 1551–1565). Hershey, PA: Information Science Reference; doi:10.4018/978-1-61520-967-5.ch094

Kraemer, K., & King, J. L. (2006). Information technology and administrative reform: Will e-government be different? *International Journal of Electronic Government Research, 2*(1), 1–20. doi:10.4018/jegr.2006010101

Kraemer, K., & King, J. L. (2008). Information technology and administrative reform: Will e-government be different? In D. Norris (Ed.), *E-government research: Policy and management* (pp. 1–20). Hershey, PA: IGI Publishing; doi:10.4018/978-1-59904-913-7.ch001

Lampathaki, F., Tsiakaliaris, C., Stasis, A., & Charalabidis, Y. (2011). National interoperability frameworks: The way forward. In Y. Charalabidis (Ed.), *Interoperability in digital public services and administration: Bridging e-government and e-business* (pp. 1–24). Hershey, PA: Information Science Reference; doi:10.4018/978-1-61520-887-6.ch001

Lan, Z., & Scott, C. R. (1996). The relative importance of computer-mediated information versus conventional non-computer-mediated information in public managerial decision making. *Information Resources Management Journal, 9*(1), 27–0. doi:10.4018/irmj.1996010103

Law, W. (2004). *Public sector data management in a developing economy*. Hershey, PA: IGI Global; doi:10.4018/978-1-59140-259-6.ch034

Law, W. K. (2005). Information resources development challenges in a cross-cultural environment. In M. Khosrow-Pour (Ed.), *Encyclopedia of information science and technology* (pp. 1476–1481). Hershey, PA: Information Science Reference; doi:10.4018/978-1-59140-553-5.ch259

Law, W. K. (2009). Cross-cultural challenges for information resources management. In M. Khosrow-Pour (Ed.), *Encyclopedia of information science and technology* (2nd ed., pp. 840–846). Hershey, PA: Information Science Reference; doi:10.4018/978-1-60566-026-4.ch136

Law, W. K. (2011). Cross-cultural challenges for information resources management. In *Global business: Concepts, methodologies, tools and applications* (pp. 1924–1932). Hershey, PA: Business Science Reference; doi:10.4018/978-1-60960-587-2.ch704

Malkia, M., & Savolainen, R. (2004). eTransformation in government, politics and society: Conceptual framework and introduction. In M. Malkia, A. Anttiroiko, & R. Savolainen (Eds.), eTransformation in governance: New directions in government and politics (pp. 1-21). Hershey, PA: Idea Group Publishing. doi:10.4018/978-1-59140-130-8.ch001

Management Association. I. (2010). Information resources management: Concepts, methodologies, tools and applications (4 Volumes). Hershey, PA: IGI Global. doi:10.4018/978-1-61520-965-1

Management Association. I. (2010). Electronic services: Concepts, methodologies, tools and applications (3 Volumes). Hershey, PA: IGI Global. doi:10.4018/978-1-61520-967-5

Mandujano, S. (2011). Network manageability security. In D. Kar & M. Syed (Eds.), *Network security, administration and management: Advancing technology and practice* (pp. 158–181). Hershey, PA: Information Science Reference; doi:10.4018/978-1-60960-777-7.ch009

Marich, M. J., Schooley, B. L., & Horan, T. A. (2012). A normative enterprise architecture for guiding end-to-end emergency response decision support. In M. Jennex (Ed.), *Managing crises and disasters with emerging technologies: Advancements* (pp. 71–87). Hershey, PA: Information Science Reference; doi:10.4018/978-1-4666-0167-3.ch006

Markov, R., & Okujava, S. (2008). Costs, benefits, and risks of e-government portals. In G. Putnik & M. Cruz-Cunha (Eds.), *Encyclopedia of networked and virtual organizations* (pp. 354–363). Hershey, PA: Information Science Reference; doi:10.4018/978-1-59904-885-7.ch047

Martin, N., & Rice, J. (2013). Evaluating and designing electronic government for the future: Observations and insights from Australia. In V. Weerakkody (Ed.), *E-government services design, adoption, and evaluation* (pp. 238–258). Hershey, PA: Information Science Reference; doi:10.4018/978-1-4666-2458-0.ch014

i. Martinez, A. C. (2008). Accessing administration's information via internet in Spain. In F. Tan (Ed.), *Global information technologies: Concepts, methodologies, tools, and applications* (pp. 2558–2573). Hershey, PA: Information Science Reference; doi:10.4018/978-1-59904-939-7.ch186

Mbarika, V. W., Meso, P. N., & Musa, P. F. (2006). A disconnect in stakeholders' perceptions from emerging realities of teledensity growth in Africa's least developed countries. In M. Hunter & F. Tan (Eds.), *Advanced topics in global information management* (Vol. 5, pp. 263–282). Hershey, PA: Idea Group Publishing; doi:10.4018/978-1-59140-923-6.ch012

Mbarika, V. W., Meso, P. N., & Musa, P. F. (2008). A disconnect in stakeholders' perceptions from emerging realities of teledensity growth in Africa's least developed countries. In F. Tan (Ed.), *Global information technologies: Concepts, methodologies, tools, and applications* (pp. 2948–2962). Hershey, PA: Information Science Reference; doi:10.4018/978-1-59904-939-7.ch209

Means, T., Olson, E., & Spooner, J. (2013). Discovering ways that don't work on the road to success: Strengths and weaknesses revealed by an active learning studio classroom project. In A. Benson, J. Moore, & S. Williams van Rooij (Eds.), *Cases on educational technology planning, design, and implementation: A project management perspective* (pp. 94–113). Hershey, PA: Information Science Reference; doi:10.4018/978-1-4666-4237-9.ch006

Melitski, J., Holzer, M., Kim, S., Kim, C., & Rho, S. (2008). Digital government worldwide: An e-government assessment of municipal web sites. In G. Garson & M. Khosrow-Pour (Eds.), *Handbook of research on public information technology* (pp. 790–804). Hershey, PA: Information Science Reference; doi:10.4018/978-1-59904-857-4.ch069

Memmola, M., Palumbo, G., & Rossini, M. (2009). Web & RFID technology: New frontiers in costing and process management for rehabilitation medicine. In L. Al-Hakim & M. Memmola (Eds.), *Business web strategy: Design, alignment, and application* (pp. 145–169). Hershey, PA: Information Science Reference; doi:10.4018/978-1-60566-024-0.ch008

Meng, Z., Fahong, Z., & Lei, L. (2008). Information technology and environment. In Y. Kurihara, S. Takaya, H. Harui, & H. Kamae (Eds.), *Information technology and economic development* (pp. 201–212). Hershey, PA: Information Science Reference; doi:10.4018/978-1-59904-579-5.ch014

Mentzingen de Moraes, A. J., Ferneda, E., Costa, I., & Spinola, M. D. (2011). Practical approach for implementation of governance process in IT: Information technology areas. In N. Shi & G. Silvius (Eds.), *Enterprise IT governance, business value and performance measurement* (pp. 19–40). Hershey, PA: Information Science Reference; doi:10.4018/978-1-60566-346-3.ch002

Merwin, G. A. Jr, McDonald, J. S., & Odera, L. C. (2008). Economic development: Government's cutting edge in IT. In M. Raisinghani (Ed.), *Handbook of research on global information technology management in the digital economy* (pp. 1–37). Hershey, PA: Information Science Reference; doi:10.4018/978-1-59904-875-8.ch001

Meso, P., & Duncan, N. (2002). Can national information infrastructures enhance social development in the least developed countries? An empirical investigation. In M. Dadashzadeh (Ed.), *Information technology management in developing countries* (pp. 23–51). Hershey, PA: IRM Press; doi:10.4018/978-1-931777-03-2.ch002

Meso, P. N., & Duncan, N. B. (2002). Can national information infrastructures enhance social development in the least developed countries? In F. Tan (Ed.), *Advanced topics in global information management* (Vol. 1, pp. 207–226). Hershey, PA: Idea Group Publishing; doi:10.4018/978-1-930708-43-3.ch014

Middleton, M. (2008). Evaluation of e-government web sites. In G. Garson & M. Khosrow-Pour (Eds.), *Handbook of research on public information technology* (pp. 699–710). Hershey, PA: Information Science Reference; doi:10.4018/978-1-59904-857-4.ch063

Mingers, J. (2010). Pluralism, realism, and truth: The keys to knowledge in information systems research. In D. Paradice (Ed.), *Emerging systems approaches in information technologies: Concepts, theories, and applications* (pp. 86–98). Hershey, PA: Information Science Reference; doi:10.4018/978-1-60566-976-2.ch006

Mital, K. M. (2012). ICT, unique identity and inclusive growth: An Indian perspective. In A. Manoharan & M. Holzer (Eds.), *E-governance and civic engagement: Factors and determinants of e-democracy* (pp. 584–612). Hershey, PA: Information Science Reference; doi:10.4018/978-1-61350-083-5.ch029

Mizell, A. P. (2008). Helping close the digital divide for financially disadvantaged seniors. In F. Tan (Ed.), *Global information technologies: Concepts, methodologies, tools, and applications* (pp. 2396–2402). Hershey, PA: Information Science Reference; doi:10.4018/978-1-59904-939-7.ch173

Molinari, F., Wills, C., Koumpis, A., & Moumtzi, V. (2011). A citizen-centric platform to support networking in the area of e-democracy. In H. Rahman (Ed.), *Cases on adoption, diffusion and evaluation of global e-governance systems: Impact at the grass roots* (pp. 282–302). Hershey, PA: Information Science Reference; doi:10.4018/978-1-61692-814-8.ch014

Molinari, F., Wills, C., Koumpis, A., & Moumtzi, V. (2013). A citizen-centric platform to support networking in the area of e-democracy. In H. Rahman (Ed.), *Cases on progressions and challenges in ICT utilization for citizen-centric governance* (pp. 265–297). Hershey, PA: Information Science Reference; doi:10.4018/978-1-4666-2071-1.ch013

Monteverde, F. (2010). The process of e-government public policy inclusion in the governmental agenda: A framework for assessment and case study. In J. Cordoba-Pachon & A. Ochoa-Arias (Eds.), *Systems thinking and e-participation: ICT in the governance of society* (pp. 233–245). Hershey, PA: Information Science Reference; doi:10.4018/978-1-60566-860-4.ch015

Moodley, S. (2008). Deconstructing the South African government's ICT for development discourse. In A. Anttiroiko (Ed.), *Electronic government: Concepts, methodologies, tools, and applications* (pp. 622–631). Hershey, PA: Information Science Reference; doi:10.4018/978-1-59904-947-2.ch053

Moodley, S. (2008). Deconstructing the South African government's ICT for development discourse. In C. Van Slyke (Ed.), *Information communication technologies: Concepts, methodologies, tools, and applications* (pp. 816–825). Hershey, PA: Information Science Reference; doi:10.4018/978-1-59904-949-6.ch052

Mora, M., Cervantes-Perez, F., Gelman-Muravchik, O., Forgionne, G. A., & Mejia-Olvera, M. (2003). DMSS implementation research: A conceptual analysis of the contributions and limitations of the factor-based and stage-based streams. In G. Forgionne, J. Gupta, & M. Mora (Eds.), *Decision-making support systems: Achievements and challenges for the new decade* (pp. 331–356). Hershey, PA: Idea Group Publishing; doi:10.4018/978-1-59140-045-5.ch020

Mörtberg, C., & Elovaara, P. (2010). Attaching people and technology: Between e and government. In S. Booth, S. Goodman, & G. Kirkup (Eds.), *Gender issues in learning and working with information technology: Social constructs and cultural contexts* (pp. 83–98). Hershey, PA: Information Science Reference; doi:10.4018/978-1-61520-813-5.ch005

Murphy, J., Harper, E., Devine, E. C., Burke, L. J., & Hook, M. L. (2011). Case study: Lessons learned when embedding evidence-based knowledge in a nurse care planning and documentation system. In A. Cashin & R. Cook (Eds.), *Evidence-based practice in nursing informatics: Concepts and applications* (pp. 174–190). Hershey, PA: Medical Information Science Reference; doi:10.4018/978-1-60960-034-1.ch014

Mutula, S. M. (2013). E-government's role in poverty alleviation: Case study of South Africa. In H. Rahman (Ed.), *Cases on progressions and challenges in ICT utilization for citizen-centric governance* (pp. 44–68). Hershey, PA: Information Science Reference; doi:10.4018/978-1-4666-2071-1.ch003

Nath, R., & Angeles, R. (2005). Relationships between supply characteristics and buyer-supplier coupling in e-procurement: An empirical analysis. [IJEBR]. *International Journal of E-Business Research, 1*(2), 40–55. doi:10.4018/jebr.2005040103

Nissen, M. E. (2006). Application cases in government. In M. Nissen (Ed.), *Harnessing knowledge dynamics: Principled organizational knowing & learning* (pp. 152–181). Hershey, PA: IRM Press; doi:10.4018/978-1-59140-773-7.ch008

Norris, D. F. (2003). Leading-edge information technologies and American local governments. In G. Garson (Ed.), *Public information technology: Policy and management issues* (pp. 139–169). Hershey, PA: Idea Group Publishing; doi:10.4018/978-1-59140-060-8.ch007

Norris, D. F. (2008). Information technology among U.S. local governments. In G. Garson & M. Khosrow-Pour (Eds.), *Handbook of research on public information technology* (pp. 132–144). Hershey, PA: Information Science Reference; doi:10.4018/978-1-59904-857-4.ch013

Northrop, A. (1999). The challenge of teaching information technology in public administration graduate programs. In G. Garson (Ed.), *Information technology and computer applications in public administration: Issues and trends* (pp. 1–22). Hershey, PA: Information Science Reference; doi:10.4018/978-1-87828-952-0.ch001

Northrop, A. (2003). Information technology and public administration: The view from the profession. In G. Garson (Ed.), *Public information technology: Policy and management issues* (pp. 1–19). Hershey, PA: Idea Group Publishing; doi:10.4018/978-1-59140-060-8.ch001

Northrop, A. (2007). Lip service? How PA journals and textbooks view information technology. In G. Garson (Ed.), *Modern public information technology systems: Issues and challenges* (pp. 1–16). Hershey, PA: Idea Group Publishing; doi:10.4018/978-1-59904-051-6.ch001

Null, E. (2013). Legal and political barriers to municipal networks in the United States. In A. Abdelaal (Ed.), *Social and economic effects of community wireless networks and infrastructures* (pp. 27–56). Hershey, PA: Information Science Reference; doi:10.4018/978-1-4666-2997-4.ch003

Okunoye, A., Frolick, M., & Crable, E. (2006). ERP implementation in higher education: An account of pre-implementation and implementation phases. *Journal of Cases on Information Technology*, *8*(2), 110–132. doi:10.4018/jcit.2006040106

Olasina, G. (2012). A review of egovernment services in Nigeria. In A. Tella & A. Issa (Eds.), *Library and information science in developing countries: Contemporary issues* (pp. 205–221). Hershey, PA: Information Science Reference; doi:10.4018/978-1-61350-335-5.ch015

Orgeron, C. P. (2008). A model for reengineering IT job classes in state government. In G. Garson & M. Khosrow-Pour (Eds.), *Handbook of research on public information technology* (pp. 735–746). Hershey, PA: Information Science Reference; doi:10.4018/978-1-59904-857-4.ch066

Owsinski, J. W., & Pielak, A. M. (2011). Local authority websites in rural areas: Measuring quality and functionality, and assessing the role. In Z. Andreopoulou, B. Manos, N. Polman, & D. Viaggi (Eds.), *Agricultural and environmental informatics, governance and management: Emerging research applications* (pp. 39–60). Hershey, PA: Information Science Reference; doi:10.4018/978-1-60960-621-3.ch003

Owsiński, J. W., Pielak, A. M., Sęp, K., & Stańczak, J. (2014). Local web-based networks in rural municipalities: Extension, density, and meaning. In Z. Andreopoulou, V. Samathrakis, S. Louca, & M. Vlachopoulou (Eds.), *E-innovation for sustainable development of rural resources during global economic crisis* (pp. 126–151). Hershey, PA: Business Science Reference; doi:10.4018/978-1-4666-4550-9.ch011

Pagani, M., & Pasinetti, C. (2008). Technical and functional quality in the development of t-government services. In A. Anttiroiko (Ed.), *Electronic government: Concepts, methodologies, tools, and applications* (pp. 2943–2965). Hershey, PA: Information Science Reference; doi:10.4018/978-1-59904-947-2.ch220

Pani, A. K., & Agrahari, A. (2005). On e-markets in emerging economy: An Indian experience. In M. Khosrow-Pour (Ed.), *Advanced topics in electronic commerce* (Vol. 1, pp. 287–299). Hershey, PA: Idea Group Publishing; doi:10.4018/978-1-59140-819-2.ch015

Papadopoulos, T., Angelopoulos, S., & Kitsios, F. (2011). A strategic approach to e-health interoperability using e-government frameworks. In A. Lazakidou, K. Siassiakos, & K. Ioannou (Eds.), *Wireless technologies for ambient assisted living and healthcare: Systems and applications* (pp. 213–229). Hershey, PA: Medical Information Science Reference; doi:10.4018/978-1-61520-805-0.ch012

Papadopoulos, T., Angelopoulos, S., & Kitsios, F. (2013). A strategic approach to e-health interoperability using e-government frameworks. In *User-driven healthcare: Concepts, methodologies, tools, and applications* (pp. 791–807). Hershey, PA: Medical Information Science Reference; doi:10.4018/978-1-4666-2770-3.ch039

Papaleo, G., Chiarella, D., Aiello, M., & Caviglione, L. (2012). Analysis, development and deployment of statistical anomaly detection techniques for real e-mail traffic. In T. Chou (Ed.), *Information assurance and security technologies for risk assessment and threat management: Advances* (pp. 47–71). Hershey, PA: Information Science Reference; doi:10.4018/978-1-61350-507-6.ch003

Papp, R. (2003). Information technology & FDA compliance in the pharmaceutical industry. In M. Khosrow-Pour (Ed.), *Annals of cases on information technology* (Vol. 5, pp. 262–273). Hershey, PA: Information Science Reference; doi:10.4018/978-1-59140-061-5.ch017

Parsons, T. W. (2007). Developing a knowledge management portal. In A. Tatnall (Ed.), *Encyclopedia of portal technologies and applications* (pp. 223–227). Hershey, PA: Information Science Reference; doi:10.4018/978-1-59140-989-2.ch039

Passaris, C. E. (2007). Immigration and digital government. In A. Anttiroiko & M. Malkia (Eds.), *Encyclopedia of digital government* (pp. 988–994). Hershey, PA: Information Science Reference; doi:10.4018/978-1-59140-789-8.ch148

Pavlichev, A. (2004). The e-government challenge for public administration. In A. Pavlichev & G. Garson (Eds.), *Digital government: Principles and best practices* (pp. 276–290). Hershey, PA: Idea Group Publishing; doi:10.4018/978-1-59140-122-3.ch018

Penrod, J. I., & Harbor, A. F. (2000). Designing and implementing a learning organization-oriented information technology planning and management process. In L. Petrides (Ed.), *Case studies on information technology in higher education: Implications for policy and practice* (pp. 7–19). Hershey, PA: Idea Group Publishing; doi:10.4018/978-1-878289-74-2.ch001

Planas-Silva, M. D., & Joseph, R. C. (2011). Perspectives on the adoption of electronic resources for use in clinical trials. In M. Guah (Ed.), *Healthcare delivery reform and new technologies: Organizational initiatives* (pp. 19–28). Hershey, PA: Information Science Reference; doi:10.4018/978-1-60960-183-6.ch002

Pomazalová, N., & Rejman, S. (2013). The rationale behind implementation of new electronic tools for electronic public procurement. In N. Pomazalová (Ed.), *Public sector transformation processes and internet public procurement: Decision support systems* (pp. 85–117). Hershey, PA: Engineering Science Reference; doi:10.4018/978-1-4666-2665-2.ch006

Postorino, M. N. (2012). City competitiveness and airport: Information science perspective. In M. Bulu (Ed.), *City competitiveness and improving urban subsystems: Technologies and applications* (pp. 61–83). Hershey, PA: Information Science Reference; doi:10.4018/978-1-61350-174-0.ch004

Poupa, C. (2002). Electronic government in Switzerland: Priorities for 2001-2005 - Electronic voting and federal portal. In Å. Grönlund (Ed.), *Electronic government: Design, applications and management* (pp. 356–369). Hershey, PA: Idea Group Publishing; doi:10.4018/978-1-930708-19-8.ch017

Powell, S. R. (2010). Interdisciplinarity in telecommunications and networking. In *Networking and telecommunications: Concepts, methodologies, tools and applications* (pp. 33–40). Hershey, PA: Information Science Reference; doi:10.4018/978-1-60566-986-1.ch004

Priya, P. S., & Mathiyalagan, N. (2011). A study of the implementation status of two e-governance projects in land revenue administration in India. In M. Shareef, V. Kumar, U. Kumar, & Y. Dwivedi (Eds.), *Stakeholder adoption of e-government services: Driving and resisting factors* (pp. 214–230). Hershey, PA: Information Science Reference; doi:10.4018/978-1-60960-601-5.ch011

Prysby, C., & Prysby, N. (2000). Electronic mail, employee privacy and the workplace. In L. Janczewski (Ed.), *Internet and intranet security management: Risks and solutions* (pp. 251–270). Hershey, PA: Idea Group Publishing; doi:10.4018/978-1-878289-71-1.ch009

Prysby, C. L., & Prysby, N. D. (2003). Electronic mail in the public workplace: Issues of privacy and public disclosure. In G. Garson (Ed.), *Public information technology: Policy and management issues* (pp. 271–298). Hershey, PA: Idea Group Publishing; doi:10.4018/978-1-59140-060-8.ch012

Prysby, C. L., & Prysby, N. D. (2007). You have mail, but who is reading it? Issues of e-mail in the public workplace. In G. Garson (Ed.), *Modern public information technology systems: Issues and challenges* (pp. 312–336). Hershey, PA: Idea Group Publishing; doi:10.4018/978-1-59904-051-6.ch016

Radl, A., & Chen, Y. (2005). Computer security in electronic government: A state-local education information system. *International Journal of Electronic Government Research, 1*(1), 79–99. doi:10.4018/jegr.2005010105

Rahman, H. (2008). Information dynamics in developing countries. In C. Van Slyke (Ed.), *Information communication technologies: Concepts, methodologies, tools, and applications* (pp. 104–114). Hershey, PA: Information Science Reference; doi:10.4018/978-1-59904-949-6.ch008

Ramanathan, J. (2009). Adaptive IT architecture as a catalyst for network capability in government. In P. Saha (Ed.), *Advances in government enterprise architecture* (pp. 149–172). Hershey, PA: Information Science Reference; doi:10.4018/978-1-60566-068-4.ch007

Ramos, I., & Berry, D. M. (2006). Social construction of information technology supporting work. In M. Khosrow-Pour (Ed.), *Cases on information technology: Lessons learned* (Vol. 7, pp. 36–52). Hershey, PA: Idea Group Publishing; doi:10.4018/978-1-59140-673-0.ch003

Ray, D., Gulla, U., Gupta, M. P., & Dash, S. S. (2009). Interoperability and constituents of interoperable systems in public sector. In V. Weerakkody, M. Janssen, & Y. Dwivedi (Eds.), *Handbook of research on ICT-enabled transformational government: A global perspective* (pp. 175–195). Hershey, PA: Information Science Reference; doi:10.4018/978-1-60566-390-6.ch010

Reddick, C. G. (2007). E-government and creating a citizen-centric government: A study of federal government CIOs. In G. Garson (Ed.), *Modern public information technology systems: Issues and challenges* (pp. 143–165). Hershey, PA: Idea Group Publishing; doi:10.4018/978-1-59904-051-6.ch008

Reddick, C. G. (2010). Citizen-centric e-government. In C. Reddick (Ed.), *Homeland security preparedness and information systems: Strategies for managing public policy* (pp. 45–75). Hershey, PA: Information Science Reference; doi:10.4018/978-1-60566-834-5.ch002

Reddick, C. G. (2010). E-government and creating a citizen-centric government: A study of federal government CIOs. In C. Reddick (Ed.), *Homeland security preparedness and information systems: Strategies for managing public policy* (pp. 230–250). Hershey, PA: Information Science Reference; doi:10.4018/978-1-60566-834-5.ch012

Reddick, C. G. (2010). Perceived effectiveness of e-government and its usage in city governments: Survey evidence from information technology directors. In C. Reddick (Ed.), *Homeland security preparedness and information systems: Strategies for managing public policy* (pp. 213–229). Hershey, PA: Information Science Reference; doi:10.4018/978-1-60566-834-5.ch011

Reddick, C. G. (2012). Customer relationship management adoption in local governments in the United States. In S. Chhabra & M. Kumar (Eds.), *Strategic enterprise resource planning models for e-government: Applications and methodologies* (pp. 111–124). Hershey, PA: Information Science Reference; doi:10.4018/978-1-60960-863-7.ch008

Reeder, F. S., & Pandy, S. M. (2008). Identifying effective funding models for e-government. In A. Anttiroiko (Ed.), *Electronic government: Concepts, methodologies, tools, and applications* (pp. 1108–1138). Hershey, PA: Information Science Reference; doi:10.4018/978-1-59904-947-2.ch083

Riesco, D., Acosta, E., & Montejano, G. (2003). An extension to a UML activity graph from workflow. In L. Favre (Ed.), *UML and the unified process* (pp. 294–314). Hershey, PA: IRM Press; doi:10.4018/978-1-93177-744-5.ch015

Ritzhaupt, A. D., & Gill, T. G. (2008). A hybrid and novel approach to teaching computer programming in MIS curriculum. In S. Negash, M. Whitman, A. Woszczynski, K. Hoganson, & H. Mattord (Eds.), *Handbook of distance learning for real-time and asynchronous information technology education* (pp. 259–281). Hershey, PA: Information Science Reference; doi:10.4018/978-1-59904-964-9.ch014

Roche, E. M. (1993). International computing and the international regime. *Journal of Global Information Management*, *1*(2), 33–44. doi:10.4018/jgim.1993040103

Rocheleau, B. (2007). Politics, accountability, and information management. In G. Garson (Ed.), *Modern public information technology systems: Issues and challenges* (pp. 35–71). Hershey, PA: Idea Group Publishing; doi:10.4018/978-1-59904-051-6.ch003

Rodrigues Filho, J. (2010). E-government in Brazil: Reinforcing dominant institutions or reducing citizenship? In C. Reddick (Ed.), *Politics, democracy and e-government: Participation and service delivery* (pp. 347–362). Hershey, PA: Information Science Reference; doi:10.4018/978-1-61520-933-0.ch021

Rodriguez, S. R., & Thorp, D. A. (2013). eLearning for industry: A case study of the project management process. In A. Benson, J. Moore, & S. Williams van Rooij (Eds.), Cases on educational technology planning, design, and implementation: A project management perspective (pp. 319-342). Hershey, PA: Information Science Reference. doi:10.4018/978-1-4666-4237-9.ch017

Roman, A. V. (2013). Delineating three dimensions of e-government success: Security, functionality, and transformation. In J. Gil-Garcia (Ed.), *E-government success factors and measures: Theories, concepts, and methodologies* (pp. 171–192). Hershey, PA: Information Science Reference; doi:10.4018/978-1-4666-4058-0.ch010

Ross, S. C., Tyran, C. K., & Auer, D. J. (2008). Up in smoke: Rebuilding after an IT disaster. In H. Nemati (Ed.), *Information security and ethics: Concepts, methodologies, tools, and applications* (pp. 3659–3675). Hershey, PA: Information Science Reference; doi:10.4018/978-1-59904-937-3.ch248

Ross, S. C., Tyran, C. K., Auer, D. J., Junell, J. M., & Williams, T. G. (2005). Up in smoke: Rebuilding after an IT disaster. *Journal of Cases on Information Technology*, *7*(2), 31–49. doi:10.4018/jcit.2005040103

Roy, J. (2008). Security, sovereignty, and continental interoperability: Canada's elusive balance. In T. Loendorf & G. Garson (Eds.), *Patriotic information systems* (pp. 153–176). Hershey, PA: IGI Publishing; doi:10.4018/978-1-59904-594-8.ch007

Rubeck, R. F., & Miller, G. A. (2009). vGOV: Remote video access to government services. In A. Scupola (Ed.), Cases on managing e-services (pp. 253-268). Hershey, PA: Information Science Reference. doi:10.4018/978-1-60566-064-6.ch017

Saekow, A., & Boonmee, C. (2011). The challenges of implementing e-government interoperability in Thailand: Case of official electronic correspondence letters exchange across government departments. In Y. Charalabidis (Ed.), *Interoperability in digital public services and administration: Bridging e-government and e-business* (pp. 40–61). Hershey, PA: Information Science Reference; doi:10.4018/978-1-61520-887-6.ch003

Saekow, A., & Boonmee, C. (2012). The challenges of implementing e-government interoperability in Thailand: Case of official electronic correspondence letters exchange across government departments. In *Digital democracy: Concepts, methodologies, tools, and applications* (pp. 1883–1905). Hershey, PA: Information Science Reference; doi:10.4018/978-1-4666-1740-7.ch094

Sagsan, M., & Medeni, T. (2012). Understanding "knowledge management (KM) paradigms" from social media perspective: An empirical study on discussion group for KM at professional networking site. In M. Cruz-Cunha, P. Gonçalves, N. Lopes, E. Miranda, & G. Putnik (Eds.), *Handbook of research on business social networking: Organizational, managerial, and technological dimensions* (pp. 738–755). Hershey, PA: Business Science Reference; doi:10.4018/978-1-61350-168-9.ch039

Sahi, G., & Madan, S. (2013). Information security threats in ERP enabled e-governance: Challenges and solutions. In *Enterprise resource planning: Concepts, methodologies, tools, and applications* (pp. 825–837). Hershey, PA: Business Science Reference; doi:10.4018/978-1-4666-4153-2.ch048

Sanford, C., & Bhattacherjee, A. (2008). IT implementation in a developing country municipality: A sociocognitive analysis. *International Journal of Technology and Human Interaction*, *4*(3), 68–93. doi:10.4018/jthi.2008070104

Schelin, S. H. (2003). E-government: An overview. In G. Garson (Ed.), *Public information technology: Policy and management issues* (pp. 120–138). Hershey, PA: Idea Group Publishing; doi:10.4018/978-1-59140-060-8.ch006

Schelin, S. H. (2004). Training for digital government. In A. Pavlichev & G. Garson (Eds.), *Digital government: Principles and best practices* (pp. 263–275). Hershey, PA: Idea Group Publishing; doi:10.4018/978-1-59140-122-3.ch017

Schelin, S. H. (2007). E-government: An overview. In G. Garson (Ed.), *Modern public information technology systems: Issues and challenges* (pp. 110–126). Hershey, PA: Idea Group Publishing; doi:10.4018/978-1-59904-051-6.ch006

Schelin, S. H., & Garson, G. (2004). Theoretical justification of critical success factors. In G. Garson & S. Schelin (Eds.), *IT solutions series: Humanizing information technology: Advice from experts* (pp. 4–15). Hershey, PA: CyberTech Publishing; doi:10.4018/978-1-59140-245-9.ch002

Scime, A. (2002). Information systems and computer science model curricula: A comparative look. In M. Dadashzadeh, A. Saber, & S. Saber (Eds.), *Information technology education in the new millennium* (pp. 146–158). Hershey, PA: IRM Press; doi:10.4018/978-1-931777-05-6.ch018

Scime, A. (2009). Computing curriculum analysis and development. In M. Khosrow-Pour (Ed.), *Encyclopedia of information science and technology* (2nd ed., pp. 667–671). Hershey, PA: Information Science Reference; doi:10.4018/978-1-60566-026-4.ch108

Scime, A., & Wania, C. (2008). Computing curricula: A comparison of models. In C. Van Slyke (Ed.), *Information communication technologies: Concepts, methodologies, tools, and applications* (pp. 1270–1283). Hershey, PA: Information Science Reference; doi:10.4018/978-1-59904-949-6.ch088

Seidman, S. B. (2009). An international perspective on professional software engineering credentials. In H. Ellis, S. Demurjian, & J. Naveda (Eds.), *Software engineering: Effective teaching and learning approaches and practices* (pp. 351–361). Hershey, PA: Information Science Reference; doi:10.4018/978-1-60566-102-5.ch018

Seifert, J. W. (2007). E-government act of 2002 in the United States. In A. Anttiroiko & M. Malkia (Eds.), *Encyclopedia of digital government* (pp. 476–481). Hershey, PA: Information Science Reference; doi:10.4018/978-1-59140-789-8.ch072

Seifert, J. W., & Relyea, H. C. (2008). E-government act of 2002 in the United States. In A. Anttiroiko (Ed.), *Electronic government: Concepts, methodologies, tools, and applications* (pp. 154–161). Hershey, PA: Information Science Reference; doi:10.4018/978-1-59904-947-2.ch013

Seufert, S. (2002). E-learning business models: Framework and best practice examples. In M. Raisinghani (Ed.), *Cases on worldwide e-commerce: Theory in action* (pp. 70–94). Hershey, PA: Idea Group Publishing; doi:10.4018/978-1-930708-27-3.ch004

Shareef, M. A., & Archer, N. (2012). E-government service development. In M. Shareef, N. Archer, & S. Dutta (Eds.), *E-government service maturity and development: Cultural, organizational and technological perspectives* (pp. 1–14). Hershey, PA: Information Science Reference; doi:10.4018/978-1-60960-848-4.ch001

Shareef, M. A., & Archer, N. (2012). E-government initiatives: Review studies on different countries. In M. Shareef, N. Archer, & S. Dutta (Eds.), *E-government service maturity and development: Cultural, organizational and technological perspectives* (pp. 40–76). Hershey, PA: Information Science Reference; doi:10.4018/978-1-60960-848-4.ch003

Shareef, M. A., Kumar, U., & Kumar, V. (2011). E-government development: Performance evaluation parameters. In M. Shareef, V. Kumar, U. Kumar, & Y. Dwivedi (Eds.), *Stakeholder adoption of e-government services: Driving and resisting factors* (pp. 197–213). Hershey, PA: Information Science Reference; doi:10.4018/978-1-60960-601-5.ch010

Shareef, M. A., Kumar, U., Kumar, V., & Niktash, M. (2012). Electronic-government vision: Case studies for objectives, strategies, and initiatives. In M. Shareef, N. Archer, & S. Dutta (Eds.), *E-government service maturity and development: Cultural, organizational and technological perspectives* (pp. 15–39). Hershey, PA: Information Science Reference; doi:10.4018/978-1-60960-848-4.ch002

Shukla, P., Kumar, A., & Anu Kumar, P. B.Anu Kumar P.B. (2013). Impact of national culture on business continuity management system implementation. *International Journal of Risk and Contingency Management, 2*(3), 23–36. doi:10.4018/ijrcm.2013070102

Shulman, S. W. (2007). The federal docket management system and the prospect for digital democracy in U S rulemaking. In G. Garson (Ed.), *Modern public information technology systems: Issues and challenges* (pp. 166–184). Hershey, PA: Idea Group Publishing; doi:10.4018/978-1-59904-051-6.ch009

Simonovic, S. (2007). Problems of offline government in e-Serbia. In A. Anttiroiko & M. Malkia (Eds.), *Encyclopedia of digital government* (pp. 1342–1351). Hershey, PA: Information Science Reference; doi:10.4018/978-1-59140-789-8.ch205

Simonovic, S. (2008). Problems of offline government in e-Serbia. In A. Anttiroiko (Ed.), *Electronic government: Concepts, methodologies, tools, and applications* (pp. 2929–2942). Hershey, PA: Information Science Reference; doi:10.4018/978-1-59904-947-2.ch219

Singh, A. M. (2005). Information systems and technology in South Africa. In M. Khosrow-Pour (Ed.), *Encyclopedia of information science and technology* (pp. 1497–1502). Hershey, PA: Information Science Reference; doi:10.4018/978-1-59140-553-5.ch263

Singh, S., & Naidoo, G. (2005). Towards an e-government solution: A South African perspective. In W. Huang, K. Siau, & K. Wei (Eds.), *Electronic government strategies and implementation* (pp. 325–353). Hershey, PA: Idea Group Publishing; doi:10.4018/978-1-59140-348-7.ch014

Snoke, R., & Underwood, A. (2002). Generic attributes of IS graduates: An analysis of Australian views. In F. Tan (Ed.), *Advanced topics in global information management* (Vol. 1, pp. 370–384). Hershey, PA: Idea Group Publishing; doi:10.4018/978-1-930708-43-3.ch023

Sommer, L. (2006). Revealing unseen organizations in higher education: A study framework and application example. In A. Metcalfe (Ed.), *Knowledge management and higher education: A critical analysis* (pp. 115–146). Hershey, PA: Information Science Publishing; doi:10.4018/978-1-59140-509-2.ch007

Song, H., Kidd, T., & Owens, E. (2011). Examining technological disparities and instructional practices in English language arts classroom: Implications for school leadership and teacher training. In L. Tomei (Ed.), *Online courses and ICT in education: Emerging practices and applications* (pp. 258–274). Hershey, PA: Information Science Reference; doi:10.4018/978-1-60960-150-8.ch020

Speaker, P. J., & Kleist, V. F. (2003). Using information technology to meet electronic commerce and MIS education demands. In A. Aggarwal (Ed.), *Web-based education: Learning from experience* (pp. 280–291). Hershey, PA: Information Science Publishing; doi:10.4018/978-1-59140-102-5.ch017

Spitler, V. K. (2007). Learning to use IT in the workplace: Mechanisms and masters. In M. Mahmood (Ed.), *Contemporary issues in end user computing* (pp. 292–323). Hershey, PA: Idea Group Publishing; doi:10.4018/978-1-59140-926-7.ch013

Stellefson, M. (2011). Considerations for marketing distance education courses in health education: Five important questions to examine before development. In U. Demiray & S. Sever (Eds.), *Marketing online education programs: Frameworks for promotion and communication* (pp. 222–234). Hershey, PA: Information Science Reference; doi:10.4018/978-1-60960-074-7.ch014

Straub, D. W., & Loch, K. D. (2006). Creating and developing a program of global research. *Journal of Global Information Management, 14*(2), 1–28. doi:10.4018/jgim.2006040101

Straub, D. W., Loch, K. D., & Hill, C. E. (2002). Transfer of information technology to the Arab world: A test of cultural influence modeling. In M. Dadashzadeh (Ed.), *Information technology management in developing countries* (pp. 92–134). Hershey, PA: IRM Press; doi:10.4018/978-1-931777-03-2.ch005

Straub, D. W., Loch, K. D., & Hill, C. E. (2003). Transfer of information technology to the Arab world: A test of cultural influence modeling. In F. Tan (Ed.), *Advanced topics in global information management* (Vol. 2, pp. 141–172). Hershey, PA: Idea Group Publishing; doi:10.4018/978-1-59140-064-6.ch009

Suki, N. M., Ramayah, T., Ming, M. K., & Suki, N. M. (2013). Factors enhancing employed job seekers intentions to use social networking sites as a job search tool. In A. Mesquita (Ed.), *User perception and influencing factors of technology in everyday life* (pp. 265–281). Hershey, PA: Information Science Reference; doi:10.4018/978-1-4666-1954-8.ch018

Suomi, R. (2006). Introducing electronic patient records to hospitals: Innovation adoption paths. In T. Spil & R. Schuring (Eds.), *E-health systems diffusion and use: The innovation, the user and the use IT model* (pp. 128–146). Hershey, PA: Idea Group Publishing; doi:10.4018/978-1-59140-423-1.ch008

Swim, J., & Barker, L. (2012). Pathways into a gendered occupation: Brazilian women in IT. *International Journal of Social and Organizational Dynamics in IT, 2*(4), 34–51. doi:10.4018/ijsodit.2012100103

Tarafdar, M., & Vaidya, S. D. (2006). Adoption and implementation of IT in developing nations: Experiences from two public sector enterprises in India. In M. Khosrow-Pour (Ed.), *Cases on information technology planning, design and implementation* (pp. 208–233). Hershey, PA: Idea Group Publishing; doi:10.4018/978-1-59904-408-8.ch013

Tarafdar, M., & Vaidya, S. D. (2008). Adoption and implementation of IT in developing nations: Experiences from two public sector enterprises in India. In G. Garson & M. Khosrow-Pour (Eds.), *Handbook of research on public information technology* (pp. 905–924). Hershey, PA: Information Science Reference; doi:10.4018/978-1-59904-857-4.ch076

Thesing, Z. (2007). Zarina thesing, pumpkin patch. In M. Hunter (Ed.), *Contemporary chief information officers: Management experiences* (pp. 83–94). Hershey, PA: IGI Publishing; doi:10.4018/978-1-59904-078-3.ch007

Thomas, J. C. (2004). Public involvement in public administration in the information age: Speculations on the effects of technology. In M. Malkia, A. Anttiroiko, & R. Savolainen (Eds.), *eTransformation in governance: New directions in government and politics* (pp. 67–84). Hershey, PA: Idea Group Publishing; doi:10.4018/978-1-59140-130-8.ch004

Treiblmaier, H., & Chong, S. (2013). Trust and perceived risk of personal information as antecedents of online information disclosure: Results from three countries. In F. Tan (Ed.), *Global diffusion and adoption of technologies for knowledge and information sharing* (pp. 341–361). Hershey, PA: Information Science Reference; doi:10.4018/978-1-4666-2142-8.ch015

van Grembergen, W., & de Haes, S. (2008). IT governance in practice: Six case studies. In W. van Grembergen & S. De Haes (Eds.), *Implementing information technology governance: Models, practices and cases* (pp. 125–237). Hershey, PA: IGI Publishing; doi:10.4018/978-1-59904-924-3.ch004

van Os, G., Homburg, V., & Bekkers, V. (2013). Contingencies and convergence in European social security: ICT coordination in the back office of the welfare state. In M. Cruz-Cunha, I. Miranda, & P. Gonçalves (Eds.), *Handbook of research on ICTs and management systems for improving efficiency in healthcare and social care* (pp. 268–287). Hershey, PA: Medical Information Science Reference; doi:10.4018/978-1-4666-3990-4.ch013

Velloso, A. B., Gassenferth, W., & Machado, M. A. (2012). Evaluating IBMEC-RJ's intranet usability using fuzzy logic. In M. Cruz-Cunha, P. Gonçalves, N. Lopes, E. Miranda, & G. Putnik (Eds.), *Handbook of research on business social networking: Organizational, managerial, and technological dimensions* (pp. 185–205). Hershey, PA: Business Science Reference; doi:10.4018/978-1-61350-168-9.ch010

Villablanca, A. C., Baxi, H., & Anderson, K. (2009). Novel data interface for evaluating cardiovascular outcomes in women. In A. Dwivedi (Ed.), *Handbook of research on information technology management and clinical data administration in healthcare* (pp. 34–53). Hershey, PA: Medical Information Science Reference; doi:10.4018/978-1-60566-356-2.ch003

Villablanca, A. C., Baxi, H., & Anderson, K. (2011). Novel data interface for evaluating cardiovascular outcomes in women. In *Clinical technologies: Concepts, methodologies, tools and applications* (pp. 2094–2113). Hershey, PA: Medical Information Science Reference; doi:10.4018/978-1-60960-561-2.ch806

Virkar, S. (2011). Information and communication technologies in administrative reform for development: Exploring the case of property tax systems in Karnataka, India. In J. Steyn, J. Van Belle, & E. Mansilla (Eds.), *ICTs for global development and sustainability: Practice and applications* (pp. 127–149). Hershey, PA: Information Science Reference; doi:10.4018/978-1-61520-997-2.ch006

Virkar, S. (2013). Designing and implementing e-government projects: Actors, influences, and fields of play. In S. Saeed & C. Reddick (Eds.), *Human-centered system design for electronic governance* (pp. 88–110). Hershey, PA: Information Science Reference; doi:10.4018/978-1-4666-3640-8.ch007

Wallace, A. (2009). E-justice: An Australian perspective. In A. Martínez & P. Abat (Eds.), *E-justice: Using information communication technologies in the court system* (pp. 204–228). Hershey, PA: Information Science Reference; doi:10.4018/978-1-59904-998-4.ch014

Wang, G. (2012). E-democratic administration and bureaucratic responsiveness: A primary study of bureaucrats' perceptions of the civil service e-mail box in Taiwan. In K. Kloby & M. D'Agostino (Eds.), *Citizen 2.0: Public and governmental interaction through web 2.0 technologies* (pp. 146–173). Hershey, PA: Information Science Reference; doi:10.4018/978-1-4666-0318-9.ch009

Wangpipatwong, S., Chutimaskul, W., & Papasratorn, B. (2011). Quality enhancing the continued use of e-government web sites: Evidence from e-citizens of Thailand. In V. Weerakkody (Ed.), *Applied technology integration in governmental organizations: New e-government research* (pp. 20–36). Hershey, PA: Information Science Reference; doi:10.4018/978-1-60960-162-1.ch002

Wedemeijer, L. (2006). Long-term evolution of a conceptual schema at a life insurance company. In M. Khosrow-Pour (Ed.), *Cases on database technologies and applications* (pp. 202–226). Hershey, PA: Idea Group Publishing; doi:10.4018/978-1-59904-399-9.ch012

Whybrow, E. (2008). Digital access, ICT fluency, and the economically disadvantages: Approaches to minimize the digital divide. In F. Tan (Ed.), *Global information technologies: Concepts, methodologies, tools, and applications* (pp. 1409–1422). Hershey, PA: Information Science Reference; doi:10.4018/978-1-59904-939-7.ch102

Whybrow, E. (2008). Digital access, ICT fluency, and the economically disadvantages: Approaches to minimize the digital divide. In C. Van Slyke (Ed.), *Information communication technologies: Concepts, methodologies, tools, and applications* (pp. 764–777). Hershey, PA: Information Science Reference; doi:10.4018/978-1-59904-949-6.ch049

Wickramasinghe, N., & Geisler, E. (2010). Key considerations for the adoption and implementation of knowledge management in healthcare operations. In M. Saito, N. Wickramasinghe, M. Fuji, & E. Geisler (Eds.), *Redesigning innovative healthcare operation and the role of knowledge management* (pp. 125–142). Hershey, PA: Medical Information Science Reference; doi:10.4018/978-1-60566-284-8.ch009

Wickramasinghe, N., & Geisler, E. (2012). Key considerations for the adoption and implementation of knowledge management in healthcare operations. In *Organizational learning and knowledge: Concepts, methodologies, tools and applications* (pp. 1316–1328). Hershey, PA: Business Science Reference; doi:10.4018/978-1-60960-783-8.ch405

Wickramasinghe, N., & Goldberg, S. (2007). A framework for delivering m-health excellence. In L. Al-Hakim (Ed.), *Web mobile-based applications for healthcare management* (pp. 36–61). Hershey, PA: IRM Press; doi:10.4018/978-1-59140-658-7.ch002

Wickramasinghe, N., & Goldberg, S. (2008). Critical success factors for delivering m-health excellence. In N. Wickramasinghe & E. Geisler (Eds.), *Encyclopedia of healthcare information systems* (pp. 339–351). Hershey, PA: Medical Information Science Reference; doi:10.4018/978-1-59904-889-5.ch045

Wyld, D. (2009). Radio frequency identification (RFID) technology. In J. Symonds, J. Ayoade, & D. Parry (Eds.), *Auto-identification and ubiquitous computing applications* (pp. 279–293). Hershey, PA: Information Science Reference; doi:10.4018/978-1-60566-298-5.ch017

Yaghmaei, F. (2010). Understanding computerised information systems usage in community health. In J. Rodrigues (Ed.), *Health information systems: Concepts, methodologies, tools, and applications* (pp. 1388–1399). Hershey, PA: Medical Information Science Reference; doi:10.4018/978-1-60566-988-5.ch088

Yee, G., El-Khatib, K., Korba, L., Patrick, A. S., Song, R., & Xu, Y. (2005). Privacy and trust in e-government. In W. Huang, K. Siau, & K. Wei (Eds.), *Electronic government strategies and implementation* (pp. 145–190). Hershey, PA: Idea Group Publishing; doi:10.4018/978-1-59140-348-7.ch007

Yeh, S., & Chu, P. (2010). Evaluation of e-government services: A citizen-centric approach to citizen e-complaint services. In C. Reddick (Ed.), *Citizens and e-government: Evaluating policy and management* (pp. 400–417). Hershey, PA: Information Science Reference; doi:10.4018/978-1-61520-931-6.ch022

Young-Jin, S., & Seang-tae, K. (2008). E-government concepts, measures, and best practices. In A. Anttiroiko (Ed.), *Electronic government: Concepts, methodologies, tools, and applications* (pp. 32–57). Hershey, PA: Information Science Reference; doi:10.4018/978-1-59904-947-2.ch004

Yun, H. J., & Opheim, C. (2012). New technology communication in American state governments: The impact on citizen participation. In K. Bwalya & S. Zulu (Eds.), *Handbook of research on e-government in emerging economies: Adoption, e-participation, and legal frameworks* (pp. 573–590). Hershey, PA: Information Science Reference; doi:10.4018/978-1-4666-0324-0.ch029

Zhang, N., Guo, X., Chen, G., & Chau, P. Y. (2011). User evaluation of e-government systems: A Chinese cultural perspective. In F. Tan (Ed.), *International enterprises and global information technologies: Advancing management practices* (pp. 63–84). Hershey, PA: Information Science Reference; doi:10.4018/978-1-60960-605-3.ch004

Zuo, Y., & Hu, W. (2011). Trust-based information risk management in a supply chain network. In J. Wang (Ed.), *Supply chain optimization, management and integration: Emerging applications* (pp. 181–196). Hershey, PA: Business Science Reference; doi:10.4018/978-1-60960-135-5.ch013

Compilation of References

Abbate, J. (2001). Government, Business, and the Making of the Internet. *Business History Review*, *75*(01), 147–176. doi:10.2307/3116559

ABC News. (2012). *Occupy Philly shows national solidarity*. Retrieved September 21, 2012 from http://www.nbcphiladelphia.com/news/local/Occupy-Philly-Shows-National-Solidarity-138304044.html

Abel, F., Gao, Q., Houben, G. J., & Tao, K. (2011). Semantic enrichment of twitter posts for user profile construction on the social web. In C. Bussler, J. Davies, D. Fensel, & R. Studer (Eds.), *The Semantic Web: Research and Applications* (pp. 375–389). Springer. doi:10.1007/978-3-642-21064-8_26

About C. G. A. P. (2013). *CGAP's organization*. Retrieved March 28, 2013, from CGAP.org: http://www.cgap.org/about

Adams, N., Stubbs, V., & Woods, V. (2005). Psychological barriers to internet usage among older adults in the U.K. *Medical Informatics and the Internet in Medicine*, *30*(1), 3–17. doi:10.1080/14639230500066876 PMID:16036626

Agarwal, S. (2009). *Social networks as Internet barometers for optimizing content delivery networks*. Paper presented at the 3rd International Symposium on Advanced Networks and Telecommunication Systems (ANTS), New Dehli, India. doi:10.1109/ANTS.2009.5409895

Aghaei, S., Nematbakhsh, M. A., & Farsani, H. K. (2012). Evolution of the World Wide Web: From Web 1.0 to Web 4.0. *International Journal of Web & Semantic Technology*, *3*(1), 1–10. doi:10.5121/ijwest.2012.3101

Akella, D. (2003). *Unlearning the fifth discipline*. New Delhi, India: Response Books.

Akram, M., & Hussain, I. (2011). The role of microfinance in uplifting income level: A study of district Okara, Pakistan. *Interdisciplinary Journal of Contemporary Research in Business*, *3*, 83–90.

Alexander, B. (2008). Social Media and Higher Education – Scholarship, The Wave of the Future in the Digital Age. In R. Katz (Ed.), *From The Tower and the Cloud: Higher Education in the Age of Cloud Computing*. Educause 2008. Retrieved from http://www.educause.edu/thetowerandthecloud

Alexslemonadestand.org. (2014). Our mission & history. Retrieved from http://www.alexslemonade.org/about/our-mission-history

American Red Cross. (2012). *Superstorm Sandy: One month update*. Retrieved from http://www.redcross.org/news/press-release/Red-Cross-One-Month-Report-on-Superstorm-Sandy

Amin, G. (2001). *Whatever happened to the Egyptians? changes in Egyptian society from 1950 to the present*. Cairo: American University in Cairo Press.

Amitabh, A., & Sinha, S. (2012). The learning continuum: Formal and informal learning experiences-enabling learning and creation of new knowledge in an organization. *International Journal of advanced Corporate Learning, 5*(2), 10-14. doi:10.3991/ijac.v5i2.2111

Andis, K. (2012). *The Alphabet and the Origins of Writing. epublic GmbH.*

Andriole, S. J. (2012). Seven indisputable technology trends that will define 2015. *Communications of the Associations for Information Systems, 30*(4), 61-72. Retrieved from http://aisel.aisnet.org/cais/

Anklam, P. (2005). Social Network Analysis in the KM Toolkit. In M. Rao (Ed.), *Knowledge Management Tools and Techniques* (pp. 329–346). Burlington, MA: Elsevier. doi:10.1016/B978-0-7506-7818-6.50029-3

Anonymous, . (2013). Ahead of the Storm: Engineering for Disaster. *Institute of Transportation Engineers. ITE Journal, 83*(12), 22–24.

Arkin, D. (2013). Canada charges two in teen cyber-bullying suicide case. *NBC News*. Retrieved 20/8/2013, from http://www.bbc.co.uk/news/world-us-canada-23752923

Army, S. (2013, December 4). Day one down & four remain for world record! Retrieved from http://blog.salvationarmyusa.org/2013/12/04/day-one-down-four-remain-for-world-record/

Arquilla, J., & Ronfeldt, D. (2001). The Advent of Netwar (Revisited). In J. Arquilla & D. Ronfeldt (Eds.), Networks and Netwars: The Future of Terror, Crime, and Militancy (pp. 1-25). Santa Monica, CA: RAND Publications.

Ashta, A., & Assadi, D. (2008). *Do social cause and social technology meet?The impact of Web 2.0 technologies on peer-to-peer lending transactions.* Paper presented at the Asia Microfinance Forum, Hanoi.

Asur, S., & Huberman, B. A. (2010). *Predicting the Future with Social Media.* Paper presented at the International Conference on Web Intelligence and Intelligent Agent Technology (WI-IAT), Toronto, Canada. doi:10.1109/WI-IAT.2010.63

Auer, M. R. (2011). The Policy Sciences of Social Media. *Policy Studies Journal: the Journal of the Policy Studies Organization, 39*(4), 709–736. doi:10.1111/j.1541-0072.2011.00428.x

Aulakh, G. (2014). *Philips to re-enter Indian smartphone market.* Retrieved from http://timesofindia.indiatimes.com/tech/tech-news/Philips-to-re-enter-Indian-smartphone- market/articleshow/35614928.cms

Avdelidis, K., Dimoulas, C., Kalliris, G., Bliatsiou, C., Passias, T., Stoitsis, J., & Papanikolaou, G. (2002). *Multilingual automated digital talking character*. Paper presented at the International Broadcasting Convention, Amsterdam.

Barbieri, D., Braga, D., Ceri, S., Valle, E. D., Yi, H., Tresp, V., & Wermser, H. et al. (2010). Deductive and Inductive Stream Reasoning for Semantic Social Media Analytics. *IEEE Intelligent Systems, 25*(6), 32–41. doi:10.1109/MIS.2010.142

Barlett, C. P., Gentile, D. A., Anderson, C. A., Suzuki, K., Sakamoto, A., Yamaoka, A., & Katsura, R. (2014). Cross-Cultural Differences in Cyberbullying Behavior A Short-Term Longitudinal Study. *Journal of Cross-Cultural Psychology, 45*(2), 300–313. doi:10.1177/0022022113504622

Barlow, J. P. (1995). Thinking Locally, Acting Globally. *Time, 147*(3), 76.

Barnett, C. (2013, May 8). Top 10 crowdfunding sites for fundraising. *Forbes*. Retrieved from http://www.forbes.com/sites/chancebarnett/2013/05 /08/top-10-crowdfunding-sites-for-fundraising/

Barton, C. (2011, October 29). Occupy Auckland protest speaks with many voices. *New Zealand Herald.* Retrieved August 12, 2012 from http://www.nzherald.co.nz/nz/news/article.cfm?c_id=1&objectid=10762353

Batzios, A., Dimou, C., Symeonidis, A. L., & Mitkas, P. A. (2008). BioCrawler: An intelligent crawler for the semantic web. *Expert Systems with Applications*, *35*(1-2), 524–530. doi:10.1016/j.eswa.2007.07.054

BBC News Media Reports. (2001, April 29). China Internet Café Debate Hots Up. *BBC News Online*. Retrieved from: http://news.bbc.co.uk/1/hi/world/monitoring/media_reports/1302309.stm

BBC News. (2012b, October 3). *Teenager convicted of Facebook murder*. Retrieved fromhttp://www.bbc.co.uk/news/world-europe-19467308 retrieved 20/8/2013

BBC News. (2013, August 20). Teenager's death sparks cyber-blackmailing probe. *BBC News Scotland*. Retrieved 20/8/2013, from http://www.bbc.co.uk/news/uk-scotland-23723169

BBC.com. (2014). How do you solve a problem like Maria? Retrieved from http://www.bbc.co.uk/maria/about/index.shtml

BBC.com. (2014). I'd do anything archive. Retrieved from http://www.bbc.co.uk/oliver/archive.shtml

Behrend, F. D., & Erwee, R. (2009). Mapping knowledge flows in virtual teams with SNA. *Journal of Knowledge Management, 13*(4), 99-114. doi: 10.1108/13673270910971860

Belanger, F., & Carter, L. (2009). The impact of the digital divide on e-government use. *Communications of the ACM*, *54*(4), 132–135. doi:10.1145/1498765.1498801

Belardo, S., Karwan, K., & Wallace, W. (1984). An Investigation of System Design Considerations for Emergency Management Decision Support. *IEEE Transactions on Systems, Man, and Cybernetics*, *14*(6), 795–804. doi:10.1109/TSMC.1984.6313308

Bélisle, J. F., & Bodur, H. O. (2010). Avatars as information: Perception of consumers based on their avatars in virtual worlds. *Psychology and Marketing*, *27*(8), 741–765. doi:10.1002/mar.20354

Belsey, B. (2005). *Cyberbullying: An emerging threat to the 'always on' generation*. Retrieved March 2 2015 from http://www.inukshuk.com/pdf/Cyberbullying_Presentation_Description.pdf

Benedikt, M. (1991). Cyberspace. First Steps. Cambridge, MA: M.I.T. Press.

Benson, J., Greaves, W., O'Donnell, M., & Taglialatela, J. (2002). Evidence for Symbolic Language Processing in a Bonobo (Pan paniscus). *Journal of Consciousness Studies*, *9*(12), 33–56.

Beran, T., & Li, Q. (2005). Cyber-harassment: A study of a new method for an old behavior. *Journal of Educational Computing Research*, *32*(3), 265–277. doi:10.2190/8YQM-B04H-PG4D-BLLH

Beran, T., & Li, Q. (2008). The relationship between cyberbullying and school bullying. *The Journal of Student Wellbeing*, *1*(2), 16–33.

Bermingham, A., Conway, M., McInerney, L., O'Hare, N., & Smeaton, A. (2009). *Combining Social Network Analysis and Sentiment Analysis to Explore the Potential for Online Radicalisation*. Paper presented at the 2009 Advances in Social Network Analysis and Mining. doi:10.1109/ASONAM.2009.31

Bernal, P. A. (2010). Web 2.5: The symbiotic web. *International Review of Law Computers & Technology*, *24*(1), 25–37. doi:10.1080/13600860903570145

Bernton, H. (2011). Portland police arrest 50 protesters. *Seattle Times*. Retrieved September 1, 2012 from http://www.seattletimes.com/seattle-news/portland-police-arrest-50-protesters

Betts, S. C., & Stouder, M. D. (2004). The Network Perspective in Organization Studies: Network Organizations or Network Analysis. *Academy of Strategic Management Journal*, *3*, 1–20.

Bhanot, S. (2012). Use of Social Media by Companies to Reach their Customers. *SIES Journal of Management*, *8*(1), 47–55.

Bharosa, N., Lee, J., & Janssen, M. (2010). Challenges and Obstacles in Sharing and Coordinating Information during Multi-agency Disaster Response: Propositions from Field Exercises. *Information Systems Frontiers*, *12*(1), 49–65. doi:10.1007/s10796-009-9174-z

Billington, M. G., & Billington, P. J. (2012). Social media tools for leaders and manager. *Journal of Leadership, Accountability and Ethics*, *9*(6), 11-18. Retrieved from http://www.na-businesspress.com/jlaeopen.html

Billington, M. G., & Billington, P. J. (2012). Social Media Tools for Leaders and Managers. *Journal of Leadership. Accountability and Ethics*, *9*(6), 11–19.

Birnhack, M. D., & Elkin-Koren, N. (2003). The Invisible Handshake: The Re-emergence of the State in the Digital Environment. *The Virginia : The Journal of Law and Technology*, (Summer), 2–3.

Bishop, M., & Yocom, J. (2011). Education, Multimedia and Social Media: Are We Ready for Showtime? In *27th Annual Conference on Distance Teaching and Learning*. Retrieved from http:www/uwex.edu/disted/conference

Bitcurrent. (2011, February 4). *How to think like a cloud: Rules of the new on-demand architecture*. Retrieved from http://cloud-computing-library.cloudconnectevent.com/download-preview?asset=/1384/how-to-think-like-a-cloud

Blackwood, C. (2012). Social entrepreneurship in Egypt: Turning revolutionary ambition into tangible results. *SIT Graduate Institute, 26*. Retrieved from http://digitalcollections.sit.edu/cgi/viewcontent.cgi?article=3594&context=capstones

Blanchard, B. (2004). *Logistics and Engineering Management* (6th ed.). Upper Saddle River, NJ: Pearson Prentice Hall.

Bondebjoerg. (2009). Behind the headlines: Documentaries, the War on Terror and everyday life. *Studies in Documentary Film, 3*, 219.

Bornstein, D. (2011). Grameen Bank and the public good. *The New York Times*. Retrieved from http://opinionator.blogs.nytimes.com/2011/03/24/grameen-bank-and-the-public-good

Boronenko, V., Menshikov, V., & Marzano, G. (2013). Topicality of cyberbullying among teenagers in Russia and Latvia. *Social Sciences Bulletin*, *2*(17), 84–104.

Bothos, E., Apostolou, D., & Mentzas, G. (2010). Using Social Media to Predict Future Events with Agent-Based Markets. *IEEE Intelligent Systems*, *25*(6), 50–58. doi:10.1109/MIS.2010.152

Brentlinger, J. (2000). Revolutionizing spirituality. *Reflections on Marxism and Religion, 64*(2), 171-193.

Breslin, J., & Decker, S. (2007). The future of social networks on the internet: The need for semantics. *IEEE Internet Computing*, *11*(6), 86–90. doi:10.1109/MIC.2007.138

Briggs, A., & Burke, P. (2009). A social History of Media (3rd ed.). Academic Press.

Browning, M., & Lusardi, A. (1996). Household saving: Micro theories and micro facts. *Journal of Economic Literature*, *34*(4), 1797–1855.

Brown, L. (2004). *State of The World 2004. WorldWatch Institute Report on Progress Toward A Sustainable Society* (pp. 170–180). New York: W.W. Norton Company.

Buck, D. A., Trainor, J. E., & Aguirre, B. E. (2006). A Critical Evaluation of the Incident Command System and NIMS. *Journal of Homeland Security and Emergency Management*, *3*(3), 1252. doi:10.2202/1547-7355.1252

Bull, G., Thompson, A., Searson, M., Garofalo, J., Park, J., Young, C., & Lee, J. (2008). Connecting Informal and Formal Learning: Experiences in the Age of Participatory Media. *Contemporary Issues in Technology & Teacher Education*, *8*(2). Retrieved from http://www.citejournal.org

Bulut, S., & Gündüz, S. (2012). Exploring violence in the context of Turkish culture and schools. In S. R. Jimerson, A. B. Nickerson, M. J. Mayer, & M. J. Furlong (Eds.), *Handbook of school violence and school safety: International research and practice* (2nd ed.; pp. 165–174). New York, NY: Routledge.

Burton, L. (2014). Bring Reading Rainbow back for every child, everyone! Retrieved from https://www.kickstarter.com/projects/readingrainbow/bring-reading-rainbow-back-for-every-child-everywh

Businessweek. (2012). Occupy DC camps persist as rally planned next week. *Businessweek*. Retrieved September 5, 2012 from http://www.businessweek.com/occupydc

Caldicott, S. M. (2013). *Midnight lunch: The four phases of team collaboration success from Thomas Edison's lab*. Hoboken, NJ: John Wiley & Sons.

California Business and Professions Code. (2003). *California Online Privacy Protection Act. Sections 22575-22579*. Author.

Campbell, M. A. (2005). Cyber bullying: An old problem in a new guise? *Australian Journal of Guidance & Counselling*, *15*(1), 68–76. doi:10.1375/ajgc.15.1.68

Carlman, A. (2010). *Development 2.0? The case of Kiva.org and online social lending for development*. Stellenbosch: Stellenbosch University.

Carne Cruda 2.0. (2012). Retrieved 28/8/2013 http://youtu.be/kS709ZyZ_YU

Cassidy, W., Brown, K., & Jackson, M. (2011). Moving from cyber-bullying to cyber-kindness: What do students, educators and parents say? In E. Dunkels, G.-M. Franberg, & C. Hallgren (Eds.), *Youth culture and net culture: Online social practices* (pp. 256–277). Hershey, PA: Information Science Reference. doi:10.4018/978-1-60960-209-3.ch015

Cassidy, W., Faucher, C., & Jackson, M. (2013). Cyberbullying among youth: A comprehensive review of current international research and its implications and application to policy and practice. *School Psychology International*, *34*(6), 575–612. doi:10.1177/0143034313479697

Castells, M. (1997). The Power of Identity:*The Information Age: Economy, Society and Culture (vol. 2)*. Oxford, UK: Blackwell Press.

Centers for Disease Control and Prevention. (2013). Deaths associated with Hurricane Sandy - October-November 2012. (No. 20). Atlanta, GA: Author.

Chalmers, D. J. (1995). *Facing up to the problem of consciousness*. Retrieved 20/8/2013, from http://philpapers.org/rec/CHAFUT

Cha, M., Perez, J., & Haddadi, H. (2009). *Flash floods and ripples: the spread of media content through the blogospere*. Association for the Advancement of Artificial Intelligence.

Chang, A., Huang, F., & Wang, E. (2009). Strategies for efficiency expanding the operations of Kiva.org. *The American Economic Review*, *99*(1), 431–457. Retrieved from http://www.mcafee.cc/Classes/BEM106/Papers/2009/Kiva.pdf

Chapman, N., & Chapman, J. (2009). *Digital multimedia* (3rd ed.). Chichester, UK: John Wiley & Sons, Ltd.

Chen, R., Sharman, R., Rao, H. R., & Upadhyaya, S. (2007). Design Principles for Critical Incident Response Systems. *Information Systems and eBusiness Management, 5*(3), 201-227. doi:10.1007/s10257-007-0046-0

Chen, Y., & Li, S. (2008). Group identity and social preferences. Retrieved from http://www.jstor.org/discover/10.2307/29730190?uid=3739560&uid=2&uid=4&uid=3739256&sid=21103613955097

Chin, B., Jones, B., McNutt, M., & Pebler, L. (2014). Veronica Mars Kickstarter and crowd funding. In M. Stanfil & M. Condis (Eds.), *Fandom and/as Labor. Transformative Works and Cultures 15* [Special Issue]., doi:10.3983/twc.2014.0519

Choi, S. O., & Kim, B. (2007). Power and Cognitive Accuracy in Local Emergency Management Networks. *Public Administration Review*, *67*(S1), 198–209. doi:10.1111/j.1540-6210.2007.00828.x

Choucri, N. (2000). Introduction: CyberPolitics in International Relations. *International Political Science Review*, *21*(3), 243–263. doi:10.1177/0192512100213001

Chroust, P. (2000). Neo-Nazis and the Taliban On-Line: Anti-Modern Political Movements and the Modern Media. In P. Ferdinand (Ed.), *The Internet, Democracy and Democratisation* (pp. 102–117). London: Frank Cass. doi:10.1080/13510340008403647

Chun, S. A., & Warner, J. (2010). Finding information in an era of abundance: Towards a collaborative tagging environment in government. *Information Policy, 15*, 89-103. doi:10.3233/IP-2010-0201

Churchill, C. (2006). *Protecting the poor: A microinsurance compendium* (Vol. 1). Munich: Re Foundation.

Clements, M., de Vries, A. P., & Reinders, M. J. (2010). The influence of personalization on tag query length in social media search. *Information Processing & Management*, *46*(4), 403–412. doi:10.1016/j.ipm.2009.03.006

Cohn, J. F. (2007). Foundations of human computing: Facial expression and emotion. In J. G. Carbonell & J. Siekmann (Eds.), *Artificial Intelligence for Human Computing* (pp. 1–16). Springer. doi:10.1007/978-3-540-72348-6_1

Colombi, J. M., & Cobb, R. G. (2009). Application of Systems Engineering to Rapid Prototyping for Close Air Support. *Defense AR Journal*, *16*(3), 284–303.

Cooper, R., Dibbell, J., & Spaight, T. (2007). *Alter ego: avatars and their creators*. London: Chris Boot.

Corbalis, M. (2009). *The Evolution of Language*. The Year on Cognitive Neuroscience.

Correa, T., Hinsley, A. W., & De Zuniga, H. G. (2010). Who interacts on the Web?: The intersection of users' personality and social media use. *Computers in Human Behavior*, *26*(2), 247–253. doi:10.1016/j.chb.2009.09.003

Costa, B., & Boiney, J. (2012). *Social Radar*. McLean, VA: The MITRE Corporation. Retrieved December 5, 2012 from https://www.mitre.org/sites/default/files/pdf/12_0581.pdf

Country Profile. (2014). Un.org.eg. United Nations. Retrieved from http://www.un.org.eg/UNInner1.aspx?pageID=3

Couvillon, M. A., & Ilieva, V. (2011). Recommended practices: A review of schoolwide preventative programs and strategies on cyberbullying. *Preventing School Failure: Alternative Education for Children and Youth*, *55*(2), 96–101. doi:10.1080/1045988X.2011.539461

Cowie, H., Hutson, N., Jennifer, D., & Myers, C. A. (2008). Taking Stock of Violence in UK Schools Risk, Regulation, and Responsibility. *Education and Urban Society*, *40*(4), 494–505. doi:10.1177/0013124508316039

Cox, G. (2014, March 11). George Takei uses social media following to tout Broadway-bound tuner. *Variety*. Retrieved from http://variety.com/2014/legit/news/star-trek-veteran-george-takei-uses-his-social-media-following-to-tout-broadway-bound-tuner-1201129693/

Coyle, K. (1997). Coyle's information highway handbook: A practical file on the new information order. Chicago, IL: American Library Association.

Craig, W. M., Henderson, K., & Murphy, J. G. (2000). Prospective teachers' attitudes toward bullying and victimization. *School Psychology International*, *21*(1), 5–21. doi:10.1177/0143034300211001

Crowdfunding for Nonprofits. (2011). Greenmarketing.tv. Retrieved from http://www.greenmarketing.tv/nonprofit-crowdfunding/

Crowe, A. (2010). The Elephant in the JIC: The Fundamental Flaw of Emergency Public Information within the NIMS Framework. *Journal of Homeland Security and Emergency Management*, *7*(1), 1662. doi:10.2202/1547-7355.1662

Cucchiarelli, A., Velardi, P., D'Antonio, F., & Curzi, M. (2008). *Modeling collaborations content in Social Network Analysis*. Paper presented at the IEEE/WIC/ACM International Conference on Web Intelligence and Intelligent Agent Technology, Sydney, Australia. doi:10.1109/WIIAT.2008.119

Cuevas, A., & Kohle, F. (2010). A Case study in using youtube and factbook as social media Tools in enhancing student centred learning and engagement. In *ICERI2010 Proceedings* (pp. 3596-3601). ICERI.

Cvorovic, J. (2006). Gypsies Drown in Shallow Water: Oral Narratives among Macva Gypsies. *Journal of Folklore Research*, *43*(2), 129–148. doi:10.2979/JFR.2006.43.2.129

Cyram. (2012). *Cyram NetMiner*. Retrieved January 15, 2012 from www.netminer.com

Dabbagh, N., & Kitsantas, A. (2011). Personal Learning Environments, Social Media and Self Regulated Learning: A Natural Formula for Connecting Formal and Informal Learning. *The Internet and Higher Education*, *15*(1), 3–8. doi:10.1016/j.iheduc.2011.06.002

Danitz, T., & Strobel, W. P. (2001). Networking Dissent: Cyber Activists Use the Internet to Promote Democracy in Burma. In J. Arquilla & D. Ronfeldt (Eds.), Networks and Netwars: The Future of Crime, Terrorism and Militancy (pp. 129-169). Santa Monica, CA: RAND Publications.

David-Ferdon, C., & Hertz, M. F. (2007). Electronic media, violence, and adolescents: An emerging public health problem. *The Journal of Adolescent Health*, *41*(6), S1–S5. doi:10.1016/j.jadohealth.2007.08.020 PMID:18047940

Davis, A., & Zigurs, I. (2008). Teaching and Learning about Virtual Collaboration: What We Know and Need to Know. In *AMCIS 2008 Proceedings*. Paper 168. Retrieved from http://aisel.aisnet.org

Dawoud, K., Alhajj, R., & Rokne, J. (2010). *A Global measure for estimating the degree of organization of terrorist networks*. Paper presented at the 2010 International Conference on Advances in Social Networks Analysis and Mining, Odense, Denmark. doi:10.1109/ASONAM.2010.84

De Choudhury, M., Sundaram, H., John, A., & Seligmann, D. (2009). Social Synchrony: Predicting mimicry of user actions in online social media. In *Proceedings of the 2009 International Conference on Computational Science and Engineering*. Vancouver, Canada: IEEE. doi:10.1109/CSE.2009.439

De Choudhury, M., Sundaram, H., John, A., & Seligmann, D. (2010). Analyzing the dynamics of communication in online social networks. In Handbook of Social Network Technologies and Applications (pp. 59 – 94). Academic Press.

De Wever, S., Martens, R., & Vandenbempt, K. (2005). The Impact of Trust on Strategic Resource Acquisition through Interorganizational Networks: Towards a Conceptual Model. *Human Relations*, *58*(12), 1523–1543. doi:10.1177/0018726705061316

Deans, P. C. (2012). Integration of study abroad with social media technologies and decision-Making applications. *Decision Sciences Journal of Innovative Education*, *10*(3), 299–336. doi:10.1111/j.1540-4609.2012.00344.x

Del Rey, R., Elipe, P., & Ortega-Ruiz, R. (2012). Bullying and cyberbullying: Overlapping and predictive value of the co-occurrence. *Psicothema*, *24*(4), 608–613. PMID:23079359

Dellarocas, C. (2003). The Digitization of Word of Mouth: Promise and Challenges of Online Feedback Mechanisms. *Management Science*, *49*(10), 1407–1424. doi:10.1287/mnsc.49.10.1407.17308

Demaray, M. K., Malecki, C. K., Jenkins, L. D., & Westermann, L. D. (2012). Social support in the lives of students involved in aggressive and bullying behaviours. In Handbook of school violence and school safety: International research and practice (pp. 57-67). Academic Press.

Department of Homeland Security Lessons Learned Information System. (2011). *Emergency public information: The Harris County, Texas, Office of Homeland Security and Emergency Management's implementation of a social media protocol*. Retrieved from https://www.llis.dhs.gov/

Department of Homeland Security Lessons Learned Information System. (2013). *Social media in emergency management: The virtual operations support team concept*. Retrieved from https://www.llis.dhs.gov/

Department of Homeland Security. (2008). *National response framework*. Retrieved from http://www.fema.gov/pdf/emergency/nrf/nrf-core.pdf

Department of Homeland Security. (2010). *Quadrennial homeland security review report: A strategic framework for a secure homeland.* Retrieved from https://www.dhs.gov/xlibrary/assets/qhsr_report.pdf

Department of Homeland Security. (2013a). *DHS uses social media to enhance information sharing and mission operations, but additional oversight and guidance are needed* (No. OIG-13-115). Retrieved from http://www.oig.dhs.gov/assets/Mgmt/2013/OIG_13-115_Sep13.pdf

Department of Homeland Security. (2013b). *Disaster response and recovery*. Retrieved from https://www.dhs.gov/topic/disaster-response-and-recovery

Department of Homeland Security. (2013c). *Lessons learned: Social media and Hurricane Sandy.* Retrieved from http://www.oig.dhs.gov/assets/Mgmt/2013/OIG_13-115_Sep13.pdf

Department of Homeland Security. (2013d). *National response framework* (2nd ed.). Retrieved from http://www.fema.gov/media-library-data/20130726-1914-25045-1246/final_national_response_framework_20130501.pdf

Department of Homeland Security. (n.d.a). *If you see something, say something*. Retrieved from http://www.dhs.gov/if-you-see-something-say-something%E2%84%A2-campaign

Department of Homeland Security. (n.d.b). *Disasters overview*. Retrieved from https://www.dhs.gov/disasters-overview

Department of Homeland Security. (n.d.c). *Our mission*. Retrieved from https://www.dhs.gov/our-mission

Department of Homeland Security. (n.d.d). *Who joined DHS*. Retrieved from http://www.dhs.gov/who-joined-dhs

Derks, D., Bos, A. E., & Von Grumbkow, J. (2008). Emoticons in computer-mediated communication: Social motives and social context. *Cyberpsychology & Behavior*, *11*(1), 99–101. doi:10.1089/cpb.2007.9926 PMID:18275321

Dey, A. K., Salber, D., & Abowd, G. D. (2001). A conceptual framework and a toolkit for supporting the rapid prototyping of context-aware applications. *Human-Computer Interaction (HCI) Journal*, *16*, 97–166. doi:10.1207/S15327051HCI16234_02

Digital Dementia on the rise as young people increasingly rely on technology instead of their brain. (2013, August 6). *Mail Online*. Retrieved on 6/8/2013, from http://www.dailymail.co.uk/health/article-2347563/Digital-dementia-rise-young-people-increasingly-rely-technology-instead-brain.html

Dimou, C., Symeonidis, A. L., & Mitkas, P. A. (2008). Data Mining and Agent Technology: a Fruitful Symbiosis. In O. Maimon & L. Rokach (Eds.), *Soft Computing for Knowledge Discovery and Data Mining* (pp. 327–362). Springer. doi:10.1007/978-0-387-69935-6_14

Dimoulas, C., Avdelidis, A., Kalliris, G., & Papanikolaou, G. (2008). Joint Wavelet video denoising and motion activity detection in multi-modal human activity analysis: Application to video – Assisted bioacoustic/psycho-physiological monitoring. *EURASIP Journal on Advances in Signal Processing*, *2008*(1), 792028. doi:10.1155/2008/792028

Dimoulas, C., Kalliris, G., Chatzara, E., Tsipas, N., & Papanikolaou, G. (2014a). Audiovisual production, restoration-archiving and content management methods to preserve local tradition and folkloric heritage. *Journal of Cultural Heritage*, *15*(3), 234–241. doi:10.1016/j.culher.2013.05.003

Dimoulas, C., Papanikolaou, G., & Petridis, V. (2011). Pattern classificationand audiovisual content management techniques using hybrid expert systems: A video-assisted bioacoustics application in abdominal sounds pattern analysis. *Expert Systems with Applications*, *38*(10), 13082–13093. doi:10.1016/j.eswa.2011.04.115

Dimoulas, C., Veglis, A., & Kalliris, G. (2014b). Application of Mobile Cloud-Based Technologies in News Reporting: Current Trends and Future Perspectives. In J. Rodrigues, K. Lin, & J. Lloret (Eds.), *Mobile Networks and Cloud Computing Convergence for Progressive Services and Applications* (pp. 320–343). Hershey, PA: Information Science Reference; doi:10.4018/978-1-4666-4781-7.ch017

Dimoulas, C., Veglis, A., & Kalliris, G. (2015). Audiovisual hypermedia in the semantic Web. In M. Khosrow-Pour (Ed.), *Encyclopedia of Information Science and Technology* (3rd ed.; pp. 7594–7604). IGI global. doi:10.4018/978-1-4666-5888-2.ch748

Dinakar, K., Reichart, R., & Lieberman, H. (2011, July). Modeling the Detection of Textual Cyberbullying. In *The Social Mobile Web*. Retrieved from http://web.media.mit.edu/~kdinakar/3841-16937-1-PB.pdf

Diou, C., Stephanopoulos, G., Panagiotopoulos, P., Papachristou, C., Dimitriou, N., & Delopoulos, A. (2010). Large-scale concept detection in multimedia data using small training sets and cross-domain concept fusion. *IEEE Transactions on Circuits and Systems for Video Technology*, *20*(12), 1808–1821. doi:10.1109/TCSVT.2010.2087814

Dixon, B. E. (2010). Towards e-government 2.0: An assessment of where e-government 2.0 is and where it is headed. *Public Administration & Management, 15*(2), 418-454. Retrieved from http://www.spaef.com/pam.php

Doclab, I. D. F. A. (n.d.). Retrieved from http://www.doclab.org/

Doclab, T. L. (n.d.). Retrieved 20/8/2013, www.film-and-television.com

Dooley, J. J., Pyżalski, J., & Cross, D. (2009). Cyberbullying versus face-to-face bullying. *Zeitschrift für Psychologie. The Journal of Psychology*, *217*(4), 182–188.

Dormehl, L. (2014). *Everything you need to know about Facebook's new app accelerator.* Retrieved from http://www.fastcolabs.com/3030615/everything-you-need-to-know-about-facebooks-new-app-accelerator

Dowla, A. (2006). In credit we trust: Building social capital by Grameen Bank in Bangladesh. *Journal of Socio-Economics*, *35*(1), 102–122. doi:10.1016/j.socec.2005.12.006

Drezner, D. W. (2002). *Who Rules? The Regulation of Globalisation.* Paper presented at the 2002 Midwestern Political Science Association Annual Meeting. Retrieved from: http://apsaproceedings.cup.org/Site/abstracts/017/017012DreznerDan.htm

Drolet, J. (2005). Women and Microcredit: Towards an understanding of women's experiences in Cairo, Egypt. The School of Social Work, McGill University.

Drossos, K., Floros, A., & Kanellopoulos, N. G. (2012). Affective acoustic ecology: towards emotionally enhanced sound events. In *Proceedings of the 7th Audio Mostly Conference: A Conference on Interaction with Sound* (pp. 109-116). ACM. doi:10.1145/2371456.2371474

Drucker, P. F. (1997). The Global Economy and the Nation State. *Foreign Affairs*, *76*(5), 159–171. doi:10.2307/20048206

Dunbar. (2003). The social brain: Mind, language, and society in evolutionary perspective. *Annual Review of Anthropology, 32*, 163-181.

Dupree, S. (2012). Crowdfunding 101: Pros and Cons. Center of entrepreneurial studies, Stanford Business. Retrieved from, http://www.gsb.stanford.edu/ces/crowdfunding-101

El-Baz, S. (1997). *The impact of social and economic factors on women's group formation in Egypt*. Oxford, UK: Berg Press.

Ellison, N. B., Steinfield, C., & Lampe, C. (2007). The benefits of Facebook "friends": Social capital and college students' use of online social network sites. *Journal of Computer-Mediated Communication*, *12*(4), article 1. Retrieved from http://jcmc.indiana.edu

Emergency Management Institute. (n.d.). *IS-42: Social media in emergency management course*. Retrieved from http://training.fema.gov/is/courseoverview.aspx?code=is-42

Ensor, J. (2013). Family of Skype suicide teen calls on David Cameron to tackle cyber bullying. *The Telegraph*. Retrieved 20/8/2013, from http://www.telegraph.co.uk/technology/news/10248058/Family-of-skype-suicide-teen-calls-on-David-Cameron-to-tackle-cyber-bullying.html

Espuny, C., Gonzalez, J., Lleixa, M., & Gisbert, M. (2011). University Students' Attitudes Towards and Expectations of the Educational Use of Social Networks. *Revista de Universidad y Sociedad del Conocimiento, 8*(1), 186-199. Retrieved from http://rusc.uoc.edu/

Evans, M. M., & Wensley, A. K. P. (2009). Predicting the influence of network structure on trust in knowledge communities: Addressing the interconnectedness of four network principles and trust. *Electronic Journal of Knowledge Management*, *7*(1), 41–54. Retrieved from www.ejkm.com/issue/download.html?idArticle=158

Facebook. (n.d.). Retrieved from https://www.facebook.com/tinylittledoclab

Facebook.com. (2014). Wil Wheaton. Retrieved from https://www.facebook.com/itswilwheaton?fref=ts

Federal Emergency Management Agency. (2003). *Terrorism incident annex*. Retrieved from http://www.fas.org/irp/offdocs/pdd39_frp.htm

Federal Emergency Management Agency. (2007a). *ICS all-hazard core competencies*. Retrieved from http://www.fema.gov/media-library-data/20130726-1617-20490-3897/ics_core_competencies_sep07.pdf

Federal Emergency Management Agency. (2007b). *Target capabilities list: A companion to the national preparedness guidelines*. Retrieved from http://www.fema.gov/pdf/government/training/tcl.pdf

Federal Emergency Management Agency. (2008a). *Emergency support function annexes: Introduction*. Retrieved from http://www.fema.gov/media-library-data/20130726-1825-25045-0604/emergency_support_function_annexes_introduction_2008_.pdf

Federal Emergency Management Agency. (2008b). *National incident management system*. Retrieved from http://www.fema.gov/pdf/emergency/nims/NIMS_core.pdf

Federal Emergency Management Agency. (2010). *FEMA independent study course IS208: State disaster management.* Retrieved from http://training.fema.gov/is/courseoverview.aspx?code=IS-208.a

Federal Emergency Management Agency. (2011). *A whole community approach to emergency management: Principles, themes, and pathways for action.* Retrieved from http://www.fema.gov/media-library-data/20130726-1813-25045-3330/whole_community_dec2011__2_.pdf

Federal Emergency Management Agency. (2013). *Hurricane sandy FEMA after-action report.* Retrieved from http://www.fema.gov/media-library-data/20130726-1923-25045-7442/sandy_fema_aar.pdf

Federal Emergency Management Agency. (2014). *Hurricane Sandy: Timeline.* Retrieved from https://www.fema.gov/hurricane-sandy-timeline

Federal Emergency Management Agency. (n.d.). *ICS Organization.* Retrieved from https://training.fema.gov/emiweb/is/icsresource/assets/icsorganization.pdf

Federal Trade Commission. (1999). *New rule will protect privacy of children online. Effective April 2000 certain web sites must obtain parental consent before collecting personal information from children.* Retrieved from http://www.ftc.gov/news-events/press-releases/1999/10/new-rule-will-protect-privacy-children-online

Federal Trade Commission. (2007). *COPPA protects children but challenges lie ahead.* Retrieved from http://www.ftc.gov/news-events/press-releases/2007/02/coppa-protects-children-challenges-lie-ahead

Federal Trade Commission. (2010). *FTC testifies on protecting teen privacy.* Retrieved from http://www.ftc.gov/news-events/press-releases/2010/07/ftc-testifies-protecting-teen-privacy

Federal Trade Commission. (2013). *FTC grants approval for new COPPA verifiable parental consent method.* Retrieved from http://www.ftc.gov/news-events/press-releases/2013/12/ftc-grants-approval-new-coppa-verifiable-parental-consent-method

Federal Trade Commission. (2014). *FTC's 'net cetera' advises parents on how to talk to their kids about internet use newly released update includes advice on mobile apps, Wi-Fi, text message spam and more.* Retrieved from http://www.ftc.gov/news-events/press-releases/2014/01/ftcs-net-cetera-advises-parents-how-talk-their-kids-about-0

Feldman, A. (2013, November 2). Indiegogo vs. RocketHub: Which crowdfunding site should you choose? Retrieved from http://www.crowdfundinsider.com/2013/11/25715-indiegogo-vs-rockethub-crowdfunding-site-choose/

Feldon, D. F., & Kafai, Y. B. (2008). Mixed methods for mixed reality: Understanding users' avatar activities in virtual worlds. *Educational Technology Research and Development*, *56*(5-6), 575–593. doi:10.1007/s11423-007-9081-2

Finextra. (2011). *Social media spreads Bank Transfer Day message.* Retrieved August 15, 2012 from www.finextra.com/news/fullstory.aspx?newsitemid=23133

Finkelhor, D., Vanderminden, J., Turner, H., Shattuck, A., & Hamby, S. (2014). Youth exposure to violence prevention programs in a national sample. *Child Abuse & Neglect*, *38*(4), 677–686. doi:10.1016/j.chiabu.2014.01.010 PMID:24630440

FireDogLake.com. (2011). *List of active Occupy encampments.* Retrieved September 21, 2012 from http://fdlaction.firedoglake.com/2011/12/20/list-of-active-occupy-encampments-across-the-country-now-at-61

Flannery, M. (2007). Kiva and the birth of person-to-person microfinace. Retrieved from http://www.mitpressjournals.org/doi/pdf/10.1162/itgg.2007.2.1-2.31

Flannery, M., & Shah, P. (May 2009). *Kiva and Web 2.0.* Presentation given at Kiva Fellows Training.

Flashmob oficina paro. (n.d.). Retrieved from http://www.npr.org/blogs/thetwo-way/2013/01/14/169345343/flash-mob-brings-some-sunshine-into-spanish-unemployment-office

Foroughi, A. (2011). A Research Framework for Evaluating the Effectiveness of Implementations of Social Media in Higher Education. *Online Journal for Workforce Education and Development, 5*(1). Retrieved from http://opensiuc.lib.siu.edu

Fortune, M., Spielman, M., & Pangelinan, D. (2011). Student's Perception of Online and Face to Face Learning and Social Media in Hospitality, Recreation and Tourism. *Merlot Journal of Online Learning and Teaching, 7*(1). Retrieved from http://jolt.merlot.org

Franco, L. R., Bellerín, M., Borrego, J., Díaz, F., & Molleda, C. (2012). Tolerance Towards Dating Violence in Adolescents. *Psicothema, 24*(2), 236–242. PMID:22420351

Friedman, T. L. (2001). *The Lexus and the Olive Tree: Understanding Globalization*. New York: Anchor Books.

Fugate, C. (2011a). *Evolution of emergency management and communication*. Hearing before the Subcommittee on Homeland Security on the U.S. Committee on Appropriations. House of Representatives. Retrieved from https://www.fema.gov/pdf/about/programs/legislative/testimony/2011/6_8_2011_evolution_of_emergency_management_and_communication.pdf

Fugate, C. (2011b). *Improving the nation's response to catastrophic disasters: How to minimize costs and streamline our emergency management programs*. Hearing before the Subcommittee on Economic Development, Public Buildings and Emergency Management on the Transportation and Infrastructure Committee. Retrieved from https://www.fema.gov/pdf/about/programs/legislative/testimony/2011/3_30_2011_improving_the_nations_response_to_catastrophic_disasters.pdf

Fugate, C. (2011c). *Understanding the power of social media as a communication tool in the aftermath of disasters*. Hearing before the Subcommittee on Disaster Recovery and Intergovernmental Affairs, Senate Committee on Homeland Security and Governmental Affairs. Retrieved from https://www.fema.gov/pdf/about/programs/legislative/testimony/2011/5_5_2011_use_of_ social_media_in_disasters.pdf

Fundable.com. (2014). Crowdfunding statistics. Retrieved from http://www.fundable.com/crowdfunding101/crowdfunding-statistics

G8 Summit. (2004, June 10). *G8 action plan: Applying the power of entrepreneurship to the eradication of poverty*. Retrieved from http://www.canadainternational.gc.ca/g8/summit-sommet/2004/action-plan-action_090604.aspx

Gabel, D. (2007). Broadband and universal service. *Telecommunications Policy, 31*(6/7), 347–358.

Gadkari, P. (2013). Ask.fm unveils changes to safety policy. *BBC News*. Retrieved 20/8/2013, from http://www.bbc.co.uk/news/world-us-canada-23752923

Gamel, H. (August, 2013). What's going on in Egypt: A brief explanation. *Huffington Post*. Retrieved from http://www.huffingtonpost.com/2013/08/14/whats-going-on-in-egypt_n_3758890.html

Garaigordobil, M. (2011). Prevalencia y consecuencias del cyberbullying: Una revisión. *International Journal of Psychology & Psychological Therapy, 11*(2), 233–254.

Garaigordobil, M., & Berrueco, L. (2011). Effects of a play program on creative thinking of preschool children. *The Spanish Journal of Psychology, 14*(2), 608–618. doi:10.5209/rev_SJOP.2011.v14.n2.9 PMID:22059307

Gasser, U. & Ernst, S. (2006). *From Shakespeare to DJ Danger Mouse: A Quick Look at Copyright and User Creativity in the Digital Age*. The Berkman Centre for Internet and Society Research Publication No. 2006-2005.

Gawande, K., & Wheeler, T. (1999). Measures of Effectiveness for Governmental Organizations. *Management Science*, *45*(1), 42–58. doi:10.1287/mnsc.45.1.42

Giugale, M., & Mubarak, H. (1996). *Private sector development in Egypt.* Cairo: American University in Cairo Press.

Gladwell, M. (2000). *The tipping point, how little things can make a big difference.* Retrieved from http://ebookbrowsee.net/the-tipping-point-how-little-things-can-malcolm-gladwell-pdf-d493731844

Gloor, P., Krauss, J., Nann, S., Fischbach, K., & Schoder, D. (2009). *Web Science 2.0: Identifying trends through semantic Social Network Analysis.* Paper presented at the 2009 International Conference on Computational Science and Engineering, Vancouver, Canada. doi:10.1109/CSE.2009.186

Golder, S. A., & Huberman, B. A. (2006). Usage patterns of collaborative tagging systems. *Journal of Information Science*, *32*(2), 198–208. doi:10.1177/0165551506062337

Goldwasser, O. (2010). How the Alphabet was born from Hieroglyphs. *Biblical Archealogy Review, 36*(2).

Goodwin, B. (2013, September). Macmillan drives staff collaboration with cloud-based social media tools. *Computer Weekly.com,* 24-30. Retrieved from http://www.computerweekly.com/

Gorham-Oscilowski, U., & Jaeger, P. (2008). National Security Letters, the USA PATRIOT Act, and the Constitution: The tensions between national security and civil rights. *Government Information Quarterly*, *25*(4), 625–644. doi:10.1016/j.giq.2008.02.001

Grow, K. (2014, July 31). Rob Zombie talks insane clowns and crowdfunding his new film: 31. *Rolling Stone.* Retrieved from http://www.rollingstone.com/music/news/rob-zombie-talks-insane-clowns-and-crowdfunding-his-new-film-31-20140731

Grunewald, D., & Baron, P. (2011). A tale of two banks: Grameen Bank and SKS microfinance. *Journal of Leadership, Accountability & Ethics.* Retrieved from http://Web.a.ebscohost.com/ehost/pdfviewer/pdfviewer?sid=48b97718-0676-44a2-b169-9690e62a7ef7%40sessionmgr4003&vid=2&hid=4107

Gueyie, J., Manos, R., & Yaron, J. (2013). *Microfinance in developing countries: Issues, policies and performance evaluation.* Palgrave Macmillan. Retrieved fromhttp://www.palgraveconnect.com/pc/doifinder/10.1057/9781137301925.0014

Gunts, S. (2011). *Sustainability and profitability of microfinance institutions.* Center for applied international finance and development. Retrieved from, http://www.th-nuernberg.de/fileadmin/Fachbereiche/bw/studienschwerpunkte/international_business/Master/CAIFD/ResearchPapers/SustainabilityAndProfitabilityOfMicrofinanceInstitutions_Guntz.pdf

Haas, P. (1992). Introduction: Epistemic Communities and International Policy Coordination. *International Organization*, *46*(1), 1–35. doi:10.1017/S0020818300001442

Habermas, J., & Lennox, F. (1974). Article. *New German Critique*, (3), 49-55.

Hafheinz, A. (2007). Arab internet use. In Arab media and political renewal (pp. 56–79). Academic Press.

Hamblin, M. (2010, February 13). Text 'Haiti' earthquake relief effort raises $32M for Red Cross. *Computerworld.* Retrieved from http://www.computerworld.com/s/article/9156678/Text_Haiti_earthquake_relief_effort_raises_32M_for_Red_Cross

Handoussa, H. (2010). *Situation analysis: key development challenges facing Egypt.* United Nations. Retrieved from http://www.un.org.eg/docs/101100%20SA%20Report%20final%20pdf%20version.pdf

Hansen, D. L., Shneiderman, B., & Smith, M. A. (2011). *Analyzing social media networks with NodeXL: insights from a connected world.* Burlington, MA: Morgan Kaufmann.

Hartley, S. (March, 2010). *KIVA.org: Crowd-Sourced microfinance and cooperation in group lending*. Retrieved from http://dash.harvard.edu/bitstream/handle/1/3757699/Hartley_Kiva_DASH.pdf?sequence=2

Hatzchristou, C., Polychroni, F., Issari, P., & Yfanti, T. (2012). A synthetic approach for the study of aggression and violence in Greek schools. In S. R. Jimerson, A. B. Nickerson, M. J. Mayer, & M. J. Furlong (Eds.), *Handbook of school violence and school safety: International research and practice* (2nd ed., pp. 141–152). New York, NY: Routledge.

Heirman, W., & Walrave, M. (2008). Assessing Concerns and Issues about the Mediation of Technology in Cyberbullying. *Cyberpsychology: Journal of Psychosocial Research on Cyberspace, 2*(2), article 1. Retrieved February 12 2015 from http://cyberpsychology.eu/view.php?cisloclanku=2008111401&article=1

Heirman, W., & Walrave, M. (2012). Predicting adolescent perpetration in cyberbullying: An application of the theory of planned behavior. *Psicothema*, *24*(4), 614–620. PMID:23079360

Helms, B. (2006). *Access for all: building inclusive financial systems*. Consultative Group To Assist The Poor.

Hemmi, A., Bayne, S. & Land, R. (2009). The Appropriation and Repurposing of Social Technologies in Higher Education. *Journal of Computer Assisted Learning, 2*, 19–30. doi: 10.1111/j.1365-2729.2008.00306.x

Hémond, Y., & Benoît, R. (2012). Preparedness: The State of the Art and Future Prospects. *Disaster Prevention and Management*, *21*(4), 404–417. doi:10.1108/09653561211256125

Hernandez, B. (2013). What the w00t?: Drew Curtis/Wil Wheaton/Greg Koch Stone Farking Wheaton w00tstout. Retrieved from http://blog.stonebrew.com/index.php/w00tstout/

Herzog, W. (2012). *Cave of Forgotten Dreams*. Retrieved 20/8/2013, from http://www.imdb.com/title/tt1664894/

Hill, K. A., & Hughes, J. E. (1998). *Cyberpolitics: Citizen Activism in the Age of the Internet*. Oxford, UK: Rowman and Littlefield Publishers.

Hinduja, S., & Patchin, J. W. (2009). *Preventing cyberbullying: Top ten tips for educators*. Retrieved December 28. 2013 from http://www.nyasp.biz/conf_2013_files/Kelly_Top_Ten_Tips_Educators_Cyberbullying_Prevention.pdf

Hinduja, S., & Patchin, J. W. (2012). Cyberbullying: Neither an epidemic nor a rarity. *European Journal of Developmental Psychology*, *9*, 539–543, doi:.(2012).706448.10.1080/17405629

Hinduja, S., & Patchin, J. (2006). Bullies move beyond the schoolyard: A preliminary look at cyberbullying. *Youth Violence and Juvenile Justice*, *4*(2), 148–169. doi:10.1177/1541204006286288

Hinduja, S., & Patchin, J. (2007). Offline consequences of online victimization: School violence and delinquency. *Journal of School Violence*, *6*(3), 89–112. doi:10.1300/J202v06n03_06

Hinduja, S., & Patchin, J. (2008). Cyberbullying: An exploratory analysis of factors related to offending and victimization. *Deviant Behavior*, *29*(2), 129–156. doi:10.1080/01639620701457816

History of Crowdfunding. (2013). *Crowdfunding World's Summit*. Retrieved from http://www.crowdfundingworldsummit.com/crowdfunding/

History of Crowdfunding. (2014). Retrieved from http://www.fundable.com/crowdfunding101/history-of-crowdfunding

History, C., &Toys, T. (n.d.). Retrieved 20/7/2012, from http://www.conflicthistory.com/#/period/0700BC-0588BC

History. (2013). *KIVA.org*. Retrieved from http://www.kiva.org/about/historyhttp://www.kiva.org/about/history

Hofstede, G., & Hofstede, G. J. (2005). *Cultures and Organizations: Software of the Mind: Revised and expanded* (2nd ed.). New York: McGraw-Hill.

Holmes, D. (2013, October 14). Infographic: Kickstarter vs. Indiegogo. *Pando Daily*. Retrieved from http://pandodaily.com/2013/10/14/infographic-kickstarter-vs-indiegogo/

Horsley, J. S., & Barker, R. T. (2002). Toward a Synthesis Model for Crisis Communication in the Public Sector. *Journal of Business and Technical Communication*, *16*(4), 406–440. doi:10.1177/105065102236525

Hossain, D. (January, 2013). *Social capital and microfinance: The case of Grameen Bank*. University of Dhaka.

How Kiva Works. (2014). Retrieved from http://www.kiva.org/about/how

Howes, L. (November, 2013). *5 Misconceptions about crowdfunding*. Analytical Ones. Retrieved from http://www.analyticalones.com/5-misconceptions-about-crowdfunding/

Huang, A. H., Yen, D. C., & Zhang, X. (2008). Exploring the potential effects of emoticons. *Information & Management*, *45*(7), 466–473. doi:10.1016/j.im.2008.07.001

Huberman, B., Romero, D. M., & Wu, F. (2008). Social networks that matter: Twitter under the microscope. *First Monday*. Retrieved September 21, 2012 from http://firstmonday.org/htbin/cgiwrap/bin/ojs/index.php/fm/article/view/2317/2063

Hughes, A. L., & Palen, L. (2012). The Evolving Role of the Public Information Officer: An Examination of Social Media in Emergency Management. *Journal of Homeland Security and Emergency Management*, *9*(1). doi:10.1515/1547-7355.1976

Hutter, C., Hartmann, B., Bohm, K., Heistermann, T., Kohlmeyer, K., Reckling, R., . . . Parra, D. S. (2010). *SONAR: Towards User-Centric Social Network Analysis and Visualization*. Paper presented at the ACM International Conference on Web Intelligence and Intelligent Agent Technology, Los Alamitos, CA. doi:10.1109/WI-IAT.2010.67

IBDB.com. (2104). The Producers. Retrieved June 30, 2014: http://www.ibdb.com/show.php?id=10510

INCOSE. (2006). *Systems Engineering handbook, version 3* (C. Haskins, Ed.). Seattle, WA: INCOSE.

Indiegogo. (n.d.). Retrieved from http://www.indiegogo.com/

Indiegogo.com. (2014). "Join MILES AHEAD – A Don Cheadle film." Retrieved July 5, 2014: https://www.indiegogo.com/projects/join-miles-ahead-a-don-cheadle-film

International Council on Systems Engineering. (2004). *What is systems engineering?* Retrieved from http://www.incose.org/practice/whatissystemseng.aspx

Iskander, E. (2011). Connecting the national and the virtual: Can facebook activism remain relevant after Egypt's January 25 uprising? *International Journal of Communication*, 13–15.

Jacobsson, M., Rost, M., & Holmquist, L. E. (2006). When Media Gets Wise: collaborative filtering with mobile media agents. In *Proceedings of the 11th international conference on Intelligent user interfaces* (pp. 291-293). ACM. doi:10.1145/1111449.1111514

Jafari, M., Akhavan, P., Bourouni, A., & Amiri, R. H. (2009). A framework for the selection of knowledge mapping techniques. *Journal of Knowledge Management Practice*, *10*(1), 40–53.

Jefferson, T. L. (2006). Evaluating the Role of Information Technology in Crisis and Emergency Management. *Vine*, *36*(3), 261–264. doi:10.1108/03055720610703542

Jenkins, H. (2006). *Convergence culture: Where old and new media collide*. New York, NY: New York University.

Jensen, J. (1998). Interactivity – Tracking a new concept in media and communication studies. *Nordicom Review*, *19*(1), 185–204.

Johnson, T., & Kaye, B. (2004). Wag the blog: How reliance on traditional media and the internet influence credibility perceptions of weblogs among blog users. *Journalism & Mass Communication Quarterly*, *81*(3), 622–642. doi:10.1177/107769900408100310

Jones, C., Berry, L., & Stevens, C. (2007). Synthesized speech intelligibility and persuasion: Speech rate and non-native listeners. *Computer Speech & Language*, *21*(4), 641–651. doi:10.1016/j.csl.2007.03.001

Jones, S. (2003). *Encyclopedia of new media: an essential reference to communication and technology*. Thousand Oaks, CA: Sage Publications. doi:10.4135/9781412950657

Jordan, T. (1999). *Cyberpower: The Culture and Politics of Cyberspace and the Internet*. London: Routledge Press. doi:10.4324/9780203448632

Kalal, Z., Mikolajczyk, K., & Matas, J. (2012). Tracking-learning-detection. *Pattern Analysis and Machine Intelligence. IEEE Transactions on*, *34*(7), 1409–1422.

Kalliris, G. M., Dimoulas, C. A., Papanikolaou, G. V., Avdelidis, K., Passias, T. N., & Stoitsis, J. S. (2002). Phoneme recognition for 3d modeled digital character talking emulation. In Audio Engineering Society Convention 112 (paper no. 5626).

Kanazi, R. (2014, July 31). "Tweet and delete: On Gaza, celebrity courage – and cowardice – over social media." *Salon*. Retrieved July 31, 2014: http://www.salon.com/2014/07/31/tweet_and_delete_on_gaza_celebrity_courage_and_cowardice_over_social_media/

Kaplan, A., & Haenlein, M. (2010). Users of the World unite! The challenges and opportunities of social media. *Business Horizons*, *53*(1), 59–68. doi:10.1016/j.bushor.2009.09.003

Kapucu, N., Arslan, T., & Demiroz, F. (2010). Collaborative Emergency Management and National Emergency Management Network. *Disaster Prevention and Management*, *19*(4), 452–468. doi:10.1108/09653561011070376

Katims, L. (2013). *Virtual operations support teams harness social media: The teams of emergency managers and disaster volunteers lend virtual support to those on the site of a disaster*. Retrieved from http://www.emergencymgmt.com/disaster/Virtual-Operations-Support-Teams-Social-Media.html

KCLU.org. (2014, July 23). "Woman struck, killed by passenger train on South Coast." Retrieved July 23, 2014: http://www.kclu.org/2014/07/23/woman-struck-killed-by-passenger-train-on-south-coast/

Kennedy, M. (2014, February 24). Backers of 'Allegiance' embrace innovation. *Associated Press*. Retrieved from http://news.yahoo.com/backers-39-allegiance-39-embrace-innovation-121731702.html

Keohane, R. O., & Nye, J. S. Jr. (2002). Power and Interdependence in the Information Age. In E. C. Kamarck & J. S. Nye Jr., (Eds.), *Governance.com: Democracy in the Information Age* (pp. 161–177). Washington, DC: Brookings Institution Press.

Kerr, A., Whitson, J., Harvey, A., Shepherd, T., & O'Donnell, C. (2013). Strategies and tactics for promoting indie game design. *Selected Papers of Internet Research*, North America, 3. Retrieved from http://spir.aoir.org/index.php/spir/article/view/874

Kerr, O. S. (2009). Vagueness Challenges to the Computer Fraud and Abuse Act. *Minnesota Law Review*, *94*, 1561–1587.

Keveney, B. (2012). 5 reasons 'American Idol' may go on nearly forever – or not. Retrieved from http://usatoday30.usatoday.com/life/television/news/story/2012-01-14/American-Idol-endurance/52625742/1

Khan, M. (2008, December 4). Red Cross attracts $190k in pledges via Text 2HELP program. *Mobile Marketer*. Available at: http://www.mobilemarketer.com/ cms/news/messaging/2226.html

Khandker, S., Khalily, B., & Khan, Z. (1995). *Grameen Bank: performance and sustainability*. Retrieved from http://books.google.com/books?id=rNN3Dfxcc3oC&printsec=frontcover#v=onepage&q&f=false

Khandker, S. (1996). Grameen Bank: Impact, costs, and program sustainability. *Asian Development Review*, 65–85. Retrieved from http://www.microfinancegateway.org/gm/document-1.9.28425/1835.pdf

Khonsari, K. K., Nayeri, Z. A., Fathalian, A., & Fathalian, L. (2010, 9-11 Aug. 2010). *Social Network Analysis of Iran's Green Movement opposition groups using Twitter.* Paper presented at the 2010 International Conference on Advances in Social Networks Analysis and Mining (ASONAM), Odense, Denmark. doi:10.1109/ASONAM.2010.75

Kickstarter. (n.d.). Retrieved from http://www.kickstarter.com/

Kickstarter.com. (2014). Wojciechowski s. Retrieved from https://www.kickstarter.com/help/stats

King, I., Li, J., & Chan, K. T. (2009). *A brief survey of computational approaches in social computing*. Paper presented at the International Joint Conference on Neural Networks, Atlanta, GA. doi:10.1109/IJCNN.2009.5178967

Kirkpatrick, D. (2011). The Facebook effect. New York: Simon & Schuster.

Kirkpatrick, D., & Myers, S. (2011). Egypt raids offices of nonprofits, 3 backed By US. *The New York Times*. Retrieved from http://www.nytimes.com/2011/12/30/world/middleeast/egypts-forces-raid-offices-of-us-and-other-civil-groups.html?pagewanted=all&_r=0

Kitchens, R., & Torrence, P. D. (2012). The JOBS Act-crowdfunding and beyond. *Economic Development Journal*, *11*(4), 42–47. doi:10.5755/j01.em.18.1.3713

Klamma, R., Spaniol, M., Cao, Y., & Jarke, M. (2006). Pattern-Based cross media Social Network Analysis for technology enhanced learning in Europe. In W. Nejdl & K. Tochtermann (Eds.), *EC-TEL 2006* (pp. 242–256). Berlin, Germany: Springer-Verlag. doi:10.1007/11876663_20

Knowledge, C. (2012). *4th Annual Nonprofit Social Network Benchmark Report*. Retrieved from http://www.nten.org/sites/default/files/2012_nonprofit_social_networking_benchmark_report_final.pdf

Kochenderfer-Ladd, B., & Pelletier, M. E. (2008). Teachers' views and beliefs about bullying: Influences on classroom management strategies and students' coping with peer victimization. *Journal of School Psychology*, *46*(4), 431–453. doi:10.1016/j.jsp.2007.07.005 PMID:19083367

Kohle, F. (2012). *The Arab Spring and the Wall Street Movement: Challenges and Implications for Documentary Filmmakers and Social Media*. Retrieved from http://www.thelondonfilmandmediaconference.com/registered-speakers-2012-a-to-k/

Kohle, F., & Cuevas, A. (2010). A case study in using youtube and facebook as social media tools in enhancing student centered learning and engagement. In *Proceedings of 3rd International Conference of Education, Research and Innovation*. Madrid, Spain: ICERI2010.

Kohle, F., & Cuevas, A. (2012). Social media: Changing the way we teach and changing the way we learn. In *Proceedings of 6th International Technology, Education and Development Conference*. Valencia, Spain: INTED2012.

Kong, D., Gray, D., & Tao, H. (2006). *A viewpoint invariant approach for crowd counting*. Paper presented at the 18th International Conference on Pattern Recognition, Hong Kong, China. doi:10.1109/ICPR.2006.197

König, A., Gollwitzer, M., & Steffgen, G. (2010). Cyberbullying as an act of revenge? *Australian Journal of Guidance & Counselling*, *20*(02), 210–224. doi:10.1375/ajgc.20.2.210

Kontostathis, A., Reynolds, K., Garron, A., & Edwards, L. (2013, May). Detecting cyberbullying: query terms and techniques. In *Proceedings of the 5th Annual ACM Web Science Conference* (pp. 195-204). ACM. doi:10.1145/2464464.2464499

Kotsakis, R., Kalliris, G., & Dimoulas, C. (2012). Investigation of broadcast-audio semantic analysis scenarios employing radio-programme-adaptive pattern classification. *Speech Communication*, *54*(6), 743–762. doi:10.1016/j.specom.2012.01.004

Kowalski, R. M., Giumetti, G. W., Schroeder, A. N., & Lattanner, M. R. (2014, February10). Bullying in the Digital Age: A Critical Review and Meta-Analysis of Cyberbullying Research Among Youth. *Psychological Bulletin*, *140*(4), 1073–1137. doi:10.1037/a0035618 PMID:24512111

Kowalski, R. M., & Limber, S. P. (2007). Electronic bullying among middle school students. *The Journal of Adolescent Health*, *41*(6), S22–S30. doi:10.1016/j.jadohealth.2007.08.017 PMID:18047942

Kowalski, R. M., Limber, S. P., & Agatston, P. W. (2008). *Cyber bullying*. Malden, MA: Blackwell. doi:10.1002/9780470694176

Kowalski, R. M., Limber, S. P., & Agatston, P. W. (2012a). *Cyberbullying: bullying in the digital age* (2nd ed.). Malden, MA: Wiley-Blackwell.

Kowalski, R. M., Morgan, C. A., & Limber, S. P. (2012b). Traditional bullying as a potential warning sign of cyberbullying. *School Psychology International*, *33*(5), 505–519. doi:10.1177/0143034312445244

Krueger, G. P. (2013). Human Systems Integration (HSI): Psychological Influences in Design Produce Exceptional Operator Performance. *Journal of the Washington Academy of Sciences*, *99*(1), 1.

Kuppuswamy, V., & Bayus, B. (2014, January 29). Crowdfunding creative ideas: The dynamics of project backers in Kickstarter. UNC Kenan-Flagler Research Paper No. 2013-15. Retrieved from http://ssrn.com/abstract=2234765 or10.2139/ssrn.2234765

Lasica, J. (2012, February 15). The 7 elements of a strategic social media plan. Retrieved from http://www.socialbrite.org/2012/02/15/the-7-elements-of-a-social-media-strategic-plan/

Lauinger, J. (2011). An Excavated Dossier of Cuneiform Tablets from Level VII Alalah?American Schools of Oriental Research, 362, 21-64.

Law, D. M., Shapka, J. D., Hymel, S., Olson, B. F., & Waterhouse, T. (2012). The changing face of bullying: An empirical comparison between traditional and internet bullying and victimization. *Computers in Human Behavior*, *28*(1), 226–232. doi:10.1016/j.chb.2011.09.004

Lea, B. L., Clayton, M., Draude, B., & Barlow, S. (2001). The Impact of Technology on Teaching and Learning Faculty Attitudes. *EDUCAUSE Quarterly*, *2*(2), 69–71. Retrieved from http://net.educause.edu/

Lee, J. G., Moon, S., & Salamatian, K. (2010). *An approach to model and predict the popularity of online contents with explanatory factors*. Paper presented at the 2010 IEEE/WIC/ACM International Conference on Web Intelligence and Intelligent Agent Technology, Los Alamitos, CA. doi:10.1109/WI-IAT.2010.209

Lee, H. W., Choi, J.-S., Shin, Y.-C., Lee, J.-Y., Jung, H. Y., & Kwon, J. S. (2012, July). Impulsivity in Internet Addiction: A Comparison with Pathological Gambling. *Cyberpsychology, Behavior, and Social Networking*, *15*(7), 373–377. doi:10.1089/cyber.2012.0063

Legrain, P. (2002). *Open World: The Truth About Globalisation*. London: Abacus Press.

Lending Teams. (2014). Retrieved from http://www.kiva.org/teams

Lenhart, A. (2010). Cyberbullying 2010: what the research tells us. *Youth Online Safety Working Group*. Retrieved March 1 2015, from http://www.pewinternet.org/2010/05/06/cyberbullying-2010-what-the-research-tells-us/

Lenk, K. (1997). The Challenge of Cyberspatial Forms of Human Interaction to Territorial Governance and Policing. In B. D. Loader (Ed.), *The Governance of Cyberspace: Politics, Technology and Restructuring* (pp. 126–135). London: Routledge Press. doi:10.4324/9780203360408_chapter_8

Leskovec, J. (2011). Social media analytics: tracking, modeling and predicting the flow of information through networks. In *Proceedings of the 20th international conference companion on World wide web* (pp. 277-278). ACM. doi:10.1145/1963192.1963309

Liebowitz, J. (2005). Linking Social Network Analysis with the Analytic Hierarchy Process for Knowledge Mapping in Organizations. *Journal of Knowledge Management*, *9*(1), 76–86. doi:10.1108/13673270510582974

Limber, S. P. (2013). Best Practices in the Prevention of Bullying: An Example of the Olweus Bullying Prevention Program. In M. G. Masiello & D. Schroeder, (Eds.), A Public Health Approach to Bullying Prevention (pp. 127-148). Washington, DC: American Public Health Association.

Lin, Q. (2011). The Role of Web-Based Activities in Mediating Student Interaction and Engagement in Four Teacher Education Classes. *Merlot Journal of Online Teaching and Learning, 7*(1). Retrieved from http://jolt.merlot.org

Lin, S. F., Chen, J. Y., & Chao, H. X. (2001). Estimation of number of people in crowded scenes using perspective transformation. *IEEE Transactions on Systems, Man, and Cybernetics. Part A, Systems and Humans*, *31*(6), 645–654. doi:10.1109/3468.983420

Li, Q., Smith, P. K., & Cross, D. (2012). Research into cyberbullying: Context. In Q. Li, D. Cross, & P. K. Smith (Eds.), *Cyberbullying in the global playground: Research from international perspectives* (pp. 3–12). West Sussex, UK: Wiley-Blackwell. doi:10.1002/9781119954484.ch1

Livingstone, S., Haddon, L., Görzig, A., & Ólafsson, K. (2010). Risks and safety on the internet. In *The perspective of European children: Final findings from the EU Kids online survey of* (pp. 9-16). Academic Press.

Livingstone, S., & Brake, D. R. (2010). On the rapid rise of social networking sites: New findings and policy implications. *Children & Society*, *24*(1), 75–83. doi:10.1111/j.1099-0860.2009.00243.x

Loader, B. D. (1997). The Governance of Cyberspace: Politics, Technology and Restructuring. In B. D. Loader (Ed.), *The Governance of Cyberspace: Politics, Technology and Restructuring* (pp. 1–22). London: Routledge Press. doi:10.4324/9780203360408_chapter_1

Locke, J., & Bogin, B. (2006). Language and Life History: A new perspective on the development and evolution of human language. *Behavioral and Brain Sciences*, 267. PMID:17214017

Lockheed Martin. (2010). *TPS-79 Tactical Military Air Traffic Control Radar*. Lockheed Martin Corporation. Retrieved September 1, 2012 from http://www.radartutorial.eu/19.kartei/pubs/TPS-79_productcard.pdf

Los Angeles Times. (2012). Timeline Occupy Wall Street. *LA Times*. Retrieved January 2, 2012 from http://timelines.latimes.com/occupy-wall-street-movement

Lou, J., Lin, S., Chen, K., & Lei, C. (2010). *What can the temporal social behavior tell us? An estimation of vertex-betweenness using dynamic social information*. Paper presented at the 2010 International Conference on Advances in Social Networks Analysis and Mining, Odense, Denmark. doi:10.1109/ASONAM.2010.46

Lu, Y., Luo, X., Polgar, M., & Cao, Y. (2010). Social network analysis of a criminal hacker community. *The Journal of Computer Information Systems, 51*(2), 31-31-41.

Lybio.net. (2014). Reading Rainbow theme song lyrics. Retrieved from http://lybio.net/tag/reading-rainbow-theme-song-lyrics/

Ma, R., Li, L., Huang, W., & Tian, Q. (2004). *On pixel count based crowd density estimation for visual surveillance*. Paper presented at the 2004 IEEE Conference on Cybernetics and Intelligent Systems, Singapore.

Mack, E., & Grubesic, T. (2009). Forecasting broadband provision. *Information Economics and Policy*, *21*(1), 57–67.

Mainsah, E., Heuer, R., Karla, R., & Zhang, Q. (2004, Spring). Grameen Bank: taking capitalism to the Poor. *Chazen Web Journal of International Business*, 1–28.

Malaby, T. (2009). *Making virtual worlds: Linden Lab and Second Life*. Ithaca, NY: Cornell University Press.

Malhorta, D. (2010). The desire to win: The effects of competitive arousal on motivation and behavior. *Organizational Behavior and Human Decision Processes*, *111*(2), 139–146. doi:10.1016/j.obhdp.2009.11.005

Malhorta, D. (2010, April). (When) are religious people nicer? Religious salience and the "Sunday Effect" on pro-social behavior. *Judgment and Decision Making*, *5*(2), 138–143.

Mansell, R. (1996). Innovation in Telecommunication Regulation: Realising National Policy Goals in a Global Marketplace. In W. H. Dutton (Ed.), *Information and Communication Technologies: Visions and Reality* (pp. 371–386). Oxford, UK: Oxford University Press.

Marin, A., & Wellman, B. (2011). Social network analysis: An introduction. In J. Scott & P. Carrington (Eds.), *The Sage Handbook of Social Network Analysis* (pp. 11–25). Thousand Oaks, CA: SAGE Publications, Inc.

Marinescu, A. (2009). Microfinance-a way to confront poverty and stimulate entrepreneurship worldwide. *Young Economists Journal*, 65–71.

Marwick, A. E., & Miller, R. W. (2014). Online Harassment, Defamation, and Hateful Speech: A Primer of the Legal Landscape. *Fordham Center on Law and Information Policy Report*, (2). Retrieved February 2 2015, from http://ir.lawnet.fordham.edu/cgi/viewcontent.cgi?article=1002&context=clip

Marx, & Engels. (1969). The Manifesto of the Communist Party. In *Marx/Engels Selected Works* (vol. 1, pp. 98–137). Moscow: Progress Publishers.

Mateas, M., & Sengers, P. (1999). Narrative intelligence. In *Proceedings AAAI Fall Symposium on Narrative Intelligence* (pp. 1-10). AAAI.

Maticka-Tyndale, E., & Barnett, J. P. (2010). Peer-led interventions to reduce HIV risk of youth: A review. *Evaluation and Program Planning*, *33*(2), 98–112. doi:10.1016/j.evalprogplan.2009.07.001 PMID:19647874

Maybury, M. (2010). *Social Radar for smart power*. McLean, VA: The MITRE Corporation. Retrieved September 15, 2012 from http://www.mitre.org/publications/technical-papers/social-radar-for-smart-power

May, C. (2002). *The Information Society: A Sceptical View*. Cambridge, MA: Polity Press.

Mc Guckin, C., Perren, S., Corcoran, L., Cowie, H., Dehue, F., Ševčíková, A., . . . Völlink, T. (2014). Coping with cyberbullying: How can we prevent cyberbullying and how victims can cope with it. In P. K. Smith & G. Steffgen (Eds.), Cyberbullying through the new media: Findings from an international network. New York, NY: Psychology Press.

McCarthy, C. (2009). Nonprofits next to test Facebook payment platform. *CNet*. Retrieved from, http://news.cnet.com/8301-17939_109-10320315-2.html

McCoy, J., Treanor, M., Samuel, B., Tearse, B., Mateas, M., & Wardrip-Fruin, N. (2010). Authoring game-based interactive narrative using social games and comme il faut. In *Proceedings of the 4th International Conference & Festival of the Electronic Literature Organization: Archive & Innovate*. Academic Press.

McKenna, K. Y., & Green, A. S. (2002). Virtual group dynamics. *Group Dynamics*, *6*(1), 116–127. doi:10.1037/1089-2699.6.1.116

McLuhan, M. (1994). *Understanding media: The extensions of man*. MIT Press.

Meadows, D. (2003). Digital storytelling: Research-based practice in new media. *Visual Communication*, *2*(2), 189–193. doi:10.1177/1470357203002002004

Mechling, J. (2002). Information Age Governance: Just the Start of Something Big? In E. C. Kamarck & J. S. Nye Jr., (Eds.), *Governance.com: Democracy in the Information Age* (pp. 141–160). Washington, DC: Brookings Institution Press.

Meeker, M., & Wu, L. (2013, May 19). *Internet trends: D11 conference*. Retrieved from http://www.theguardian.com/technology/blog/2013/may/30/mary-meeker-2013-internet-trends

Meister, J. (2014, March 26). Social media: Moving from danger to brand building opportunity. *Forbes Magazine*. Retrieved from http://www.forbes.com/sites/jeannemeister/2014/03/26/social-media-moving-from-danger-to-brand-building-opportunity/

Mendonca, D. J., & Wallace, W. (2007). A Cognitive Model of Improvisation in Emergency Management. *IEEE Transactions on Systems, Man, and Cybernetics. Part A, Systems and Humans*, *37*(4), 547–561. doi:10.1109/TSMCA.2007.897581

Menesini, E., Nocentini, A., Palladino, B. E., Frisén, A., Berne, S., Ortega-Ruiz, R., & Smith, P. K. et al. (2012). Cyberbullying definition among adolescents: A comparison across six European countries. *Cyberpsychology, Behavior, and Social Networking*, *15*(9), 455–463. doi:10.1089/cyber.2012.0040 PMID:22817693

Meredith, J. P. (2010). Combating cyberbullying: Emphasizing education over criminalization. *Federal Communications Law Journal*, *63*(1), 311–340.

Merola, M. (2007). *Archeaology, Messages of the Dead*. Retrieved 29/8/2013, from www.archive.archaeology.org/0701/abstracts/qatna.html

Messinger, P. R., Ge, X., Stroulia, E., Lyons, K., Smirnov, K., & Bone, M. (2008). On the relationship between my avatar and myself. *Journal of Virtual Worlds Research*, *1*(2).

Mestyán, M., Yasseri, T., & Kertész, J. (2013, August21). Early prediction of movie box office success based on Wikipedia activity big data. *PLoS ONE*, *8*(8), e71226. doi:10.1371/journal.pone.0071226 PMID:23990938

Mickahail, B., & Andrews, K. (2015). Collaborative innovative leadership and emerging social media technologies. In C. Luntz (Ed.), The Refractive Thinker® – An Anthology of Higher Learning: Effective Business Practices in Leadership and Emerging Technologies (vol. 9). Las Vegas, NV: The Refractor's Press.

Microcredit Summit Campaign. (2005, January 24). *Global microcredit summit coming to Halifax, Canada*. Message posted to Microcredit Summit mailing list. Retrieved from http://www.microcreditsummit.org

Miles, J. (2000, January 29). Can Governments Control the Internet? *BBC News Asia Pacific*. Retrieved from: http://news.bbc.co.uk/1/hi/world/asia-pacific/623339.stm

Milicevic, A. K., Nanopoulos, A., & Ivanovic, M. (2010). Social tagging in recommender systems: A survey of the state-of-the-art and possible extensions. *Artificial Intelligence Review*, *33*(3), 187–209. doi:10.1007/s10462-009-9153-2

Mooney, C., & Lee, M. (1995). Legislating morality in the American states: The case of pre-Roe abortion regulation reform. *American Journal of Political Science*, *39*(3), 599–627. doi:10.2307/2111646

Moore, M. J., Nakano, T., Suda, T., & Enomoto, A. (2013). Social interaction and automated detection: Tools in Cyberbullying. In A. Merlo, M. Coccoli, & L. Caviglione (Eds.), Social Network Engineering for Secure Web Data and Services (pp. 67-87). IGI Global.

Mortensen, P. (2014). *The future of technology is contextual.* Retrieved from http://www.fastcodesign.com/1672531/the-future-of-technology-isnt-mobile-its-contextual

Mossberger, K., Tolbert, C., & McNeal, R. (2007). *Digital citizenship: the internet, society and participation.* Cambridge, MA: MIT Press.

Mossberger, K., Tolbert, C., & Stansbury, M. (2003). *Virtual inequality: beyond the digital divide.* Washington, DC: Georgetown Press.

Mowery, D. C. (2001). Is the Internet a US invention?—An economic and technological history of computer networking. Timothy Simcoe Haas School of Business, University of California at Berkeley.

Moynihan, D. P. (2008). Learning Under Uncertainty: Networks in Crisis Management. *Public Administration Review*, *68*(2), 350–365. doi:10.1111/j.1540-6210.2007.00867.x

Mura, G., Topcu, C., Erdur-Baker, O., & Diamantini, D. (2011). An international study of cyberbullying perception and diffusions among adolescents. *Social and Behavioral Sciences*, *15*(2), 3805–3809.

MyLifetime.com. (2014). Vote for your favorite designer! Retrieved from http://www.mylifetime.com/shows/project-runway/season-13/fan-favorite

Nadkarni, A., & Hofmann, S. (2012). Why do people use Facebook? *Personality and Individual Differences*, *52*(3), 243–249. doi:10.1016/j.paid.2011.11.007 PMID:22544987

Nadrljanski, M., & Batinica, V. (2010). Intelligent media agents. *International Journal of Intelligent Defence Support Systems*, *3*(1), 128–138. doi:10.1504/IJIDSS.2010.033682

National Conference of State Legislatures. (2011a). *Privacy laws of government websites. Notice of monitoring of employee e-mail communications and internet access.* Retrieved from http://www.ncsl.org/issues-research/telecommunications-information-technology/state-laws-related-to-internet-privacy.aspx#govpolicies

National Conference of State Legislatures. (2011b). *Privacy policies: Government websites.* Retrieved from http://www.delcode.delaware.gov/title29/c090c/sc02/index.shtml

National Conference of State Legislatures. (2013). *State spyware laws.* Retrieved from http://www.ncsl.org/research/telecommunications-and-information-technology/state-spyware-laws.aspx

National Conference of State Legislatures. (2014). *Employer access to social media usernames and passwords 2013.* Retrieved from http://www.ncsl.org/research/telecommunications-and-information-technology/employer-access-to-social-media-passwords-2013.aspx

National Conference of State Legislatures. (2015). *Data disposal laws.* Retrieved from http://www.ncsl.org/research/telecommunications-and-information-technology/data-disposal-laws.aspx

National Endowment for the Arts. (2014). National Endowment for the Arts Appropriations. Retrieved from http://arts.gov/open-government/national-endowment-arts-appropriations-history

National Telecommunications and Information Administration. (2000). *Falling through the net: Toward digital inclusion. A report on Americans access to technology tools*. Retrieved from http://www.ntia.doc.gov/files/ntia/publications/fttn00.pdf

National Telecommunications and Information Administration. (2014). *Exploring the digital nation: Embracing the mobile internet*. Retrieved from http://www.ntia.doc.gov/files/ntia/publications/exploring_the_digital_nation_embracing_the_mobile_internet_10162014.pdf

Naylor, P., & Cowie, H. (1999). The effectiveness of peer support systems in challenging school bullying: The perspectives and experiences of teachers and pupils. *Journal of Adolescence*, *22*(4), 467–479. doi:10.1006/jado.1999.0241 PMID:10469511

News, B. B. C. (2012c, October 22). *Facebook party invite sparks riot in Haren*. Retrieved from http://www.bbc.co.uk/news/world-europe-1968470 retrieved 20/8/2013

News, N. B. C. (2012). Occupy Oakland: 400 arrested near Oakland city hall. *NBC News*. Retrieved September 15, 2012 from http://usnews.nbcnews.com/_news/2012/01/30/10268080-occupy-oakland-400-arrested-after-violent-protest

News-record.com. (2011). Black Friday occupations. *News & Record*. Retrieved September 11, 2012 from www.news-record.com/2011/black-friday-occupations

NHTV. (n.d.). Retrieved 20/8/2013, from www.nhtv.nl

Nikolic, V., Savic, S., & Stankovic, M. (2007). Designing a Multimedia Platform for Emergency Management. *Management of Environmental Quality*, *18*(2), 198–210. doi:10.1108/14777830710725858

NPR. (2012). Occupy movement has captured the nation's attention, if not its cities. *National Public Radio*. Retrieved October 1, 2012 from http://www.npr.org/2012/02/01/146205425/occupying-the-nations-attention-if-not-its-cities

NUA Internet Surveys. (2013). *How Many Online?* Retrieved February 1, 2014, from: http://www.nua.ie/surveys/how_many_online/

O'Reilly, T. (2005). *Web 2.0: compact definition?* Retrieved from http://radar.oreilly.com/archives/2005/10/Web-20-compact-definition.html

Ochs, M., Pelachaud, C., & Sadek, D. (2008). An empathic virtual dialog agent to improve human-machine interaction. In *Proceedings of the 7th international joint conference on Autonomous agents and multiagent systems-Volume 1* (pp. 89-96). International Foundation for Autonomous Agents and Multiagent Systems.

Office of Personnel and Management. (2014). Combined Federal Campaign. Retrieved http://www.opm.gov/combined-federal-campaign/

Olweus, D. (1993). *Bullying at school: What we know and what we can do*. Malden, MA: Wiley-Blackwell.

Olweus, D. (1999). Norway. In P. K. Smith, Y. Morita, J. Junger-Tas, D. Olweus, R. Catalano, & P. Slee (Eds.), *The nature of school bullying: A cross-national perspective* (pp. 28–48). London: Routledge.

Olweus, D. (2012a). Cyberbullying: An overrated phenomenon? *European Journal of Developmental Psychology*, *9*(5), 520–538. doi:10.1080/17405629.2012.682358

Olweus, D. (2012b). Comments on cyberbullying article: A rejoinder. *European Journal of Developmental Psychology*, *9*(5), 559–568. doi:10.1080/17405629.2012.705086

Ortega, R., Mora-Merchán, J. A., & Jäger, T. (2007). *Acting against school bullying and violence. The role of media, local authorities and the internet*. Landau: Verlag Empirische Pädagogik. Retrieved January 40 2015, from http://iamnotscared.pixel-online.org/data/database/publications/618_Acting_against_school_bullying_and_violence.pdf

Ortega-Ruiz, R., & Nunez, J. C. (2012). Bullying and cyberbullying: Research and intervention at school and social contexts. *Psicothema*, *24*(4), 603–607. PMID:23079358

Osman, T. (2010). *Egypt on the brink: from the rise of Nasser to the fall of Mubarak*. Retrieved from http://books.google.com/books?hl=en&lr=&id=MWReGwoyjcsC&oi=fnd&pg=PP2&dq=political+scene+in+Egypt+after+2011&ots=lXb93gfxfc&sig=OUPXAJF8szmv6ZNzn-FQW5UMt4c#v=onepage&q=political%20scene%20in%20Egypt%20after%202011&f=false

Ownby, D. (2008). *Falun Gong and the Future of China*. New York: Oxford University Press. doi:10.1093/acprof:oso/9780195329056.001.0001

Ozkan, S., Cakir, M., & Bilgen, S. (2008). A Maturity Based Qualitative Information Systems Effectiveness Evaluation of a Public Organization in Turkey. *Journal of Cases on Information Technology*, *10*(3), 58–71. doi:10.4018/jcit.2008070106

Page, D. (2012). Using electronic portfolios. *Performance Improvement*, *51*(8), 34–36. doi:10.1002/pfi.21296

Pagel, M. (n.d.). *University of Reading*. Retrieved 20/8/2013, from http://www.ted.com/talks/mark_pagel_how_language_transformed_humanity.html

Palladino, B. E., Nocentini, A., & Menesini, E. (2012). Online and Offline peer led models against bullying and cyberbullying. *Psicothema*, *24*(4), 634–639. PMID:23079363

Pallis, G., Zeinalipour-Yazti, D., & Dikaiakos, M. (2011). Online social networks: Status and trends. In New Directions in Web Data Management (pp. 213 - 234). Berlin: Springer-Verlag.

Panitch, L. (1996). Rethinking the Role of the State. In J. Mittelman (Ed.), *Globalisation: Critical Reflections. In International Political Economy Yearbook* (Vol. 9, pp. 83–113). London: Lynne Rienner.

Papadopoulos, S., Kompatsiaris, Y., Vakali, A., & Spyridonos, P. (2012). Community detection in social media. *Data Mining and Knowledge Discovery*, *24*(3), 515–554. doi:10.1007/s10618-011-0224-z

Park, J., Shin, Y., Kim, K., & Chung, B.-S. (2010). Searching Social Media Streams on the Web. *IEEE Intelligent Systems*, *25*(6), 24–31. doi:10.1109/MIS.2010.150

Paul, S., Smith, P. K., & Blumberg, H. H. (2010). Addressing cyberbullying in school using the quality circle approach. *Australian Journal of Guidance & Counselling*, *20*(02), 157–168. doi:10.1375/ajgc.20.2.157

Paul, S., Smith, P. K., & Blumberg, H. H. (2012). Investigating legal aspects of cyberbullying. *Psicothema*, *24*(4), 640–645. PMID:23079364

Pellissier, R. (2011). The Implementation of Resilience Engineering to Enhance Organizational Innovation in a Complex Environment. *International Journal of Business and Management*, *6*(1), 145–164.

Perren, S., Corcoran, L., Cowie, H., Dehue, F., Garcia, D. J., Mc Guckin, C., & Völlink, T. et al. (2012). Tackling cyberbullying: Review of empirical evidence regarding successful responses by students, parents, and schools. *International Journal of Conflict and Violence*, *6*(2), 283–292.

Perren, S., Dooley, J., Shaw, T., & Cross, D. (2010). Bullying in school and cyberspace: Associations with depressive symptoms in Swiss and Australian adolescents. *Child and Adolescent Psychiatry and Mental Health*, *4*(28), 1–10. PMID:21092266

Perritt, H. H. (1998). The Internet as a Threat to Sovereignty? Thoughts on the Internet's Role in Strengthening National and Global Governance. *Indiana Journal of Global Legal Studies*, *5*(2), 431–344.

Perry, R. W. (2003). Incident Management Systems in Disaster Management. *Disaster Prevention and Management, 12*(5), 405–412. doi:10.1108/09653560310507226

Petit, M. (2012). Retrieved 20/8/2013, from http://www.imdb.com/title/tt1838723/

Petrakou, A. (2010). Interacting through avatars: Virtual worlds as a context for online education. *Computers & Education, 54*(4), 1020–1027. doi:10.1016/j.compedu.2009.10.007

Pew Internet & American Life Project. (2012a). *Parents, teens, and online privacy.* Retrieved from http://www.pewinternet.org/files/oldmedia/Files/Reports/2012/PIP_ParentsTeensAndPrivacy.pdf

Pew Internet & American Life Project. (2012b). *Mobile health 2012.* Retrieved from http://www.pewinternet.org/Reports/2012/Mobile-Health.aspx

Pew Internet & American Life Project. (2012c). *Twitter users 2012.* Retrieved from http://www.pewinternet.org/files/old-media//Files/Reports/2012/PIP_Twitter_Use_2012.pdf

Pew Internet & American Life Project. (2013a). *The demographics of social media users – 2012.* Retrieved from http://pewinternet.org/Reports/2013/Social-media-users.aspx

Pew Internet & American Life Project. (2013b). *Anonymity, privacy, and security online.* Retrieved from http://pewinternet.org/Reports/2013/Anonymity-online.aspx

Pew Internet & American Life Project. (2013c). *Teens and mobile apps privacy.* Retrieved from http://www.pewinternet.org/Reports/2013/Teens-and-Mobile-Apps-Privacy.aspx

Pew Internet & American Life Project. (2013d). *72% of online adults are social networking site users.* Retrieved from http://www.pewinternet.org/files/oldmedia/Files/Reports/2013/PIP_Social_networking_sites_update_PDF.pdf

Pew Internet & American Life Project. (2013e). *Teens, social media, and privacy.* Retrieved from http://www.pewinternet.org/2013/05/21/teens-social-media-and-privacy

Pew Internet & American Life Project. (2013f). *51% of U.S. adults bank online 32% of adults transact bank business on their mobile phones.* Retrieved from http://www.pewinternet.org/2013/08/07/51-of-u-s-adults-bank-online

Pike, H. (2012). U-Series Dating of Paleolithic Art in 11 Caves in Spain. *Science, 336*(6087), 1409-1413. DOI: 10.1126/science.1219957

PlanetFinance. (2008). *National impact survey of microfinance.* Retrieved from http://www.sfdegypt.org/c/document_library/get_file?uuid=d11593f0-46b6-49d3-b64d-2eab2c0e6f12&groupId=10136

Popovic´-C´ itic´, B., Djuric´, S., & Cvetkovic´, V. (2011). The prevalence of cyberbullying among adolescents: A case study of middle schools in Serbia. *School Psychology International, 32*, 412–424.

Population. (2013). *Central Agency for Public Mobilization and Statistics.* Retrieved from http://www.capmas.gov.eg/pdf/Electronic%20Static%20Book2013/english/population/untitled1/pop.aspx

Privacy Rights Clearing House. (2014b). *Fact sheet 7: Workplace privacy and employee monitoring.* Retrieved from https://www.privacyrights.org/workplace-privacy-and-employee-monitoring#socialmedia

Privacy Rights Clearinghouse. (2014a). *Fact sheet 35: Social networking privacy: How to be safe, secure and social.* Retrieved from https://www.privacyrights.org/social-networking-privacy-how-be-safe-secure-and-social#user-shares

Prive, T. (2012). What is crowdfunding and how does it benefit the economy. *Forbes.com.* Retrieved from http://www.forbes.com/sites/tanyaprive/2012/11/27/what-is-crowdfunding-and-how-does-it-benefit-the-economy/

Public Law 106-102. (1999). Gramm-Leach-Bliley Act. 106th Congress.

Public Law 107-56. (2001). Uniting and Strengthening America by Providing Appropriate Tools Required to Intercept and Obstruct Terrorism (USA PATRIOT ACT) of 2001. 107th Congress.

Public Law 99-508. (1986). The Electronic Communications Privacy Act of 1986. 99th Congress.

Puttnam, D. (2013). *In his guest speaker engagement at NHTV*. University of Applied Sciences, Netherlands. Retrieved 20/8/2013, http://youtu.be/HOL14ONdzIU

PWC. (2012). *Media and Entertainment Outlook 2012-2016*. Retrieved 20/2/2013, from http://www.pwc.nl/nl/entertainment-media-outlook/2012site/index.jhtml

Qiu, L., & Benbasat, I. (2005). An investigation into the effects of Text-To-Speech voice and 3D avatars on the perception of presence and flow of live help in electronic commerce. *ACM Transactions on Computer-Human Interaction*, *12*(4), 329–355. doi:10.1145/1121112.1121113

Ravilious, K. (2010). The writing on the cave wall. *New Scientist Magazine*, (2748).

Reed, D. (2001, February). The law of the pack. *Harvard Business Review*. Retrieved from http://hbr.org/2001/02/the-law-of-the-pack/ar/1

Renninger, B. J. (2013, September 19). Why Getting on Upworthy is the New Goal for Kickstarter Campaigns. Retrieved from http://www.indiewire.com/article/why-getting-on-upworthy-is-the-new-goal-for-all-kickstarter-campaigns

Reynolds, K., Kontostathis, A., & Edwards, L. (2011, December). Using machine learning to detect cyberbullying. In *Machine Learning and Applications and Workshops (ICMLA), 2011 10th International Conference on* (Vol. 2, pp. 241-244). IEEE. doi:10.1109/ICMLA.2011.152

Ricciardelli, L. (2009). Documentary Filmmaking in the Postmodern Age: Errol Morris & The Fog of Truth. *Studies in Documentary Film*, *3*, 219.

Rideout, J., Foehr, U., & Roberts, F. (2010). Generation M: Media in the lives of 8-18 year olds. *Kaiser Family Foundation*. Retrieved 25/8/2013, from http://kaiserfamilyfoundation.files.wordpress.com/2013/01/8010.pdf

Riebel, J., Jaeger, R. S., & Fischer, U. C. (2009). Cyberbullying in Germany–an exploration of prevalence, overlapping with real life bullying and coping strategies. *Psychology Science Quarterly*, *51*(3), 298–314.

Riedl, M. O., & Stern, A. (2006). Believable agents and intelligent story adaptation for interactive storytelling. In *Technologies for Interactive Digital Storytelling and Entertainment* (pp. 1–12). Springer. doi:10.1007/11944577_1

Riedl, M. O., & Young, R. M. (2010). Narrative planning: Balancing plot and character. *Journal of Artificial Intelligence Research*, *39*(1), 217–268.

Rivers, I., & Noret, N. (2010). 'I h8 u': Findings from a five-year study of text and email bullying. *British Educational Research Journal*, *36*(4), 643–671. doi:10.1080/01411920903071918

RMHC.org. (2014). Tax returns & audited financials. Retrieved from http://www.rmhc.org/tax-returns-and-audited-financials

Robbins, A. (2001). The loss of personal privacy and the consequences for social research. *Journal of Government Information*, *28*(4), 493–527. doi:10.1016/S1352-0237(02)00343-X

Robin, B. R. (2008). Digital storytelling: A powerful technology tool for the 21st century classroom. *Theory into Practice*, *47*(3), 220–228. doi:10.1080/00405840802153916

Rogers, K. (2012). Small businesses weigh in: should crowdfunding be legal? *Fox Business Journal.* Retrieved from http://smallbusiness.foxbusiness.com/technology-web/2012/02/17/small-businesses-weigh-in-should-crowdfunding-be-legal/

Rose, M. (1988). The Author as Proprietor: Donaldson v. Becket and the Genealogy of Modern Authorship. *Representations (Berkeley, Calif.),*(23), 51–85. doi:10.2307/2928566

Ryall, J. (2012). Surge in digital dementia. *The Telegraph.* Retrieved on 6/8/2012, from http://www.telegraph.co.uk/news/worldnews/asia/southkorea/10138403/Surge-in-digital-dementia.html

Ryan, T., Kariuki, M., & Yilmaz, H. (2011). A Comparative Analysis of Cyberbullying Perceptions of Preservice Educators: Canada and Turkey. *Turkish Online Journal of Educational Technology-TOJET, 10*(3), 1–12.

Sage, A., & Rouse, W. (2009). *Handbook of Systems Engineering and Management* (2nd ed.). Hoboken, NJ: Wiley.

Sahin, M. (2010). Teachers' Perceptions of Bullying in High Schools: A Turkish Study. *Social Behavior and Personality: an International Journal, 38*(1), 127–142. doi:10.2224/sbp.2010.38.1.127

Santos, E., Pan, L., Arendt, D., & Pittkin, M. (2006). *An effective anytime anywhere parallel approach for centrality measurements in Social Network Analysis.* Paper presented at the 2006 IEEE International Conference on Systems, Man, and Cybernetics, Taipei, Taiwan. doi:10.1109/ICSMC.2006.385045

Sassen, S. (1997). *Losing Control: Sovereignty in an Age of Globalisation.* New York: Columbia University Press.

Satoh, I. (2010). *Context-aware Media Agent for Public Spaces.* ECAI.

Saulais, P., & Ermine, J. (2012). Creativity and Knowledge Management. *Vine, 42*(3/4), 416–438. doi:10.1108/03055721211267521

Savage, K. (2014, March 2). Fans pledge their allegiance to Allegiance (a new musical). Retrieved from http://blog-ken-savage.com/facebook/fans-pledge-their-allegiance-to-allegiance-a-new-musical/

Scholte, J. A. (1997). Global Capitalism and the State. *International Affairs, 73*(3), 427–452. doi:10.2307/2624266

Schurgin O'Keefe, G., & Clarke-Pearson, K. (2011, April1). American Academy of Pediatrics. *Pediatrics, 127*(4), 800–804. PMID:21444588

Secretary of Defense. (2011). *Sociocultural behavior research and engineering in the Department of Defense context.* Retrieved August 15, 2013 from http://www.dtic.mil/dtic/tr/fulltext/u2/a549230.pdf

Selke, K. (2007). *Understanding social networks - A key requirement for systems engineers.* Retrieved from http://www.dtic.mil/ndia/2007systems/Thursday/AM/Track8/UnderstandingSocialNetworks_ BriefingFINAL.pdf

Selwyn, N. (2010). *The Educational Significance of Social Media – A Critical Perspective.* Keynote debate at Ed-Media Conference 2010, Toronto, 28th June to 2nd July. Toronto: Selwyn. Retrieved from http://www.scribd.com

Selwyn, N. (2004). Reconsidering political and popular understandings of the digital divide. *New Media & Society, 6*(3), 341–362. doi:10.1177/1461444804042519

Senge, P. (2013, June 14). *Real collaboration takes more than meetings and powerpoints.* Retrieved from http://nbs.net/real-collaboration-takes-more-than-meetings-and-powerpoints/

Senge, P. (1990). *The fifth discipline: The art and practice of the learning organization.* New York, NY: Doubleday.

Serino, R. (2011). *Accountability at FEMA: Is Quality Job #1?* Hearing before the Ad Hoc Subcommittee on Disaster Recovery and Intergovernmental Affairs. U. S. Senate Committee on Homeland Security and Government Affairs. Retrieved from https://www.fema.gov/pdf/about/programs/legislative/testimony/2011/10_20_2011_accountability_at_fema.pdf

Shaluf, I. M., Ahmadun, F., & Said, A. M. (2003). A Review of Disaster and Crisis. *Disaster Prevention and Management, 12*(1), 24–32. doi:10.1108/09653560310463829

Sharma, R. (2014, March 13). How nonprofits use social media to engage with their communities. *Nonprofit Quarterly*. Retrieved from https://nonprofitquarterly.org/management/23837-how-nonprofits-use-social-media-to-engage-with-their-communities.html

Shaughnessy, H. (2012, April 10). *Forbes Magazine*. Retrieved 20/8/2013, from http://www.forbes.com/sites/haydn-shaughnessy/2012/10/04/facebooks-1-billion-users-why-the-sky-is-still-the-limit/

Siemens, G., & Weller, M. (Eds.). (2011). The Impact of Social Networks on Teaching and Learning. *Revisita de Universidad y Sociedad del Conocimiento, 8*(1), 164-170. Retrieved from http://oro.open.ac.uk

Silius, K., Miilumäki, T., Huhtamäki, J., Tebest, T., Meriläinen, J., & Pohjolainen, S. (2010). Social Media Enhanced Learning in Higher Education. *IEEE Educon Education Engineering 2010 – The Future of Global Learning Engineering Education*. Retrieved from http://www.doaj.org

Sklair, L. (2002). *Globalization: Capitalism and Its Alternatives*. Oxford, UK: Oxford University Press.

Slate.com. (2011). *Portland police arrest 50*. Retrieved September 10, 2012 from www.slate.com/blogs/2011/portland_police_arrest_50.html

Slaughter, C. (2014). May the force be with you: Converging innovations will empower entrepreneurs in the developing world. *The Guardian*. Retrieved from http://www.theguardian.com/global-development-professionals-network/partner-zone-path/entrepreneurs-innovation-developing-world

Smith, P. K., Mahdavi, J., Carvalho, M., & Tippett, N. (2006). *An investigation into cyberbullying, its forms, awareness and impact, and the relationship between age and gender in cyberbullying*. Research Brief No. RBX03-06. London: DfES.

Smith, P. K., Cowie, H., Olafsson, R., & Liefooghe, A. (2002). Definitions of bullying: A comparison of terms used, and age and gender differences, in a fourteen-country international comparison. *Child Development, 73*(4), 1119–1133. doi:10.1111/1467-8624.00461 PMID:12146737

Smith, P. K., Mahdavi, J., Carvalho, M., Fisher, S., Russell, S., & Tippett, N. (2008). Cyberbullying: Its nature and impact in secondary school pupils. *Journal of Child Psychology and Psychiatry, and Allied Disciplines, 49*(4), 376–385. doi:10.1111/j.1469-7610.2007.01846.x PMID:18363945

Smith, P. K., & Slonje, R. (2010). Cyberbullying: The nature and extent of a new kind of bullying, in and out of school. In S. R. Jimerson, S. M. Swearer, & D. L. Espelage (Eds.), *Handbook of bullying in schools: An international perspective* (pp. 249–262). New York, NY: Routledge.

Snakenborg, J. A., Van Acker, R., & Gable, R. A. (2011). Cyberbullying: Prevention and intervention to protect our children and youth. *Preventing School Failure, 55*(2), 88–95. doi:10.1080/1045988X.2011.539454

Sørensen, I. (2012, September). Crowdsourcing and outsourcing: The impact of online funding and distribution on the documentary film industry. *Media Culture & Society, 34*(6), 726–743. doi:10.1177/0163443712449499

Soto, C. M., Campos, J. C., & Morales, L. B. (2012). Bullying in Peru: A code of silence. In S. Jimerson, A. Nickerson, M. J. Mayer, & M. J. Furlong (Eds.), Handbook of school violence and school safety: International research and practice. New York: Routledge.

Sourander, A., Brunstein Klomek, A., Ikonen, M., Lindroos, J., Luntamo, T., Koskelainen, M., & Helenius, H. et al. (2010). Psychosocial risk factors associated with cyberbullying among adolescents: A population-based study. *Archives of General Psychiatry, 67*(7), 720. doi:10.1001/archgenpsychiatry.2010.79 PMID:20603453

Spar, D. L. (1999). Lost in (Cyber) Space: The Private Rules of Online Commerce. In C. A. Cutler, V. Haufler, & T. Porter (Eds.), *Private Authority and International Affairs* (pp. 31–49). Albany, NY: SUNY Press.

Spencer, J. W. (2003). Global gatekeeping, representation, and network structure: A longitudinal analysis of regional and global knowledge-diffusion networks. *Journal of International Business Studies*, *34*(5), 428–442. doi:10.1057/palgrave.jibs.8400039

Spitzer, M. (2012). Digitale Demenz. Droemer Verlag.

Sproles, N. (2000). Coming to Grips with Measures of Effectiveness. *Systems Engineering*, *3*(1), 50–58. doi:10.1002/(SICI)1520-6858(2000)3:1<50::AID-SYS4>3.0.CO;2-U

Spyridou, P.-L., & Veglis, A. (2008). Exploring Structural Interactivity in Online Newspapers: A Look at the Greek Web Landscape. *First Monday*, *13*(5). doi:10.5210/fm.v13i5.2164

St. Denis, L. A., Palen, L., & Hughes, A. (2012). Trial by fire: The Deployment of Trusted Digital Volunteer in the 2011 Shadow Lake Fire. In *Proceedings of the 9th International ISCRAM Conference*. Vancouver, Canada: ISCRAM.

Standish, P. (2008). Preface. *Journal of Philosophy of Education*, *42*(3-4), 3–4, 349–353. doi:10.1111/j.1467-9752.2008.00659.x

Stanley, L. D. (2003). Beyond access: Psychosocial barrier to computer literacy. *The Information Society*, *19*(5), 407–416. doi:10.1080/715720560

Statista.com. (2014). Most active crowdfunding categories of 2012. Retrieved from http://www.statista.com/statistics/269975/most-active-crowdfunding-categories-of-2012/

Statistics. (2014). Retrieved from http://www.kiva.org/about/stats

Strasburger, Jordan, & Donnerstein. (2010). Article. *Pediatrics*, *125*(756). DOI: 10.1542/peds.2009-2563

Sun, Z., Yang, Y., Zhou, Y., & Cruickshank, H. (2014). Agent-Based Resource Management for Mobile Cloud. In J. Rodrigues, K. Lin, & J. Lloret (Eds.), *Mobile Networks and Cloud Computing Convergence for Progressive Services and Applications* (pp. 118–134). Hershey, PA: Information Science Reference; doi:10.4018/978-1-4666-4781-7.ch007

Sutton, R. I. (2007). *The no asshole rule: Building a civilized workplace and surviving one that isn't*. New York: Business Plus.

Su, Y. S., Wardell, C., & Thorkildsen, Z. (2012). *Social media in the emergency management field (No. IPP-2013-U-004984)*. National Emergency Management Association.

SyFy.com. (2104). The Wil Wheaton Project. Retrieved from http://www.syfy.com/wilwheaton/about

Symeonidis, A. L., & Mitkas, P. A. (2005). Agent intelligence through data mining. Springer.

Takei, G. (2013, March 28). *My Broadway debut at age 75*. TEDxBroadway Presentation delivered to TEDxBroadway. Retrieved from http://blog.ted.com/2013/03/28/george-takei-on-star-trek-musicals-and-japanese-american-internment/

Tang, L., & Huan, L. (2010). Understanding group structures and properties in social media. In P. S. Yu (Ed.), Link Mining: Models, Algortithms, and Applications (pp. 163 - 185). Berlin, Germany: Springer Science+Business Media, LLC.

Taylor, J. (1999). The Information Polity. In W. H. Dutton (Ed.), *Society on the Line: Information Politics in the Digital Age* (pp. 127–128). Oxford, UK: Oxford University Press.

Telegraph. (2012). Police Clash with Occupy Wall Street Protesters. *Telegraph*. Retrieved September 12, 2012 from http://www.telegraph.co.uk/finance/8811627/Police-clash-with-Occupy-Wall-Street-protesters.html

Thales Raytheon Systems. (2011). *Thales awarded tactical control radar contract.* Retrieved September 1, 2012 from http://www.thalesgroup.com/Press_Releases/Countries/Canada/2011/Tactical_Control_Radar_Contract

The Economist. (2001). Special: The Internet and the Law: Stop Signs on the Web. *The Economist, 358*(8204), 21-23.

Thomas, R. (2103). The Veronica Mars movie project. Retrieved from https://www.kickstarter.com/projects/559914737/the-veronica-mars-movie-project

Thomas, H. J., Connor, J. P., & Scott, J. G. (2014). Integrating Traditional Bullying and Cyberbullying: Challenges of Definition and Measurement in Adolescents–a Review. *Educational Psychology Review, 27*(1), 135–152. doi:10.1007/s10648-014-9261-7

Thussu, D. K. (2000). *International Communication: Continuity and Change.* London: Arnold Publishers.

Tsai, J., & Chi, C. S. F. (2012). Cultural Influence on the Implementation of Incident Command System for Emergency Management of Natural Disasters. *Journal of Homeland Security and Emergency Management, 9*(1). doi:10.1515/1547-7355.1970

Tsipas, N., Dimoulas, C., Kalliris, G., & Papanikolaou, G. (2013). Collaborative annotation platform for audio semantics. In *134th Audio Engineering Society Convention 2013*, (pp. 218-222).

Tsirelis, T., & Delopoulos, A. (2011). Automatic ground-truth image generation from user tags. In *12th International Workshop on Image Analysis for Multimedia Interactive Services (WIAMIS 2011).* Academic Press.

Tulving, E. (2002). Episodic Memory: From Mind to Brain. *Annual Review of Psychology, 53*(1), 1–25. doi:10.1146/annurev.psych.53.100901.135114 PMID:11752477

Turner, G., & Shepherd, J. (1999). A method in search of a theory: Peer education and health promotion. *Health Education Research, 14*(2), 235–247. doi:10.1093/her/14.2.235 PMID:10387503

Turoff, M., Chumer, M., Hiltz, S. R., & Klashner, R. (2004b). Assuring Homeland Security: Continuous Monitoring, Control & Assurance of Emergency Preparedness. *Journal of Information Technology Theory and Application, 6*(3), 1–24.

Turoff, M., Chumer, M., Van de Walle, B., & Yao, X. (2004a). The Design of a Dynamic Emergency Response Management Information System (dermis). *Journal of Information Technology Theory and Application, 5*(4), 1–35.

Twitter.com. (2014). @midnight. Retrieved from https://twitter.com/midnight

Twitter.com. (2014). Anne Wheaton. Retrieved from https://twitter.com/AnneWheaton/status/492020385487917056

Twitter.com. (2014). The Wil Wheaton Project. Retrieved from https://twitter.com/wilwproject

Twitter.com. (2014). Wil Wheaton. Retrieved from https://twitter.com/wilw

U.S Department of Homeland Security. (2013). *Fact sheet: Executive order on cybersecurity / presidential policy directive on critical infrastructure security and resilience.* Retrieved from http://www.dhs.gov/news/2013/02/13/fact-sheet-executive-order-cybersecurity-presidential-policy-directive-critical

Uganda: A Little Goes A Long Way. (2009). *Pbs.org, Front Line World.* Retrieved from http://www.pbs.org/frontlineworld/stories/uganda601/video_index.html

UNICEF.org. (2012, September 12). Milestones: Year by year through UNICEF's history. Retrieved from http://www.unicef.org/about/history/index_milestones.html

Valkenburg, P. M., & Peter, J. (2011). Online communication among adolescents: An integrated model of its attraction, opportunities, and risks. *The Journal of Adolescent Health, 48*(2), 121–127. doi:10.1016/j.jadohealth.2010.08.020 PMID:21257109

Van de Walle, B., & Turoff, M. (2008). Decision Support for Emergency Situations. *Information Systems and eBusiness Management, 6*(3), 295-316. doi:10.1007/s10257-008-0087-z

Van Dijk, J. (1999). *The network society, social aspects of new media*. London: Sage.

Van Royen, K., Poels, K., Daelemans, W., & Vandebosch, H. (2015). Automatic monitoring of cyberbullying on social networking sites: From technological feasibility to desirability. *Telematics and Informatics, 32*(1), 89–97. doi:10.1016/j.tele.2014.04.002

Vandebosch, H., Beirens, L., D'Haese, W., Wegge, D., & Pabian, S. (2012). Police actions with regard to cyberbullying: The belgian case. *Psicothema, 24*(4), 646–652. PMID:23079365

Vandebosch, H., & Van Cleemput, K. (2008). Defining cyberbullying: A qualitative research into the perceptions of youngsters. *Cyberpsychology & Behavior, 11*(4), 499–503. doi:10.1089/cpb.2007.0042 PMID:18721100

Varlamis, I., Eirinaki, M., & Louta, M. (2010). *A study on social network metrics and their application in trust networks*. Paper presented at the 2010 International Conference on Advances in Social Networks Analysis and Mining (ASONAM), Odense, Denmark. doi:10.1109/ASONAM.2010.40

Vasalou, A., & Joinson, A. N. (2009). Me, myself and I: The role of interactional context on self-presentation through avatars. *Computers in Human Behavior, 25*(2), 510–520. doi:10.1016/j.chb.2008.11.007

Vashist, A., Mau, S., Poylisher, A., Chadha, R., & Gosh, A. (2011). *Leveraging social network for predicting demand and estimating available resources for communication network management*. Paper presented at the IEEE IM 2011: Mini Conference, Dublin, Ireland. doi:10.1109/INM.2011.5990558

Veglis, A., Dimoulas, C., & Kalliris, G. (2016). Towards intelligent cross-media publishing: media practices and technology convergence perspectives. In A. Lugmayr & C. Dal Zotto (Eds.), Media convergence Handbook (Vol. 2). Germany: Springer-Verlag..

Virkar, S. (2013). Designing and Implementing e-Government Projects: Actors, Influences, and Fields of Play. In S. Saeed & C. G. Reddick (Eds.), Human-Centered Design for Electronic Government (pp. 88-110). Hershey, PA: IGI Global Inc.

Virkar, S. (2014). Consulting the British Public in the Digital Age: Emerging Synergies and Tensions in the Government 2.0 Landscape. In L. G. Anthopoulos & C. G. Reddick (Eds.), *Government e-Strategic Planning and Management: Practices, Patterns, and Roadmaps* (pp. 185–203). New York: Springer Link. doi:10.1007/978-1-4614-8462-2_10

Viskontas, I., & Mooney, C. (2014, January 3). Mark Ruffalo wants you to imagine a 100 percent clean energy future. *Mother Jones*. Retrieved from http://www.motherjones.com/environment/2014/01/inquiring-minds-mark-ruffalo-fracking-solar

Von der Puetten, A. M., Krämer, N. C., Gratch, J., & Kang, S. H. (2010). "It doesn't matter what you are!" explaining social effects of agents and avatars. *Computers in Human Behavior, 26*(6), 1641–1650. doi:10.1016/j.chb.2010.06.012

von Mare´ es, N., & Petermann, F. (2012). Cyberbullying: An increasing challenge for schools. *School Psychology International, 33*(5), 467–476.

w00stock.net. (2014). w00tstock: A night of geeks + music. Retrieved from http://w00tstock.net/

Wachs, S., Wolf, K. D., & Pan, Ch.-Ch. (2012). Cybergrooming: Risk factors, coping strategies and associations with cyberbullying. *Psicothema, 24*(4), 628–633. PMID:23079362

Wade, R. (1996). Globalisation and its Limits: Reports of the Death of the National Economy are Greatly Exaggerated. In S. Berger & R. Dore (Eds.), *National Diversity and Global Capitalism* (pp. 60–88). Ithaca, NY: Cornell University Press.

Walker, M. (2014). *The History of Social Networking - WebMaster View*. Available at: http://www.webmasterview.com/2011/08/social-networking-history/

Wall Street Journal. (2011). *Thousands occupy mall*. Retrieved August 5, 2012 from http://blogs.wsj.com/washwire/2011/11/15/occupy-dc-allowed-to-keep-on-occupying-d-c-park

Waly, G. (n.d.). Microfinance in Egypt: An Overview. *The Egyptian Financial Supervisory Authority*. Retrieved from https://www.efsa.gov.eg/content/efsa2_ar/pool_extra_efsa/UG18915UG18916.pdf

Wang, F. Y., Carley, K. M., Zeng, D., & Mao, W. (2007). Social computing: From social informatics to social intelligence. *IEEE Intelligent Systems*, *22*(2), 79–83. doi:10.1109/MIS.2007.41

Warburton, S. (2009). Second Life in higher education: Assessing the potential for and the barriers to deploying virtual worlds in learning and teaching. *British Journal of Educational Technology*, *40*(3), 414–426. doi:10.1111/j.1467-8535.2009.00952.x

Ward, R., Wamsley, G., Schroeder, A., & Robins, D. B. (2000). Network Organizational Development in the Public Sector: A Case Study of the Federal Emergency Management Administration (FEMA). *Journal of the American Society for Information Science*, *51*(11), 1018–1032. doi:10.1002/1097-4571(2000)9999:9999<::AID-ASI1004>3.0.CO;2-5

Warren, L., & Quigley, R. (2013). Police confirm 12 year old girls suicide note said she was being cyber-bullied as her sister reveals she knew about abuse but was sworn to secrecy. *Daily Mail*. Retrieved 20/8/2013, from http://www.dailymail.co.uk/news/article-2331670/Gabrielle-Molina-Police-confirm-12-year-old-girls-suicide-note-said-cyber-bullied-sister-reveals-knew-abuse-sworn-secrecy.html

Wartenkin, C. (2001). *Reshaping World Politics: NGOs, the Internet and Global Civil Society*. Oxford, UK: Rowman and Littlefield Publishers.

Watkins, T. (2010). New light on Neolithic revolution in south-west Asia. *Antiquity*, *84*(325), 621–634. doi:10.1017/S0003598X00100122

Webb, C. (2008). Measuring social capital and knowledge networks. *Journal of Knowledge Management, 12*(5), 65-78. doi: 10.1108/13673270810902948

Webster, F. (2014). *Theories of the Information Society* (4th ed.). London: Routledge Press.

Weiss, S. (2009). Privacy threat model for data portability in social network applications. *International Journal of Information Management*, *29*(4), 249–254. doi:10.1016/j.ijinfomgt.2009.03.007

Weller, M. J., & Dalziel, J. (2007). On-line Teaching: Suggestions for Instructors. In L. Cameron & J. Dalziel (Eds.), *Proceedings of the 2nd International LAMS Conference 2007: Practical Benefits of Learning Design* (pp. 76-82). Sydney: LAMS Foundation. Retrieved from http://lamsfoundation.org

West, D., & Miller, E. (2006). The digital divide in public e-health: Barriers to accessibility and privacy in state health department websites. *Journal of Health Care for the Poor and Underserved*, *17*(3), 652–666. doi:10.1353/hpu.2006.0115 PMID:16960328

White House Office of Management and Budget. (2009). *Open government directive*. Retrieved from https://www.whitehouse.gov/omb/inforeg_infocoll

White House Office of Management and Budget. (2010a). *Guidance for agency use of third-party websites and applications*. Retrieved from https://www.whitehouse.gov/omb/inforeg_infocoll

White House Office of Management and Budget. (2010b). *Guidance for online use of web measurement and customization technologies*. Retrieved from https://www.whitehouse.gov/omb/inforeg_infocoll

White House Office of Management and Budget. (2010c). *Social media, web-based interactive technologies, and the paperwork reduction act*. Retrieved from https://www.whitehouse.gov/omb/inforeg_infocoll

White House. (2003). *Homeland Security Presidential Directive-5: Management of domestic incidents*. Retrieved from http://www.dhs.gov/sites/default/files/publications/Homeland%20Security%20Presidential%20Directive%205.pdf

White House. (2003). *Homeland Security Presidential Directive-8: National preparedness*. Retrieved from http://www.fas.org/irp/offdocs/nspd/hspd-8.html

Willard, N. E. (2007). *Cyberbullying and cyberthreats: Responding to the challenge of online social aggression, threats, and distress*. Champaign, IL: Research Press.

Williams, M. D., & Smart, A. (2010). Patient Safety: A Casualty of Target Success? *International Journal of Public Sector Management*, *23*(5), 416–430. doi:10.1108/09513551011058457

Wills, D., & Reeves, S. (2009). Facebook as a political weapon: Information in social networks. *British Politics*, *4*(2), 265-281. doi: 10.1057/bp.2009.3

Wojciechowski, A. (2009). Models of charity donations and project funding in social networks. In R. Meersman, P. Herrero, & T. Dillon (Eds.), *On the Move to Meaningful Internet Systems:Proceedings of OTM 2009 Workshops* (pp. 454 – 463). doi:10.1007/978-3-642-05290-3_58

Wolff, C. (2009). *Requirements for Radar Receivers*. Retrieved September 1, 2012 from http://www.radartutorial.eu/09.receivers/rx04.en.html

Wolf, M. (2004). *The Role of the State in an Era of Globalisation. The Sir Ronald Trotter Lecture, 2004*. Wellington, New Zealand: New Zealand Business Roundtable.

Woolcock, M. (1998). *Social theory, Development Policy, and Poverty Alleviation: A Comparative Historical Analysis of Group Based Banking in Developing Economies*. (Unpublished PhD Dissertation). Brown University.

Yang, L., Prasanna, R., & King, M. (2009). On-site Information Systems Design for Emergency First Responders. *Journal of Information Technology Theory and Application*, *10*(1), 5–27.

Yang, X., & Zhang, S. (2010). Use, Challenges and Potential of Social Media in Higher Education: Evidence from a Case Study. In C. Wankel (Ed.), *Cutting-edge Social Media Approaches to Business Education: Teaching with LinkedIn, Facebook, Twitter, Second Life, Blogs, Information Age* (pp. 217–240). Charlotte, NC: Publishing.

Ybarra, M. L., & Mitchell, K. J. (2004). Online aggressor/targets, aggressors, and targets: A comparison of associated youth characteristics. *Journal of Child Psychology and Psychiatry, and Allied Disciplines*, *45*(7), 1308–1316. doi:10.1111/j.1469-7610.2004.00328.x PMID:15335350

Ybarra, M., & Mitchell, K. (2007). Prevalence and frequency of Internet harassment instigation: Implications for adolescent health. *Adolesc Health*, *41*(2), 89–95. doi:10.1016/j.jadohealth.2007.03.005 PMID:17659224

Ybarra, M., Mitchell, K., Wolak, J., & Finkelhor, D. (2006). Examining characteristics and associated distress related to Internet harassment: Findings from the Second Youth Internet Safety Survey. *Pediatrics*, *118*(4), 69–77. doi:10.1542/peds.2006-0815 PMID:17015505

Youtube. (n.d.). Retrieved 20/8/2013, from https://www.youtube.com/reproductivebill

Yunus, M. (1999). The Grameen Bank: a small experiment begun in Bangladesh has turned into a major new concept in eradicating poverty. *Scientific American*. Retrieved from http://www.anderson.ucla.edu/faculty/bhagwan.chowdhry/Yunus1999%20-%20Scientific%20American.pdf

Zeng, D., Hsinchun, C., Lusch, R., & Shu-Hsing, L. (2010). Social media analytics and intelligence. *IEEE Intelligent Systems*, *25*(6), 13–16. doi:10.1109/MIS.2010.151

Zhang, Y., Yu, X., Dang, Y., & Chen, H. (2010). An integrated framework for avatar data collection from the virtual world. *IEEE Intelligent Systems*, *25*(6), 17–23. doi:10.1109/MIS.2010.138

Zheng, L., Shen, C., Tang, L., Zeng, C., Li, T., Luis, S., & Chen, S. (2013). Data Mining Meets the Needs of Disaster Information Management. *IEEE Transactions on Human - Machine Systems, 43*(5), 451-464. doi:10.1109/THMS.2013.2281762

Zhou, F., Dun, H. B.-L., & Billinghurst, M. (2008). Trends in augmented reality tracking, interaction and display: A review of ten years of ISMAR. In *Proceedings - 7th IEEE International Symposium on Mixed and Augmented Reality 2008, ISMAR 2008,* (pp. 193-202). IEEE.

Žiberna, A., & Vehovar, V. (2009). *Using social network to predict the behavior of active members of online communities*. Paper presented at the 2009 Advances in Social Network Analysis and Mining, Athens, Greece. doi:10.1109/ASONAM.2009.24

Zimmermann, K. A. (2012). *Hurricane Katrina: Facts, damage & aftermath*. Retrieved from http://www.livescience.com/22522-hurricane-katrina-facts.html

Zuchora-Walske, C. (2010). *Internet Censorship: Protecting Citizens Or Trampling Freedom?* Minneapolis, MN: Twenty-First Century Books.

About the Contributors

John P. Sahlin is an Adjunct Professor with The George Washington University in Washington, DC. He is also currently the Technical Director for Information Dominance at Engility, where he is responsible for multitenant network design and cybersecurity programs. In this capacity, he drives the strategy and vision to apply commercial technology and best practices to the U.S. Department of Defense. Dr. Sahlin is a leading technical expert in the fields of Systems Engineering, distributed computing network design, and managed hosting services (e.g., Cloud Computing). He has over 20 years' experience designing, building, and sustaining distributed computing networks for commercial and U.S. Public Sector clients, applying his expertise in the ASP and Cloud Computing industries to the government. Dr. Sahlin holds a Ph.D. in Systems Engineering from The George Washington University, where he is an Adjunct Professor. He is active on both LinkedIn (linkedin.com/in/johnpsahlin) and Twitter (@sahlinj) posting on topics concerning technology, security, and government.

* * *

Kate Andrews has a Ph.D. in Industrial/Organizational Psychology and has worked extensively in consulting firms both internally and externally. Her continued work with organizations centers around process improvement, leadership consulting, and team dynamics. She has worked in both Israel and in Romania serving organizations and individuals. Currently, she has 3 research projects in motion and time for her 6 grandchildren.

Arleen Cuevas completed her education in the Philippines and Singapore. Besides being a lecturer at NHTV, she works as an independent film producer with a focus on Asian-European co-productions. She has produced a total of 16 films which premiered in different international film festivals. Her films have received film grants from Fonds Sud, France, World Cinema Fund, Germany, Hubert Bals Fund, Netherlands and Global Film Initiative, USA. Two of the films she co-produced, "Independencia" and "Manila" were selected and premiered at Cannes Film Festival in 2009. She was line producer for the film "Amigo", by American director, John Sayles and "The Amazing Truth about Queen Raquela", by Iceland director Olaf Johanesson. She currently produced "Kalayaan" (Wildlife), which premiered at Rotterdam Film Festival in January 2013 and "How to Disappear Completely" which premiered in Locarno Film Festival in August 2013.

Charalampos A. Dimoulas received his diploma and Ph.D. from the School of Electrical and Computer Engineering, Faculty of Engineering, Aristotle University of Thessaloniki (AUTh) in 1997 and 2006, respectively. In 2008, he received scholarship on post-doctoral research at the Laboratory of Electronic Media of the School of Journalism and Mass Communications of AUTh. Both his doctoral dissertation and his post-doc research deal with advanced audio-visual processing and content management techniques for intelligent analysis of prolonged multi-channel recordings. He elected Lecturer (November 2009) and Assistant Professor (June 2014) of Electronic Media in the School of Journalism and Mass Communications, AUTh, where he currently serves. His current scientific interests include media technologies, signal processing, machine learning, multi-modal intelligent content analysis including multimedia semantics, audiovisual content description and management automation. Dr. Dimoulas is member of IEEE, EURASIP and AES.

George Kalliris, http://kalliris.blogspot.com/, was born in Nicosia - Cyprus. In 1989 he received his 5-year study Diploma in Electrical Engineering with Telecommunications from the Aristotle University of Thessaloniki (AUTh) - Greece. In 1995 he received his Ph.D. from the same School. His doctoral research was carried out at the Lab of Electroacoustic and Television Systems. During and after completing his doctoral studies he worked in several research, development and innovation projects as well as a part-time higher education teacher. In 1998 he was elected Lecturer of Electronic Media Technology at AUTh. His current position is associate professor and director of the Electronic Media Lab at the School of Journalism and Mass Communication. He has also taught and/or teaching in two Master degree programs, to the Film Studies School and as a visiting professor of the Frederick University Cyprus. His current research interests and publications include audiovisual technologies for the new media, radio and television studio design, digital audio-video processing–production–broadcasting-webcasting, multimedia content, restoration, management and retrieval.

Fritz Kohle studied at the Surrey Institute of Art and Design and as a postgraduate at the Northern Media School, Sheffield Hallam University. He currently studies for a Ph.D. at the University of Edinburgh investigating the impact of social media on documentary production. In 2009 he obtained tenure at NHTV, University of Applied Sciences, Breda, Netherlands, as a senior lecturer teaching Film & TV production, while continuing his professional production practice. Kohle began his career as a production assistant at companies such as the The Weather Channel UK, working his way up to become Location Manager for a Channel 4 UK feature film (Prometheus), Head of Productions at the Lux Centre London, and Field Producer for iBeam Europe. As a freelance Producer and Production Manager he has produced episodes for Marienhof, one of Germany's most popular soaps, with a daily viewership exceeding 6 million. In collaboration with the Ludwigsburg Film Academy in Germany, he published the media handbook Medienmacher Heute. Kohle's projects include post-production management for Studio Babelsberg features in Berlin, such as Wim Wender's Soul of a Man, and Jackie Chan's 80 Days Around the World. In addition, he was also involved in various German soaps such as Notruf 110 and Tatort. Kohle was Bigfoot Entertainment's Head of Post Production until 2007. In June 2007 he joined New York Universities Tisch School of the Arts as Assistant Director for the Production and Post-production Centre, significantly contributing to the establishment of the NYU Singapore Campus.

Jaime Lobera is the Dean for the College of Graduate Studies at Coleman University in San Diego, California. He has mentored and guided Master of Science in Information Systems Management student through their thesis process. He also serves as a Thesis Committee Member. He worked in the field of technology management in the United States government for over twenty years by upgrading legacy systems and managing secure and non-secure networks. Mr. Lobera holds a Masters in Technology Management as well as other degrees. He is currently pursuing a Ph.D. in Information Technology with Capella University.

Gilberto Marzano. Expert in computer science and social anthropology, he is prof. at Rezekne University of Applied Science (Latvia), and Head of the laboratory of pedagogical technologies of Research Institute for Regional Studies at the same university. He is member of the professor board of Ph.D. on Economy, Ecology and Preservation of Agricultural and Environmental Systems at University of Udine (Italy). President of the Ecoistituto del Friuli Venezia Giulia, in Italy, a private non-profit research institute founded in 1989, and Vice president of IPSAPA (Interregional Society for Participation in Agribusiness Landscape and Environmental Management). For many years he was contract professor at the University of Trieste and at the University of Udine. He worked as a manager in several private ICT companies and was director of an R&D software laboratory of a large ICT Italian Company. He participates, as an expert, to international projects on cyber risks and new media.

Maria Matsiola was born in 1971 in Kavala, Greece. She received her diploma from the School of Electrical and Computer Engineering, Faculty of Engineering, Aristotle University of Thessaloniki (AUTh) in 1996 and her Ph.D. on New Technologies on Journalism, from the School of Journalism and Mass Communications, Aristotle University of Thessaloniki (AUTh) in 2008. She serves as Tenured Senior Teaching Fellow and Instructor in the Electronic Media Laboratory of the School of Journalism and Mass Communications (AUTh). Her scientific interests include New Media technologies, Web Radio and TV, audiovisual content management, distance learning, audiovisual application systems in distance education.

Thomas A. Mazzuchi is Chair of the Department of Engineering Management and Systems Engineering in the School of Engineering and Applied Science, and Professor of Engineering Management and Systems Engineering, at The George Washington University. He earned his B.A. (1978) in Mathematics from Gettysburg College, Gettysburg, PA; and his M.S. (1979) and D.Sc. (1982), both in Operations Research from George Washington University, Washington, D.C. Dr. Mazzuchi served as research mathematician at Royal Dutch Shell, and has conducted research for the U.S. Air Force, Army, and Postal Service; for NASA; and for the Port Authority of New Orleans, among others. His research interests include reliability, life testing design and inference, maintenance inspection policy analysis, and expert judgment in risk analysis.

Ramona McNeal is an associate professor in the Department of Political Science at the University of Northern Iowa. Her chief research interests include e-government, telehealth, campaign finance reform and telecommunications policy. She also studies the impact of technology on participation, including its relationship to voting, elections, and public opinion. She has published work in a number of journals including Journal of Information Technology & Politics, Social Science Quarterly, Policy Studies Journal, Public Administration Review, Political Research Quarterly and State Politics and Policy Quarterly. She is a co-author of Digital Citizenship: The Internet, Society and Participation (MIT Press, 2007) with Karen Mossberger and Caroline Tolbert.

Bethany Mickahail holds several degrees: Bachelor of Arts, in Linguistics & TESL (University of Hawaii): Masters of International Affairs and Communications (School of International & Public Affairs at Columbia University); and Doctorate in Education, Leadership & Public Policy (University of Denver). Dr. Bethany has provided learning opportunities to children and adults as an ESL teacher, staff developer-trainer, district wide program coordinator, college educator and United Nations researcher. Dr. Bethany's research efforts focus on issues related to innovative community based disability program policy, micro business development, training and systems change in California. Most recently, as a School of Advanced Studies educator and dissertation advisor, she is a Fellow with the Center for Workplace Diversity Research at the University of Phoenix. At the Center, she is pursuing international research on Innovation and Design Thinking. An award winning author and presenter, Dr. Bethany frequently speaks to groups about disability, societal change, innovation, policy development and implementation.

Shahram Sarkani, Ph.D., P.E., is Professor of Engineering Management and Systems Engineering, and Director of EMSE Off-Campus Programs, at The George Washington University. He designs and administers graduate programs that enroll over 1,000 students across the U.S. and abroad. In over 150 technical publications and in sponsored research with NASA, NIST, NSF, AID, and Departments of Interior, Navy, and Transportation, his research has application to risk analysis, system safety, and reliability.

Mary Schmeida, Ph.D., is a public policy expert who has served in several key research positions. Her chief research interests include social welfare and health policy, electronic government, and environmental policy. She has published work in a number of academic journals including Government Information Quarterly, Administration and Policy in Mental Health and Mental Health Services Research, Journal of Health Care for the Poor and Underserved, and International Journal of Professional Case Management.

James ter Veen leads multiple systems engineering efforts including systems of systems architecture, project planning, design, software and hardware lifecycle maintenance, test and integration, new product demonstrations, technical documentation as well as installation, configuration, training and support of enterprise systems at customer locations. During thirty years of experience in high technology hardware/software/systems development, customer support and software quality assurance, James has acquired expert knowledge of the iterative software development lifecycle. He has a proven track record of building and managing integrated product teams working in areas including wireless sensor networks, mass data storage technologies, electronic security systems, information systems management, medical data analysis and command and control systems. He holds an active security clearance, an AS in Electronics Technology, a Bachelor's in Information Technology, a Master of Science in Systems Engineering and a Ph.D. in Systems Engineering from The George Washington University.

Andreas Veglis is Professor, head of the Media Informatics Lab at the School of Journalism & Mass Communication at the Aristotle University of Thessaloniki. He received his BSc in Physics, MSc in Electronics and Communications, and Ph.D. in Computer Science, all from Aristotle University. From November of 2010 until February of 2015 he was head of the postgraduate programme of the Department of Journalism & Mass Media Communication, Aristotle University of Thessaloniki. In March of 2012 he was elected Deputy Chairman of the Department of Journalism Mass Media Communication. His research interests include information technology in journalism, new media, course support environments, data journalism, open data and distance learning. He is the author or co-author of 12 books, he has published 50 papers on scientific journals and he has presented 74 papers in international and national Conferences. Professor Veglis has been involved in 18 national and international research projects.

Shefali Virkar is research student at the University of Oxford, UK, currently reading for a D.Phil. in Politics. Her doctoral research seeks to explore the growing use of Information and Communication Technologies (ICTs) to promote better governance in the developing world, with special focus on the political and institutional impacts of ICTs on local public administration reform in India.Shefali holds an M.A. in Globalisation, Governance and Development from the University of Warwick, UK. Her Master's dissertation analysed the concept of the Digital Divide in a globalising world, its impact developing countries and the ensuing policy implications. At Oxford, Shefali is a member of Keble College.

Brett Winkelhake is an Adjunct Professor with San Diego Mesa College in San Diego, CA, Grossmont College in El Cajon, CA, Southwestern College in Chula Vista, CA and National University in San Diego, CA. Currently Dr. Winkelhake works for Neighborhood House Association in his full-time position. The agency works with children for the Head Start Program. Dr. Winkelhake served as the Thesis Advisor/Co-Author while at Coleman University for Mr. Ismail's chapter. Dr. Winkelhake has over 15 years of professional experience in the real estate, higher education and non-profit industries. Dr. Winkelhake in February 2015 completed his Doctorate of Management in Organizational Leadership from the University of Phoenix. His research for his dissertation was entitled of Verbal Argumentativeness: A Study of Retail Store Managers.

Kimberly Young-McLear received her BS in Industrial Engineering at Florida Agricultural and Mechanical University. In 2009, she earned her MS in Industrial Administration from Purdue University. She also holds a Project Management Professional (PMP) certification. In 2005, she earned her commission in the United States Coast Guard where she has had a robust career in marine safety, industrial engineering, and enterprise project management. She has managed multi-million dollar naval engineering projects in support of restoring Coast Guard vessels. Her research focus includes Human Systems Integration and Organizational Cybernetics. She is currently completing her Ph.D. in Systems Engineering from The George Washington University and teaching full-time at the United States Coast Guard Academy.

Ahmed Youssry is a mechanical engineer with two masters degrees in Business and Information Technology. Ahmed is currently working as a quality manager for a manufacturing company in Poway. Ahmed has developed a passion for micro-finance and crowd funding after working in several nonprofit organizations in Africa and the USA for over 6 years.

Index

H

I

K

L

M

N

O

P

R

S

T

U

W

Y

www.ingramcontent.com/pod-product-compliance
Lightning Source LLC
LaVergne TN
LVHW060424300926
849162LV00006B/1520